ENVIRONMENTAL
RISK ANALYSIS

ENVIRONMENTAL RISK ANALYSIS

Ian Lerche
Evan K. Paleologos

Department of Geological Sciences
University of South Carolina
Columbia, South Carolina

McGraw-HILL
New York   Chicago   San Francisco   Lisbon   London
Madrid   Mexico City   Milan   New Delhi   San Juan
Seoul   Singapore   Sydney   Toronto

Library of Congress Cataloging-in-Publication Data

Lerche, I. (Ian)
  Environmental risk analysis / Ian, Lerche, Evan K. Paleologos
      p. cm.
    Includes index.
    ISBN 0-07-137266-0
    1. Environmental risk assessment.   I. Paleologos, Evan K.   II. Title
  GE145.L47   2001
  333.7'14—dc21   00-068379

McGraw-Hill

A Division of The McGraw·Hill Companies

1 2 3 4 5 6 7 8 9 0   DOC/DOC   0 7 6 5 4 3 2 1

ISBN 0-07-170072-2
ISBN 978-0-07-170072-6

*The sponsoring editor for this book was Kenneth McCombs, the editing supervi-
sor was David E. Fogarty, and the production supervisor was Sherri Souffrance.
It was set in the HB1 design in Times Roman by Deirdre Sheean of McGraw-
Hill's Professional Book Group composition unit, Hightstown, New Jersey.*

This book was printed on recycled, acid-free paper containing
a minimum of 50% recycled, de-inked fiber.

To my parents,
to Katrina and Demi,
and mostly to my wife, Cleo

Evan K. Paleologos

CONTENTS

Chapter 8. Corporate Involvement in Multiple Environmental Projects   *139*

Chapter 9. Apportionment of Cost Overruns to Hazardous Waste Projects   *183*

PREFACE

The disposition of chemical, biologic, radioactive, and toxic wastes is perhaps one of the quintessential environmental problems of the past century and the foreseeable future. From the technical and scientific points of view, the major concerns have to do with transport, burial, and monitoring. The aims are to transport the waste without spillage due to mishap (natural or human error), to bury the waste in such a way that there is no leakage thereafter, and to monitor the burial site in a continuous fashion for any leakage after burial. While these aims are the zenith, the nadir is spillage during transport, burial with catastrophic leakage, and a monitoring system that fails.

Scientific analyses of environmental projects are hindered by significant uncertainties. Characterization of present-day conditions at environmental sites is almost always limited because of financial constraints and because the complex interaction of physical, chemical, and biologic processes that control the transport of contaminants is usually not well understood. This situation is accentuated by the requirement that a repository site needs to provide efficient isolation of the waste from the environment and human population over long periods of time and, correspondingly, of the liability claims that an environmental company may have to face long after completion of a project. Future geologic events (e.g., earthquakes, floods, climate change, etc.), all of which can influence the ability to control and isolate the waste over time, introduce further uncertainties. Moreover, the existing technological solutions are limited in their efficiency and expensive to implement and can generate by-products that are difficult to control. The situation increases in complexity with litigation issues, where a significant expense can develop such that funds that otherwise could have been allocated for development and application of new technologies are used up in attempts to resolve legal issues.

Over the past two decades, an enormous effort has been expended (in particular for potential nuclear waste depositories) to account for the uncertainty in site conditions, and this has led to a shift in the focus of scientific studies at academic institutions, national laboratories, and most major environmental companies from the analysis of perfectly determined systems to systems that are described statistically. Combined with the lack of efficient technological developments, this has led regulatory federal and state agencies increasingly to recognize risk as an integral part of environmental projects and to require scientific and financial risk analyses.

There is another side to the problem of waste disposal, however. Eventually, a corporation (or corporations) will to have transport, bury, and monitor the waste, usually under a contract from a regulatory agency. The contract usually will contain performance criteria to be met but also will contain a price to be paid to the contractor. The question of interest to the corporation is whether the contract is profitable and under what conditions. There is clearly some sort of financial limit depending on the prior estimates of probabilities of transport (with or without spillage), burial (with or without leakage), and ongoing monitoring costs. Additionally, regulatory agencies, concerned citizen groups, and political staff attached to a lawmaker may all wish to become involved in the various components affecting the decision making at various stages. From the corporate perspective, the problem is to ascertain what associated costs make it worthwhile to accept the contract. From an environmental corporation's perspective, scientific uncertainties and limitations of the technological solutions are only a subset of the total uncertainties it faces. Changing political, financial, and regulatory conditions constitute other unpredictable components in a project's performance and financial return.

The purpose of this book is to explore from a corporate perspective how the preceding issues can be addressed when environmental projects are assessed. In light of these considerations, one may distinguish two main objectives of this monograph. The first is to provide a coherent and unified account of the most critical components entering risk analyses of environmental projects. Thus the influence of various risk factors (arising from scientific uncertainties, technological limitations, regulatory changes, and unpredictable events) on a project's performance is fully illuminated and incorporated into the risk-analysis framework. Additionally, this book aims to incorporate in the decision analyses of environmental projects some recent techniques, which, though not usually discussed in standard risk-analysis books, have become essential in the application of effective corporate environmental strategies. Topics such as innovative types of insurance coverage, partial involvement in multiple projects, corporate alternatives to changes in regulations and/or catastrophic events have seen wide application in other business operations but still have to be included in a comprehensive way in risk analyses of the environmental industry. This book aims to bring exposure to such topics and to include these alternatives as an integral part of environmental risk analyses. The second objective of the book is to provide a step-by-step approach to planning and performing an effective environmental risk assessment. The individual chapters are organized in such a way that an environmental decision-making group can, by starting from the preliminary stage of financial analysis of a project, proceed to address more complex situations and financial alternatives in a systematic manner. Thus the summaries that are provided at the end of individual chapters can operate as a checklist of an environmental project's risk-assessment progress and help in evaluating whether all available alternatives have been explored.

This book is organized as follows: Chapter 1 provides an exposition of some common risks encountered in the transport, burial and storage, and monitoring phases of waste. Chapter 2 sets the stage for a preliminary risk analysis of environmental projects by considering the simplest situation of a contract offered under fixed regulations, with no catastrophic loss events, without inclusion of either optional insurance or of corporate risk tolerance, and without the occurrence of spillage or leakage. Chapter 3 introduces three statistical measures that can be used in risk analyses of environmental projects that face the possibility of limited or catastrophic losses. Particular emphasis in this chapter is placed on the inclusion of catastrophic scenarios in risk analyses because these can alter substantially the perspective on a project and lead environmental corporations away from investments. Chapter 4 introduces the concept of corporate risk tolerance and presents a methodology that addresses risk considerations through partial project involvement. Chapter 5 develops the framework for the inclusion of insurance alternatives in environmental risk analyses and presents a procedure that allows small environmental companies to address limited-liability claims for a number of situations. Chapters 6 and 7 expand the concept of risk-adjusted value to more complex cases of environmental risk and provide procedures that can be used to determine the dominant uncertainty factors that influence a project's return and performance. Chapter 8 presents a methodology for optimizing total corporate return for a portfolio of opportunities in the face of a constrained budget, and Chapter 9 deals with the apportionment of cost overruns to various environmental projects in such a way as to maximize total corporate return. Chapters 10 and 11 provide the framework for bayesian updating of leakage scenarios and multiple transport of hazardous material, respectively. Chapters 12, 13, and 14 address issues of regulatory compliance and project monitoring strategy, the use of option payment as an alternative to insurance coverage, and the worth of collecting additional data. Chapter 15 discusses some problems of scientific uncertainty in the representation of a hydrologic system that are due to model selection and resolution of data, while Chaps. 16 and 17 provide concluding remarks on the interrelation of human, scientific, and financial considerations for environmental projects. Finally, the two appendixes provide a summary of the regulations that govern Phase I, II, and III site assessments, and also give a brief overview of the federal statutes that are appplicable to environmental problems.

The monograph is set at a level where an environmental, civil, or chemical engineer or scientist involved in environmental problems should have little difficulty not only in following the arguments presented but also in actively building on the precepts expounded here to a higher level than we present. For corporate executives, it is hoped that the work presented here will remove some of the less than objective assessments of economic worth of an environmental project that occasionally have been the lot in the past. Although this is an application-oriented book, the material is set at a graduate study level so that it can be used as a textbook in courses of decision analysis offered at engineering or science departments. Toward this objective, the book not only provides an exposition and application of decision-analysis theory to practical environmental problems but also includes topics not usually covered in standard risk-analysis textbooks that are critical in comprehensive risk assessments of environmental projects and the analysis of corporate alternatives.

We are grateful to a large body of people for their input, advice, and criticisms of the ideas developed here. And we are keenly aware that this monograph does not do justice to all economic aspects of all environmental problems. Such a development would make for a very long tome indeed and one that is well beyond our abilities. However, we will have succeeded in our endeavor if others, more able than ourselves, can bring the tools and methods presented here to as sharp a focus as possible over the years. More appropriate statements of economic evaluation can then be made which can help to guide the future toward the most important technical and scientific developments needed to integrate scientific risk and uncertainty with economic risk and uncertainty. It is with these goals in mind that the present monograph has been written.

We are particularly grateful to Donna Black for typing parts of this manuscript, Dora Avanidou and Charles Fletcher for their assistance with parts of Chapters 3 and 5, respectively, and Theo Sarris for his help with graphics and computer assistance in the editorial processing of figures, text, and equations. Again, family and friends suffer silently (and sometimes not so silently) as one endeavors to bring a book manuscript to fruition. To them, and to many others, we are grateful and extend our thanks for their understanding, appreciation, and patience.

Finally, E. K. P. would like to gratefully acknowledge funding from the U.S. Department of Energy, Office of Environmental Management, through the Center for Water Research and Policy at the University of South Carolina (contract DE-FG02-97EW09999). Particular thanks go to the Alexander von Humboldt Stiftung and the University of Halle (Saale) for their generous financial support during a sabbatical year at Halle for I. L., which enabled this book to be completed.

IAN LERCHE
EVAN PALEOLOGOS
Columbia, South Carolina

ENVIRONMENTAL RISK ANALYSIS

CHAPTER 1
INTRODUCTION TO ENVIRONMENTAL PROBLEMS: TRANSPORT, BURIAL, MONITORING, SPILLAGE, LEAKAGE, AND CLEANUP

Humanity produces copious quantities of waste of various sorts. Some of the waste is biologic; some is in the form of both household and industrial material, some consists of chemical compounds used in various processes; some is in the form of gaseous emissions that pollute the atmosphere and the earth; some is in the form of radioactive waste as by-products or end-products of nuclear-driven power plants, submarines, and decommissioned weapons; and some is in the form of raw product such as oil and its associated waste (Freeze, 1997; Lerche, 1997).

Two major problems are present. First, in the case of directly and indirectly produced waste, there are problems of disposal. Clearly, sites must be chosen that are capable of handling the various forms of waste. These sites must be sufficiently controlled and isolated from human population by a combination of natural barriers and engineering measures that the potential is minimized for escape of harmful constituents from the site. Second, there is the potential for spill during transport of the waste and leakage to the environment after burial (Wentz, 1989). Similar issues are also of concern with the shipment of raw product by air, land, or sea, as well as with the associated containment in storage vessels (on land or sea) after transport.

In a broad sense, the quintessential problems for both waste and raw product can be classified in about five main categories. The first category arises because material must be transported from a temporary storage or processing facility to a permanent site. This transport requires that decisions be made on the transport vehicles and routes to be used as well as on the material of the transporting containers. For instance, in the case of transport of radioactive waste, the very containers that are used for the transport eventually may become radioactive, and thus the containers themselves may have to be transported and buried along with the primary radioactive waste. In the case of chemical waste transport, the absorption of toxic material to container linings means that such containers may pose problems either by corrosion of the container or by becoming toxic, again requiring that the containers be disposed of properly (Magnuson, 1980).

The very act of shipping and transporting hazardous material is fraught with difficulties because of the desire to avoid population centers where possible, the need to transport the product to a receiving site with the least delay and possibility of danger, and the fact that the method of transport can itself be subject to the vagaries of weather, human error, and unforeseen problems, such as bridge collapses, train derailments, or ship breakups, to name but a few (Abramson and Finizza, 1995).

Assuming that one has transported the material successfully to the storage site, be it a waste depository or a storage site for holding product until needed for further use, the second critical issue arises from the location and construction of the site itself. The storage site must be sufficiently well constructed that deposition of the waste material either will not comprise the integrity of the site or so rapidly fill the site that another must be sought. Site integrity, arguably, is one of the most critical ingredients because external influences, such as water flow, rock fractures, earthquakes, erosion, magmatic activity, and the like, should not be sufficiently strong that site integrity is compromised (Fischer, 1986). Indeed, geologic, hydrologic, and geochemical conditions at a site preferably should act as natural barriers to potential leakage, supplementing the role of engineering measures (Freeze, 1997). The site must be sufficiently far away from population centers that potential leaks will not affect human health and at locations where future activities (e.g., mining, urban expansion, etc.) will not endanger site integrity (U.S. EPA, 1991). In addition, in the case of long-term storage, well beyond the lifetime of any known political system of government, it becomes of paramount importance that archival records be maintained to advise future generations of the hazardous nature of the material contained in a storage site (Keeney, 1980).

A third category of problem arises with the requirement of monitoring the hazardous material during both the transport and later storage stages. One must, at the least, have a satisfactory number of monitoring devices of sufficient resolution to detect rapidly any problems that do occur and have contingency plans for action based on the monitoring results. The issues here would seem to be to estimate how many monitoring devices are needed, the length of time over which monitoring should occur (with regular maintenance and replacement being incorporated in the monitoring program), the accuracy and resolution of results reported by such devices, and how measurements performed at different scales can be used (Cushman, 1986; Christakos and Hristopulos, 1998).

The most critical problems are associated with spillage and leakage, which are treated separately here as two distinct categories. *Spillage* is usually referred to when, during the acts of transport or after storage, there is a major loss of material from the transporting system or from the storage containers that has an immediate and significant effect on the surrounding environment. Such types of spillage problems are perhaps exemplified by the breakup of a ship carrying crude oil with the immediate loss of most of the cargo, resulting in severe, immediate environmental damage, or perhaps by the Chernobyl disaster, where huge amounts of radioactive material were released to both the atmosphere and the surrounding ground, thereby causing major contamination problems that required immediate resolution. On the other hand, *leakage* usually refers to slow loss of material that, while not so hazardous in the short term, would become so unless action is taken to control and modify the transport or storage system and stop the leakage. Thus slow leakage from a corroding container filled with toxic chemical waste presents minimal immediate danger but, unless controlled, eventually will pose a significant long-term hazard (Asante-Duan, 1996).

From the scientific point of view, there are several problems that need to be addressed in relation to the transport of hazardous material, the processing of such waste, or the storage of the material in a way that minimizes the potential for spillage or leakage. First, one needs to assess the chemical, biologic, or radioactive contamination that could occur should waste be released and to provide remediation solutions to cover such contingencies (Van der Heijde et al., 1986; Nyer, 1992; Grasso, 1993; Flach and Harris, 1997). Second, one needs to plan the transport and storage in such a way that the short- or long-term (or both, if possible) potential for release is minimized. Thus, in the case of the potential storage site at Yucca Mountain, Nevada, for spent radioactive material, one needs to perform major site surveys to check out the current rock fracture pattern, the likelihood of fault and earthquake disturbances that would compromise the integrity of the storage system, the potential for groundwater contamination, the geochemical conditions that can accelerate or impede such contamination, and a host of ancillary scientific issues that then provide the best available scientific assessment of the capability of the repository for secure long-term storage. This sort of problem has been the subject of major scientific investigation over the years (Hunter and Mann, 1992; U.S. DOE, 1994; Sandia, 1994). In other cases one is perhaps not quite so fortunate to have a large body of scientific investigation at hand. The consequence is that for both short- and long-term problems one is more likely to have overlooked or misestimated a potential hazard, with the consequent increase in uncertainty on the capability of the storage system to truly maintain a secure condition (Keeney, 1980).

Despite scientific analyses, there are always situations that are difficult to guard against, control, or remediate. Perhaps the most obvious of such problems is the transport of oil by tankers. Here there is little that can be done to safeguard against major storms close to shore that can (and do) easily shipwreck tankers, with the consequent spillage of oil because of disruption of ship integrity. Or again, it would seem difficult to guard against human error as represented by the *Exxon Valdez* disaster, where a fully laden tanker ripped apart on a submarine reef as a result of human negligence, spilling its oil contents to the ocean and shore around Valdez, Alaska (Alaska Fish and Game, 1989). In other instances one has to deal with old, abandoned hazardous landfills and other sites, where leakage has occurred from corroding drums and containers that have been labeled improperly or burned incompletely and hence there is very little knowledge of the quantity and nature of the chemicals involved (Magnuson, 1980; Malle 1996). The situation is exacerbated by the poor performance of certain commonly used waste-site remediation approaches (Canter and Knox, 1986). For example, a National Research Council committee review of the performance of pump-and-treat systems at 77 contaminated sites reported that groundwater cleanup goals had been achieved at only 8 sites (MacDonald and Kavanaugh, 1995). The U.S. Environmental Protection Agency (EPA) reported that only 14 of 263 Superfund source control projects for which systematic site-remediation solutions were applied reached completion (Powell, 1994).

Thus, from both the storage and transport points of view, not only must hazard assessments be made and contingency plans developed in response to estimated hazards, but also there is a need for assessment of the resolution and accuracy of the estimates and of catastrophic cleanup problems should such eventualities occur. The point here is that the scientific measurements (from which hazard estimates are made) are limited in number, unevenly distributed in space and time, and have intrinsic uncertainties of their own (Dooge, 1997; Christakos and Hristopulos, 1998). In addition, the scientific models are themselves uncertain, require parameters that are poorly known, or do not quite address all the problems that could occur. For instance, groundwater flow and transport models cannot fully account for the spatiotemporal variabilities and uncertainties of the parameters and variables of the problems and, in addition, suffer from the inherent simplification of the complex interaction of physical, chemical, and biologic processes that affect the transport of contaminants. The validation of such predictive models in terms of slowly evolving subsurface plumes is currently elusive (Dagan and Neuman, 1997). Thus the ability to provide accurate assessments is limited by data, models, and outside considerations over which no, or only limited, control exists (Corwin et al., 1999).

Motivated by these limitations, decision-tree diagrams have been used over the years to allow for the likelihood of different occurrences in behavior, and they are often tied to penalty/reward scientific assessments and probability of occurrence (LaValle, 1978; Bunn, 1984). Such methods do not provide a universal guarantee, of course, that one has controlled the uncertainty, but they do provide a method for identifying the most likely areas of concern or ignorance, thereby allowing determination of the areas where effort should be focused to narrow the state of uncertainty (Grayson, 1960; Raiffa and Schlaifer, 1961). These methods will be discussed in more detail later in the book, including methods that account for uncertainty in the reward and penalty assessments and in the probability of occurrence.

The largest factor, however, and the one that is often overlooked in scientific assessments, is the financial analysis attendant on any decision made or to be made. Thus costs of transport, storage, monitoring, leakage remediation, spillage cleanup, insurance, catastrophic liability, and the value of bidding on a contract for such transport and storage need to be intertwined with the scientific assessment. For example, if the cost of undertaking to outfit a tanker with double or triple hulls to offset the chance of an oil spill were to be such that this cost could never be recovered, then there would be little point in taking such action; alternative methods of transport would be searched for that would allow a profit to be made.

Only a government has essentially unlimited financial resources, so any time a contractor is involved in either a transport or storage opportunity, there is always the question from the corporate side of whether the project could prove profitable or whether the project is so risky that even with maximum insurance and available technology the likelihood of spillage and/or leakage is so large as to increase the probability that the company will lose huge amounts of capital. Under such conditions, it would be a brave, or foolhardy, company that would accept such risk without a limited-liability clause written into its contract.

The purpose of this book is to provide some quantitative methods and associated examples to address the preceding problems. Illustrations are furnished to show how one can account for the uncertainty in both the scientific and economic estimates in order to reach a rational decision of the worth of a project. In this way, the methods guide the scientific and economic components of a project to increase either the amount of scientific information—and also determine what type of scientific information to increase—or perhaps accept the available scientific information but improve resolution—and to then determine which components of the scientific information need improvement. Also, one must provide a clear set of criteria to determine what is the rationale for the required improvements in narrowing scientific and economic uncertainties while at the same time addressing the worth of collecting more information. For example, if such information would do little to help resolve the economic picture or clarify questions of scientific uncertainty, which may be limited primarily by model assumptions rather than by the quality or quantity of the available data, then an intense site characterization may not be the appropriate action (an example of such a situation will be given in Chap. 3).

The primary concern throughout this book is to view the problem of transport and storage of hazardous materials from an economic perspective, as viewed by decision-making groups motivated by the need to make a profit for their corporation (else the corporation will eventually go bankrupt).

1.1 TRANSPORT PROBLEMS

Put to one side for the moment all other problems, and concentrate on the transport component of a disposal process. The main modes of transport of hazardous material are by truck, train, pipeline, ship, and occasionally, air.

In the case of transport, there are several uncontrollable hazards depending on the mode of transport. For instance, transport by truck brings to the fore the possibility of driver error, of accident due to vehicle collision, of road integrity under continuing and relentless truck weight, of truck failure (e.g., axle shaft breakage), of weather-related problems, etc. Any one, or all, of these situations could lead to the breakage of transporting containers. In such a case, spillage of the material would require not only that the spilled material be re-collected but also that the contaminated surroundings and all the material then be transported and safely stored or buried. Such an event can make for a significant increase in costs, as well as in hazard to humanity.

In the case of train transport, allied similar hazards can arise: collision with another train, collision with a vehicle at a crossing, track failure, wheel failure, signal junction switching failure, or weather-related failure. Again, then, one has the problem of re-collecting the spilled material plus contaminated surroundings, plus train parts that are also contaminated. And all such material must now be transported and stored safely.

Likewise, in the case of ship transport, one has the problems of storm-related catastrophic failure, of submarine hazards causing failure of ship integrity, of internal ship failure (e.g., explosion, fire, etc.), or of ship age causing structural failure. Once more, the problems of environmental cleanup have to be resolved.

In the case of pipeline transport, pipeline leakage problems are particularly severe as a result of pipeline age without proper maintenance and/or pipeline "tapping" to divert the material illegally.

1.2 BURIAL/STORAGE PROBLEMS

Concentrate now exclusively on storage problems. Two characteristic storage/burial conditions are operative: Either the material is being stored in a temporary container for later processing and/or use (such as a tank farm for oil or a buried container for gasoline) or the material is to be stored/buried indefinitely to minimize hazard potential. In the case of temporary storage, the material is delivered and removed from the temporary system, so not only must the holding system itself be secure but also so must the delivery and removal mechanisms. Clearly, the potential for spillage is greater in such a case than for a one-time delivery because the multiple acts of delivery and removal increase the chances that spillage will take place during such events. In addition, the corresponding variable stress with time on the holding system can weaken structural components beyond the point of failure, thereby causing loss of storage system integrity and consequent leakage or spillage. Thus a holding tank for oil at a refinery or at a gas station is filled, emptied, and refilled many times. Each time the oil level is varied, there is a corresponding change in stress on the tank walls and tank base. This variable stress can lead to loss of integrity, much as occurs with multiple flexing of airplane wings.

In the case of indefinite storage systems, the ideal system would be one in which material is brought to the storage system, made secure, and thereafter is not disturbed by further human activity. However, such long-term storage systems are subject to hydrologic and geologic disturbances, so the "pristine" conditions of initial storage are altered as time goes by. These variable conditions in time can then lead to corrosion of buried containers, to contaminated water efflux, or to upheaval or disruption of the storage system by, say, an earthquake. Thus once more one has to deal with the problems of loss of storage integrity and consequent remedial action.

1.3 MONITORING PROBLEMS

Focus now exclusively on monitoring problems. It is almost mandatory that some form of monitoring take place in both transport and storage because one wishes to know where, when, and how much a system has spilled or leaked. For instance, in the case of container storage of hazardous radioactive material or chemicals, the storage system usually has an array of monitoring devices to check whether a container or group of containers has corroded or burst, thereby spilling hazardous material into the natural system. In addition, the storage system periphery usually is ringed by a set of monitors to check loss of contaminant with time from the storage system as a whole. Several issues with the monitoring devices themselves then arise. Do the characteristics of the devices change with time so that one is no longer sure what is being monitored? How does one know (rather than guess) that the lateral and vertical distribution of monitoring devices is adequate to truly provide a safeguard measure of contaminant leakage either from the storage site as a whole or from local regions of the site? What are the quality, quantity, accuracy, precision, resolution, and uniqueness of monitor data? Is it the case that, for instance, intrusive groundwater flow would provide a "masking" (perhaps by contaminant dilution) of a hazardous situation? How often should monitor maintenance and/or replacement be carried out to improve results? How does one know when to increase (decrease) the number of monitoring devices, and what criteria does one use to make such a determination?

Clearly, these questions are of significance for determining long-term integrity. However, from an economic decision-making perspective, the goal is to *minimize* the number of monitoring devices to save costs, whereas from the scientific perspective, the goal is to *maximize* the number of monitoring devices to ensure the highest and most rapid control of potential leakage or spillage signatures. Somewhere between these two extremes a practical compromise has to be sought.

1.4 SPILLAGE PROBLEMS

Put to one side all considerations except that of spillage. The main concern here, of course, is that one does not know until after a spill has occurred what sort of remediation needs to be implemented nor the cost of that remediation. It is true that corporations plan for possible disaster scenarios prior to potential spillage occurrences and that they execute simulations to prepare their teams to handle spillage disasters. However, the reality of an actual spill is often far removed from the scenarios envisioned prior to the event, and the preparedness of a team to handle such a situation is often not as good as one would estimate based on prior simulations. Nevertheless, spills do occur, and the main concern is to mitigate, as rapidly as possible, the implied and/or overt threat to the environment. Once this aspect is brought under control, then one has the secondary concern of disposing of the contaminant— whether it is oil, chemicals, gas, radioactive material, or any other substance that can lead to toxic damage to the human population or the environment. Depending on the severity of the spillage problem, the costs of containment and disposal can truly reach astronomical proportions—yet another reason that a company is often reluctant to take on a project without some form of limited-liability clause incorporated into its contract or without an insurance policy covering catastrophic liability costs. Of course, the costs of a liability clause or of an insurance contract have to be included in the cost of the total project. Accordingly, a corporation must allow for these costs when assessing the worth of a contract and also must determine the amount it wishes to bid in the case of a government tender to perform a specific piece of work.

1.5 LEAKAGE PROBLEMS

Concentrate now on system leakage problems. Two concerns are relevant: Is the leakage internal to the storage system as a whole? Has the leakage spread beyond the confines of the storage system? Whatever the cause of the leakage, the leak must be contained to prevent further, and likely more widespread, contamination. In addition, leaked contaminant also must be brought under control and reconfined in the storage system or in another storage system. Depending on the type of contaminant involved, the leakage mechanism, and the later transport of contaminant, control of the leakage and containment of the leaked contaminant can be undertaken in a variety of ways. For instance, consider burial of a container that undergoes corrosion, thereby producing a slow leak of toxic chemicals. Once the monitoring devices recognize the leakage, the question then arises as to how to contain the leakage. Does one disinter the container, make it safe, and rebury it? Does one pump sealing compound around the container, creating an encasement, and so make the container safe? And what does one do to the material that has leaked to make it safe?

1.6 SUMMARY

In short, within the framework of environmental projects, as viewed from the corporate perspective, the objective lies in the optimal allocation of resources. In attempting to evaluate this aspect, a corporation must address the possibilities of the transport, burial, monitoring, and remediation costs due to leakage and spillage potentials within the total scheme of determining whether the corporation should venture into a particular project. It is not at all clear that sufficient scientific information or effective technological solutions are present to provide a well-constrained determination for all projects under all conditions—and often it is clear that such information and technology are not available. Nevertheless, a corporation must still be able to estimate the probability of a profit based on incomplete and uncertain information. In addition, the corporation must be able to evaluate different projects with the same risk-assessment methods so that a direct comparison of projects is possible. In this way, a corporation can then make rational decisions about the risk of each project and the fractional involvement desired.

The purpose of this book is to show how measures for such risk evaluations of environmental projects can be used by a corporation to at least quantify the likelihood of making a profit, the conditions that have to be attached to the probability of profit making as a consequence of the uncertainty of information, and the worth of obtaining additional information in attempting to narrow uncertainty. In addition, the potential for reevaluation and renegotiation of a contract, as a consequence of unknown conditions prior to contract involvement, also must be part of the overall strategy of a corporation. The alternative is that the company could indeed go bankrupt if truly adverse conditions, unsuspected at the time of assessing the original contract conditions for involvement, were to occur later. For instance, in constructing a storage containment for chemical or radioactive waste, it may be that the best prior advice was that there was no significant groundwater flow through the site. However, an undetermined underground fracture or fault system could be encountered during site preparation, in which case site preparation would be considerably more expensive than originally anticipated. It is even possible that the potential site would have to be abandoned and another found at considerable expense of purchase and cost of new site testing. Such considerations can, unless allowed for in contract rediscussions, seriously and adversely influence the ability of a corporation to produce a profit.

Thus the major point to be made is that both scientific and financial evaluations of environmental problems are uncertain. These twin aspects act in concert to create a risk to a corporation such that the profitability of an environmental project may be compromised. This book shows how to allow for these uncertainties in a quantitative fashion and how to provide a rational measure of potential profitability despite such difficulties.

The chapters that follow are arranged so that individual facets of the various decision-making aspects of environmental projects are illuminated as sharply as possible. In this way one can see how the various components of environmental problems fit together within the framework of the general requirement of a corporation to be profitable. It is toward these ends that the overall thrust of this book is aimed.

CHAPTER 2
CONTRACTS AND DECISIONS

In many areas of the world, regulatory control of hazardous waste disposal is held by local, state, or federal authorities. Conventionally, such a government body or agency offers a contract for open bid requiring either transport or burial of waste. A corporation must then decide whether it wishes to bid on all or parts of the contract. Equally, many major corporations (such as chemical, mining, and oil firms) are becoming increasingly sensitive to the long-term effects of liabilities that may arise from past activities or currently operating projects (Ness, 1992; Telego, 1998). In order to control environmental damage, the corporations either perform disposal of the waste themselves or offer contracts to specialist firms to perform the work. In either event, there is transfer of capital for waste disposal, and the corporate division (or outside firm) responsible for the task has to decide if the project can prove profitable.

This chapter considers the simplest situation of a contract being offered under fixed regulations, with a single contract offered per project, with no catastrophic loss events, without inclusion of either optional insurance or of a corporate risk tolerance, and without leakage or spillage occurring. In this respect, the concepts of this chapter can be used at the preliminary stage of the financial analysis of a project and to help expose risk-management groups to the uncertainty factors in different components of a project. As subsequent chapters unfold, more complex situations will be addressed gradually, and each of the preceding factors will be added to the general method discussed here. To set the stage for such developments, the basic situation considered in this chapter will serve as an underpinning framework.

2.1 GENERAL DEVELOPMENT

Consider that a contract is offered at a fixed price G for the transport and long-term storage of hazardous waste, including requirements on storage monitoring for a period of t years after emplacement of the waste material. The contract includes, additionally, storage-site preparation. A corporation must then determine whether the fixed-price contract can return a profit before it bids on the contract. The corporation must, of course, quantify all the factors listed in the previous paragraph; these will be dealt with later in this book and are ignored in this chapter.

The corporation assesses the cost of transportation on a unit of waste. This unit may be a volume or weight measure, or a fixed number of bequerels in the case of radioactive waste, or some other measure depending on the material being shipped. Let the transport cost per unit be C_T so that in transporting n units the total transport cost is nC_T. In the disposal component of the project, there are the fixed costs of site preparation F_s and the disposal costs D_c per unit transported for a total disposal cost of $D_T \equiv F_s + nD_c$. In addition, the contract requires monitoring after storage. This monitoring requirement includes two components: the fixed cost of all the monitoring devices M_1 and the ongoing costs of monitoring, maintenance, and replacement of failed devices. Let these latter costs be reckoned on a per-year basis at m so that after t years the total monitoring cost is $M \equiv M_1 + mt$.

A simple calculation of raw potential profit P to the corporation after t years is then

$$P_{\text{raw}} = G - nC_T - D_T - M \tag{2.1}$$

where all amounts are calculated in fixed-year dollars. There is also a discount factor that needs to be included because of corporate salaries, taxes, and corporate overhead for buildings, etc. Take the salaries to be S per year and the total corporate overhead per year to be H. Then, after t years, the raw potential profit is reduced by $(S + H)t$, yielding a potential profit of

$$P = G - nC_T - (F_s + nD_c) - (M_1 + mt) - (S + H)t \tag{2.2}$$

If the profit P is positive, then taxes must be paid. For simplicity, let the tax rate be a fixed fraction u of the profit P. Also assume that the n units of waste are transported and buried at a fixed rate r per year so that one can write $n = rt$. Then the net profit NPV to the corporation is given by

$$\text{NPV} = (1-u)\,\{G - (F_s + M_1) - t[r\,(C_T + D_c) + m + S + H]\} \tag{2.3}$$

provided that NPV > 0.

Inspection of Eq. (2.3) shows that two factors control NPV and whether it is positive or not. First, note that the total fixed costs of preparing and monitoring the site $F_s + M_1$ must be less than the contract price G, or there is no hope of making a profit. Second, note that even when $F_s + M_1 < G$, the ongoing per-year costs of unit transportation and storage, monitor maintenance, and salaries plus overhead eventually will drive the NPV negative after a time t_*, given by

$$t_* = \frac{G - (F_s + M_1)}{r\,(C_T + D_c) + m + S + H} \tag{2.4}$$

Thus there is a critical project lifetime t_* that returns NPV = 0 beyond which the project will result in a net loss. Clearly, in order for the project lifetime to extend longer than the monitoring time t_M, which is mandated in the contract, two factors must be considered: G must exceed the fixed costs $F_s + M_1$, and when such is the case, the costs of transport and storage per year plus the cost of yearly maintenance monitoring plus the cost of salaries and corporate overhead must be kept sufficiently small so that $t_* > t_M$. If such is not the case, then no profit can accrue, and the corporation loses money on the contract.

2.2 PARAMETER UNCERTAINTIES

Now the problem with this simplistic estimate, notwithstanding the fact that one has ignored all more involved factors, is that the fixed costs and yearly costs are not known ahead of their occurrence. Thus, for instance, until the project is completed, one does not know the yearly maintenance cost of monitoring. Or again, until after the disposal site has been prepared, one does not know (as opposed to estimate) the fixed costs of site preparation. For each component entering Eqs. (2.3) and (2.4), the same uncertainty on costs prevails. Generally, one has some relatively good idea of the *range* of potential costs, but the exact value of these costs is not known with precision (Yeo, 1990). Thus, in 1997, the average cost for private-sector environmental remediation projects in the United States was 25 to 50% over the initial budget (Al-Bahar and Crandell, 1990; Diekmann and Featherman, 1998).

Accordingly, there is an associated uncertainty on both the NPV and the critical project lifetime t_*. In the case where the NPV is strongly positive and t_* is much greater than the mandated monitoring time t_M, it is likely that NPV and the difference $t_* - t_M$ will be positive even when the ranges of uncertainty on the parameters are included. However, for situations where NPV and $t_* - t_M$ are only slightly positive, inclusion of the uncertainty in the parameters can end up providing cases where there are negative values for NPV and $t_* - t_M$. Thus the question devolves into one of estimating the probability of obtaining a positive worth NPV and a positive $t_* - t_M$ so that the corporation can assess the risk of making a profit on the environmental project.

To illustrate the influence of uncertainties on the NPV and the value of t_*, suppose then that the fixed component F_s of the site-preparation costs is unknown but is estimated to range from $F_{s,\min}$ to $F_{s,\max}$ with a most likely value of F_L. For the purposes of the illustration, take it that all other parameters are statistically sharp with no uncertainty; i.e., all other costs are known precisely in advance. The actual statistical distribution of F_s is, in general, unknown, so one has to make some assessment of the statistical uncertainty of F_s based on the three values $F_{s,\min}$, $F_{s,\max}$, and F_L. Appendix 2A presents a discussion of how one can perform this sort of estimation together with how one also can generate equivalent statements on means and uncertainties for parameters that occur as sums (differences) and products (ratios). The results and methods in App. 2A will be used throughout this book.

From Eqs. (2.27) and (2.28) of App. 2A it is then possible to provide an estimate of the mean value of F_s, that is, $<F>$, and of the variance of F_s, that is, $<\delta F^2>$. Note that $<\delta F^2> = <F^2> - <F>^2$, where angular brackets are reserved for statistical averages. Inspection of Eq. (2.3) shows that NPV is linear in F_s, so one can immediately write the expected value of NPV as

$$<NPV> = (1 - u)\{G - M_1 - t[r(C_T + D_c) + m + S + H]\} - (1 - u)<F> \qquad (2.5)$$

and the variance in the expected NPV is

$$<\delta NPV^2> = (1 - u)^2 <\delta F^2> \qquad (2.6)$$

Based solely on the mean $<NPV>$ and variance $<\delta NPV^2>$, one can write an equivalent gaussian cumulative probability of obtaining an NPV in excess of a preset value V as

$$P(NPV > V) = (2\pi <\delta NPV^2>)^{-1/2} \int_V^\infty \exp\left(\frac{-(x - <NPV>)^2}{2 <\delta NPV^2>} \right) dx \qquad (2.7)$$

If one asks for the cumulative probability $P(0)$ of making a profit greater than zero, performing the substitution $x = <NPV> + (2<\delta NPV^2>)^{1/2}y$ allows Eq. (2.7) to be written in the simpler form

$$P(0) = \pi^{-1/2} \int_{-b}^\infty \exp(-y^2) \, dy \qquad (2.8)$$

where

$$b = \frac{<NPV>}{(2 <\delta NPV^2>)^{1/2}} \equiv \frac{1}{(\sqrt{2} \, v)}$$

Thus the probability of making some profit is controlled not only by the expected NPV but also by the standard error of the expected value. A large positive value of b in Eq. (2.8) means that the ratio of $<NPV>$ to the standard error $<\delta NPV^2>^{1/2}$ is large and positive, so there is a high probability of making some profit, whereas a small value of b ($b \ll 1$) implies that the uncertainty on the expected NPV is large and that the probability $P(0)$ is close to 0.5; i.e., there is only about a 50% chance of profitability. The risk of the project is then measured by the volatility v. If v is large (positive or negative), then there is considerable uncertainty on $P(0)$, whereas a small value of v implies a relatively sharply defined value of $P(0)$, i.e., one with less risk.

Most corporations set a minimum internal risk factor in the form of a minimum acceptable chance MAC to make a profit. If the value of $P(0)$ for a project falls below a predetermined value of MAC, then the corporation usually will not invest in a project without some compelling reason. Thus $P(0)$ is an extremely valuable measure of worth of a project to a corporation, as is the volatility v of each project.

The particular illustration here has considered only variations in the range of F_s in order to illuminate the general procedure succinctly. In a more pragmatic situation, one should, and does, include all variations of all parameters. The same sense of measure can be applied to the lifetime t_* to zero profit versus the mandated monitoring time t_M. Here, considering F_s again to be the only uncertain parameter yields the expression

$$<t_*> = \frac{G - M_1 - <F>}{r\,(C_T + D_c) + m + S + H} \tag{2.9}$$

for the average value of t_*, whereas the variance $<\delta t_*^2>$ around the mean $<t_*>$ is given by

$$<\delta t_*^2> = \frac{<\delta F^2>}{[r\,(C_T + D_c) + m + S + H]^2} \tag{2.10}$$

Thus the cumulative probability that the project will have a profitable lifetime in excess of t_M is

$$P(t > t_M) = (2\pi <\delta t_*^2>)^{-1/2} \int_{t_M}^{\infty} \exp\left(\frac{-(x - <t_*>)^2}{2 <\delta t_*^2>}\right) dx \tag{2.11}$$

Writing $x = <t_*> + [2<\delta t_*^2>]^{1/2} y$ enables Eq. (2.11) to be written in the form

$$P(t > t_M) = \pi^{-1/2} \int_{-B}^{\infty} \exp\left(-y^2\right) dy \tag{2.12}$$

where

$$B = \frac{<t_*> - t_M}{[2\,(\delta t_*)^2]^{1/2}} \tag{2.13}$$

Thus, if $<t_*>$ is greater than t_M, then the cumulative probability that the profitable lifetime will exceed the mandated monitoring time t_M is greater than 50% (and may be much greater if $B \gg 1$). Equally, from a corporate perspective, one could set a minimum chance P_{min} that the lifetime should exceed t_M. In this case, one writes

$$P_{min} = \pi^{-1/2} \int_{-B}^{\infty} \exp\left(-y^2\right) dy \tag{2.14}$$

and then uses the preset value of P_{min} to determine B from Eq. (2.14). In this way one comes up with an inequality relationship between $<t_*>$ and $<\delta t_*^2>$ such that one can indeed satisfy the minimum chance P_{min}. Hence one can set limits on the allowed range of fluctuations in F_s so that the corporate objectives can be achieved.

The point about these simple illustrations is to show how the basic tools can be developed and used to assess the worth to a corporation of becoming involved in a project. Of particular concern are two major components: (1) to provide some assessment of the project and the NPV subject to the constraints imposed (in the simple illustration the constraint was that monitoring be carried out for a specific, mandated time after storage), (2) to illustrate how to assess the probability of making at least some profit as well as to determine the probability of honoring the mandated constraint when there is uncertainty in parameters entering the cost assessment. In order to show these aspects as simply as possible, all other considerations involved in environmental projects were ignored. In addition, only one parameter (the fixed cost of storage-site preparation) was allowed to be uncertain. In general, however, one cannot ignore problems of potential spillage, multiple projects, limited-liability insurance, corporate risk aversion, fixed budgets, etc., nor can one ignore the problems of multiparameter uncertainty, which may be correlated or independent, in influencing the corporate decision on project involvement. The later chapters in this volume will deal progressively with such concerns and will steadily and systematically include multiple factors influencing decisions.

2.3   NUMERICAL EXAMPLE

To show how the mathematical expressions of this chapter can be used, consider the situation where a government offers a contract at a fixed price of $\$10^9$ to transport and store safely radioactive waste with ongoing monitoring required for a 20-year period.

A corporation estimates total transportation and storage costs for the radioactive waste of $100 million, monitoring equipment costs of $50 million, yearly maintenance of the monitoring program at 10% of the equipment costs, corporate overhead and salaries for the 20-year period at $10 million per year, and storage-site preparation costs ranging between $100 million and $500 million, with a most likely estimate of about $200 million. The corporate decision makers have to decide whether the project is worthwhile.

From the preceding information, the parameters take the following values:

$$G = \$10^9$$

$$tr(C_T + D_c) = \$0.1 \times 10^9$$

$$m = \$5 \times 10^6$$

$$M_1 = \$5 \times 10^7$$

$$t(S + H) = \$10^7 t$$

where t is time (in years).

From App. 2A, the mean cost of site preparation is estimated to be

$$<F> = \tfrac{1}{3}(100 + 200 + 500) \times 10^6 = \$2.67 \times 10^8$$

and the standard error around the mean is

$$<\delta F^2>^{1/2} = \$8.54 \times 10^7$$

If one were to use *only* the mean site-preparation cost, then after t years the mean NPV to the corporation before taxes [Eq. (2.5) with $u = 0$] is given by

$$<\text{NPV}> = \$(5.83 \times 10^8 - 1.5 \times 10^7 \times t)$$

After 20 years of monitoring ($t = 20$), the expected NPV becomes

$$<\text{NPV}> = \$2.83 \times 10^8$$

so that, on average, the project appears profitable to the corporation. But consider now the influence of fluctuations in NPV brought about by uncertainty in the site-preparation costs. Then the variance in NPV around the mean [Eq. (2.6) with $u = 0$] is given by

$$<\delta\text{NPV}^2> = <\delta F^2> = 7.3 \times 10^{15} \qquad \text{(in dollars squared)}$$

Thus one has the ratio

$$\frac{<\text{NPV}>}{[2 <\delta\text{NPV}^2>]}{}^{1/2} = 2.36$$

so the volatility v is just $v = 0.30$.

The probability $P(0)$ of the corporation making a profit greater than zero is then

$$P(0) = \pi^{-1/2} \int_{-2.36}^{\infty} \exp(-y^2)\, dy = 0.95$$

Thus there is an extremely high probability (95%) of a profit even when allowance is made for the large range of uncertainty in site-preparation costs.

2.4 DISCUSSION

Despite the high probability of the project proving successful, the corporate decision makers would view askance any such calculation presented to them. One reason, of course, is that the complications due to potential spillage and leakage are not included. In addition, no mention of corporate liability is included, nor, for that matter, is any assessment given of the likely cost of insurance coverage. Further, it is a rare situation when a corporation is assessing only one project at a time so that no comparison is available of relative worth and relative risk of multiple projects. Two further matters complicate the issue. A corporation has only a fixed budget in any given year. If the contract funding is *not* awarded ahead of project commencement (a common situation), then the corporation must commit its own funds to the project, with partial contract payment as each agreed milestone is met. The problem for the corporation is to apportion its budget between various projects in order to succeed. This apportionment may mean that the corporation cannot take 100% interest in each and every project; the corporation must settle for a smaller working interest. Of concern to the decision makers is how one assesses the best fractional involvement in each project in order to maximize profit. The second complicating issue is the risk-aversion philosophy of the corporation. If the corporate outlay for a project is a significant fraction of the total corporate worth, or if the liability cost of potential spillage and leakage operations could bankrupt the corporation, or if the government changes rules for transport and storage requirements in the middle of a project (thereby involving the corporation in major unanticipated costs that are, again, a significant fraction of corporate worth), or if liability for cleanup of a prior project is later mandated, then under such conditions the corporation generally would limit its risk of financial ruin. This risk limitation normally is accomplished in one of two main ways within a corporation: a risk-tolerance monetary value is set as an internal limit such that projects that are likely to involve costs exceeding the risk tolerance are either not considered or only a fractional working interest is taken in such projects, or liability insurance is taken out to limit the risk to the corporation, or a combination of both.

The calculations in this chapter can serve as a preliminary assessment of the worth of projects and for the development of some form of ranking order from which a more detailed analysis of each project can be carried through. Thus Eqs. (2.3) and (2.4) can provide an initial evaluation of the net profit of a project NPV in the absence of uncertainties and of the critical project lifetime t_* over which a project that includes monitoring costs remains positive. Section 2.2 and Eqs. (2.5) to (2.13) introduce a preliminary uncertainty evaluation of the net profit and the critical lifetime of a project, respectively, by considering the situation where the costs of a component of a project are not known in precision but may range significantly. The detailed analyses that will be developed in the following chapters *will* allow for all the complications and uncertainties that have been intentionally ignored in this chapter.

APPENDIX 2A: SOME PROPERTIES OF A LOG-NORMAL DISTRIBUTION

2A.1 Exact Statements

Consider the log-normal probability distribution (Mood et al., 1974; Aitchison and Brown, 1999)

$$p(x|a,\ \mu)\ dx \propto \exp\left[\frac{-(\ln x - a)^2}{2\mu^2}\right]\frac{dx}{x} \tag{2.15}$$

where a and μ are fixed parameters, the variable x ranges in $0 \le x \le \infty$, and $p(x|a,\ \mu)$ is the differential probability of finding the value x in the range x to $x + dx$.

   With the normalization

$$\int_0^\infty p(x|a,\ \mu)\ dx = 1 \tag{2.16}$$

it follows that

$$p(x|a,\ \mu)\ dx = \frac{1}{2}\ \pi^{-1/2}\mu^{-1} \exp\left[\frac{-(\ln x - a)^2}{2\mu^2}\right]\frac{dx}{x} \tag{2.17}$$

The mean value of x^n, that is, $E_n(x)$, is given by

$$E_n(x) = \int_0^\infty x^n p(x|a,\ \mu)\ dx \equiv \exp\frac{an + n^2\mu^2}{2} \tag{2.18}$$

Thus

$$E_1(x) = \exp\frac{a + \mu^2}{2} \tag{2.19a}$$

$$E_2(x) = \exp(2a + 2\mu^2) \tag{2.19b}$$

Hence

$$\mu^2 = \ln\frac{E_2(x)}{E_1(x)^2} \tag{2.20}$$

It also follows that the value $x_{1/2}$ at which

$$\int_0^{x_{1/2}} p(x|a,\ \mu)\ dx = 0.5 \tag{2.21}$$

is given by

$$a = \ln x_{1/2} \tag{2.22}$$

so that

$$E_1(x) = x_{1/2}\ \exp\frac{\mu^2}{2} > x_{1/2} \tag{2.23}$$

Consider the cumulative probability

$$P(x|a, \mu) \equiv \int_0^x p(u|a, \mu)\, du \tag{2.24}$$

which provides the probability of obtaining a value less than or equal to x. On $x_1 = x_{1/2}\exp(\mu)$, $P(x_1|a, \mu)$ is given through

$$P(x_1|a, \mu) = \tfrac{1}{2} + (2\pi)^{-1/2}\int_0^{2^{-1/2}} \exp(-s^2)\, ds \cong 0.84 \equiv P(84) \tag{2.25a}$$

which is independent of a and μ, whereas on $x_2 = x_{1/2}\exp(-\mu)$, $P(x_2|a, \mu)$ is given through

$$P(x_2|a, \mu) = 1 - P(x_1|a, \mu) \cong 0.16 \equiv P(16) \tag{2.25b}$$

which is also independent of a and μ.

Thus, on $P(16)$, one has

$$x = E_1(x)\exp(-\mu - \mu^2/2) < x_{1/2} \tag{2.26a}$$

On $P(84)$, one has

$$x = E_1(x)\exp(\mu - \mu^2/2) > x_{1/2} \tag{2.26b}$$

On $P(50)$, one has

$$x = x_{1/2} = E_1(x)\exp(-\mu^2/2) < E_1(x) \tag{2.26c}$$

Note that for $\mu < 2$ [that is, $E_2(x) < e^4 E_1(x)^2$], $E_1(x)$ occurs between $P(50)$ and $P(84)$, whereas for $\mu > 2$, $E_1(x)$ occurs at greater than $P(84)$. For $\mu \lesssim 2$, a good pragmatic approximation is that on $x = E_1(x)$, $P[E_1(x)|a, \mu]$ occurs at about 68% [i.e., about midway between $P(50)$ and $P(84)$]. A slightly better approximation is that E_1 occurs at a cumulative probability of about $(50 + 17\mu)\%$.

2A.2 Approximate Statements

Empirically, it is often difficult, if not impossible, to obtain enough information to determine the precise shape of the frequency distribution of a particular parameter or variable. Indeed, quite often it is considered a fairly good achievement to be able to estimate a likely minimum x_{min}, a likely maximum x_{max}, and a likely most probable value x_p for a parameter. A rough idea of relevant mean and variance can then be obtained from Simpson's triangular rule (Press et al., 1987). We then have

$$E_1(x) \cong \tfrac{1}{3}(x_{min} + x_p + x_{max}) \tag{2.27a}$$

$$E_2(x) = E_1(x)^2 + \sigma^2 \tag{2.27b}$$

with

$$\sigma^2 \cong \tfrac{1}{2}E_1(x)^2 - \tfrac{1}{6}[x_{min}x_{max} + x_p(x_{min} + x_{max})] > 0$$

$$\equiv \tfrac{1}{18}\{[x_p - \tfrac{1}{2}(x_{min} + x_{max})]^2 + \tfrac{3}{4}(x_{max} - x_{min})^2\} \tag{2.27c}$$

If it is further assumed that the variable is approximately log normally distributed, then

$$\mu \cong \left[\ln \frac{1 + \sigma^2}{E_1(x)^2}\right]^{1/2} \tag{2.28}$$

so that an equivalent log-normal approximate distribution can be constructed using Eqs. (2.26) together with the estimates of $E_1(x)$ from Eq. (2.27a) and μ from Eq. (2.28).

2A.3 Multiple-Parameter Distributions

In assessments of exploration economic objectives and scientific basin analysis outputs, many parameters occur, either alone or in combination with other parameters, and each of the parameters has its own uncertainty. We need to have available practical procedures for estimating the combined effects of uncertainty of parameters on an exploration project.

Two sorts of fundamental parameter combinations seem to be prevalent: sums of parameters and products or rates of parameters. Consider each in turn.

Sums of Parameters.   It is well known that two or more independent random variables A and B, both with *normal* distributions, combine to give a sum ($A \pm B$) that is also precisely normally distributed with mean value

$$E_1(A \pm B) = E_1(A) \pm E_1(B) \tag{2.29a}$$

and with variance

$$\sigma(A \pm B)^2 = \sigma(A)^2 + \sigma(B)^2 \tag{2.29b}$$

Empirically, it appears that N independent random variables from any frequency distributions (not necessarily normally distributed) add to give a sum S_N ($\equiv x_1 \pm x_2 \pm x_3 \pm \cdots \pm x_N$) that is approximately normally distributed as N becomes large, with mean value

$$E_1(S_N) \cong E_1(x_1) \pm E_1(x_2) \pm \cdots \pm E_1(x_N) \tag{2.30a}$$

and with variance

$$\sigma(S_N)^2 = \sum_{i=1}^{N} \sigma(x_i)^2 \tag{2.30b}$$

Products of Parameters. It is well known that multiple independent random variables X, Y, Z,... from *log-normal* distributions combine in generic product form $X^a Y^b Z^c \cdots$ to give a distribution for the product that is also precisely log-normally distributed with mean value

$$E_1(X^a Y^b Z^c \cdots) = E_1(X^a) \, E_1(Y^b) \, E_1(Z^c) \cdots \tag{2.31a}$$

with second moment

$$E_2(X^a Y^b Z^c \cdots) = E_2(X^a) \, E_2(Y^b) \, E_2(Z^c) \cdots \tag{2.31b}$$

and scale factor μ given through

$$\mu^2 = \ln \frac{E_2(X^a Y^b Z^c \cdots)}{E_1(X^a Y^b Z^c \cdots)^2} \tag{2.31c}$$

Empirically, it appears that N independent random variables from any frequency distributions (not necessarily log-normally distributed) tend to combine to produce a product $P_N \, (\equiv X_1^a X_2^b X_3^c \cdots X_N^d)$ that is approximately log-normally distributed as N becomes large, with mean value

$$E_1(P_N) \cong E_1(X_1^a) \, E_1(X_2^b) \cdots \tag{2.32a}$$

and scale parameter

$$\mu^2 \cong \ln \frac{E_2(P_N)}{E_1(P_N)^2} = \sum_{i=1}^{N} \mu_i^2 \tag{2.32b}$$

where

$$\mu_i^2 = \ln \frac{E_2(X_i^p)}{E_1(X_i^p)^2} \tag{2.32c}$$

These results will be used throughout this book.

CHAPTER 3
TRANSPORT AND BURIAL HAZARDS OF RADIOACTIVE WASTE

3.1 INTRODUCTION

In most practical applications, a decision is required among different actions when unknown or unpredictable elements are involved in the analysis of the situation at hand. Such examples include construction of a dam when the magnitudes of future flood events are uncertain, choices between alternative foundation designs for a bridge when the supporting capacity of the soil is not completely known, and selection of the location and construction materials of a landfill when the hydrogeologic conditions are subject to uncertainty.

Traditionally, these kinds of problems are treated in the context of bayesian statistical decision theory. A basic element of the structure of such problems includes a set of alternative actions among which the selection is made. The choice of a particular action over another depends on the "state of nature," such as the magnitude of the maximum flood, the supporting capacity of the soil, or the hydrogeologic conditions. The various states of nature are subject to uncertainty, and probability values are assigned to them. The choice of each action leads to a different numerical measure of worth (say, dollars) for the action-state pair.

In decision analyses of engineering projects, the most prevalent criterion for reaching a rational decision has been that of the maximum expected monetary value (MEMV) (Benjamin and Cornell, 1970; Berger, 1985). Based on the concept of the "fair" price for participation in a long sequence of games, this criterion reduces the analysis of a decision-making situation to calculation of the expected monetary return of the foreseeable options and selection of that action which returns the MEMV. This procedure has been the practice in analyses of flood risk and dam failure (Goldman, 1997; Thompson et al., 1997), aquifer remediation (James et al., 1996), and geologic repositories design (Hunter and Mann, 1992). Reliance on the MEMV metric has persisted despite advances in other rules, such as those of the utility function (Krzysztofowicz, 1986; Eiser and van der Pligt, 1988; Raiffa, 1997) or of stochastic dominance (Keeney and Raiffa, 1993; Clemen, 1996). These rules seek to incorporate a decision maker's attitude toward risk (Lerche and MacKay, 1999) and provide a framework where considerations not easily measurable in monetary terms (de Souza Porto and de Freitas, 1996) can be taken into account. The need for such additional statistical criteria in the decision-making process has been noted by Burt and Stauber (1971), Loaiciga and

Marino (1986), Bouchart and Goulter (1998), and Lerche and MacKay (1999), who sought to develop procedures that reduce the variability of returns while maximizing expected revenues.

In this chapter it is shown that for environmental projects, which are frequently concerned with the long-term impact of low-probability and high-cost catastrophic events, the MEMV metric alone is not an appropriate measure of the worth of a project. The use of three simple statistical measures is a first step toward gaining insight into the decision-making analysis. Through a numerical illustration it is shown that these metrics may lead to a different selection of the optimal action than that suggested by the MEMV and that they guide one toward an analysis that accounts for the risk aversion of the decision maker (Clemen, 1996; Lerche and MacKay, 1999). A second theme of this chapter is the investigation of the impact of catastrophic scenario games on the decision-making process. The incorporation of a catastrophic event in the decision analysis can alter the perception of the profitability of projects and may lead a corporation away from potentially profitable ventures or toward actions that undermine the profitability of projects. This chapter provides a methodology for the analysis of cases where the parameters of a problem, such as probabilities of future events and estimates of benefits and costs, are not assigned statistically sharp values but rather a range of values. The procedure provides the relative contribution of each component to the overall system uncertainty. Thus one can effectively establish the set of components for which uncertainty reduction is most critical in the economic analysis of a project. These concepts can be applied equally well to uncertainty analyses of physical problems, as is illustrated with an application from the field of subsurface hydrology.

3.2   RISK AND CATASTROPHE IN THE TRANSPORT OF RADIOACTIVE WASTE

Consider the hypothetical situation of a disposal facility for hazardous or radioactive wastes where the appropriate level for a contract bid has to be selected by a consortium of contracting companies. Three main components are considered in the analysis: (1) transportation of the wastes from the location of origin (or temporary storage) to the proposed operating facility, (2) construction of the facility and burial of the transported material, and (3) operation of the facility and monitoring of the wastes for a fixed period of time. The action space A contains action a_1, which involves putting forward a bid for the construction, transportation, and burial of the wastes; and action a_2, which, additionally, includes operation of the facility and monitoring for t years. By *actions* are meant the various distinct alternatives a primary contracting company may evaluate for bidding on a project. Thus, for example, selection of a_2 over a_1 commits resources that could be used in other projects, or alternatively, if a_1 is found more profitable than a_2, a company may decide to proceed by subcontracting, purchasing insurance, or negotiating separately for the monitoring component of the project.

Of concern in the transport of wastes are two factors: whether one can (or does) transport the wastes to the facility from the originating site without a spill at some cost C_1, and if there is a spill during transport, then not only must the original material be re-collected and transported, but so must any other contaminated material. Depending on the severity of the spill, the cleanup process can increase the transportation cost to C_1' (limited spill) or task the whole project with an enormous remediation expenditure C_1'' (catastrophic spill). Designate the probabilities of transport without spill, with limited spill, and with catastrophic spill by p_s, p_f, and p_k, respectively.

Once the waste material reaches the facility, there is a direct cost C_3 for burial (C_3 also incorporates all costs for construction of the facility itself). In the case where a spill occurred during transport, the cost would be higher (at C_3' for a limited spill and C_3'' for a catastrophic spill, respectively) because material contaminated at the spill site also must be emplaced at the facility, with requirements for additional space, etc. In the case where the facility is operated and the buried material is being monitored for leakage (because of inadequately engineered barriers, corrosion of containers, unforeseen conditions, etc.) over a period of t years, there is a cost M per year in the case of only the initially transported material and higher costs M' and M'' for the cases of limited and catastrophic spills during transportation, respectively, because more material has to be monitored over the years.

These various probability branches and associated costs are depicted in Figs. 3.1 and 3.2 together with the contract bid B offered by a company. All cost values are considered in fixed year dollars. Figure 3.1 represents the case where the notion of failure of a project is viewed within the context of controlled, limited consequences where, for example, frequencies and associated costs can be defined to a satisfactory degree through the study of similar situations, historical data, etc. By contrast, Fig. 3.2 includes a catastrophic scenario where, in general, the establishment of probabilities of failure and costs cannot be predicted scientifically. Because the interest here is to study the influence of catastrophic events on the perceived profitability of projects, the probability of success p_s is kept the same ($p_s = 0.9$) in the two figures, and the probability of limited spill has been perturbed by only 0.1% to allow for the inclusion of a catastrophic spill in the second figure. This study is not meant to provide an exhaustive analysis of the hazardous and radioactive wastes transport and burial problem (such as can be found in Hunter and Mann, 1992; U.S. DOE, 1994; Sandia, 1994) but rather to illustrate the application of the preceding concepts to the decision-making process.

3.3 STATISTICAL MEASURES

Consider four statistical measures for the analysis of the optimal action. The expected value E of each branch that corresponds to an action, the variance σ^2 as a measure of the accuracy of the expected value, and the volatility v defined by

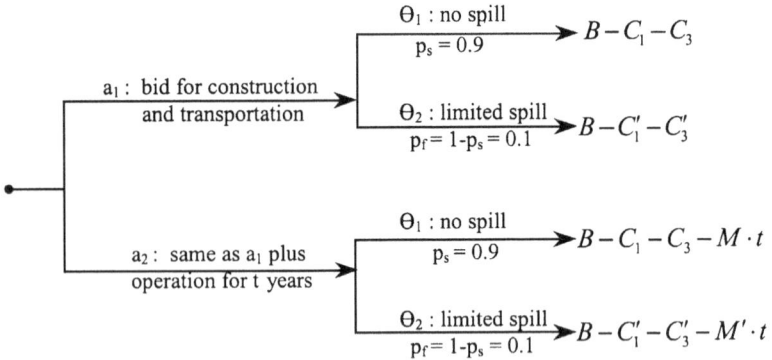

FIGURE 3.1 Decision tree: Contract choice in the presence of limited spill.

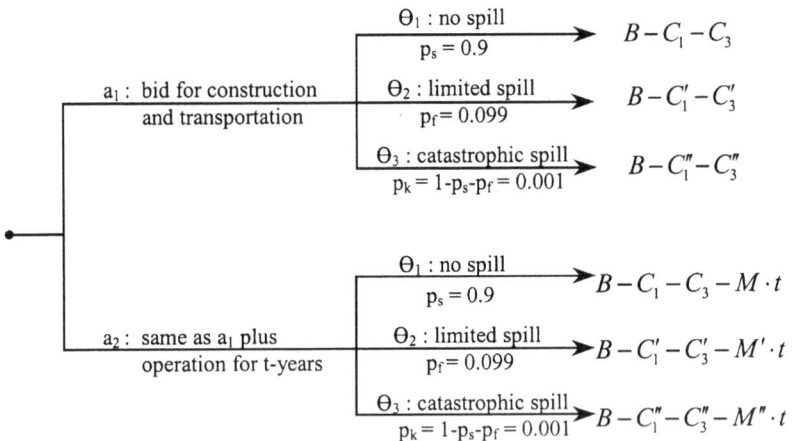

FIGURE 3.2 Decision tree: Contract choice in the presence of catastrophic spill.

$$v = \frac{\sigma}{|E|} \tag{3.1}$$

a measure of the worth of the expected value. A low value of v ($v \ll 1$) implies but little uncertainty on E, whereas $v \gg 1$ implies a large uncertainty.

Based on both the expected value E and the variance σ^2, one can write an equivalent gaussian probability value $P(V)$ of obtaining a value greater than or equal to V as

$$P(V) = 1 - \frac{1}{\sqrt{2\pi}} \int_{-\infty}^{b} \exp\left(-\frac{u^2}{2}\right) du \tag{3.2}$$

where the standardized variable u (mean 0 and standard deviation 1) and $b = (V - E)/\sigma$ are used. Note that if one wants a value V greater than or equal to the mean E (that is, $V = E$) then Eq. (3.2) returns $P(E) = \frac{1}{2}$, as it should.

Conventionally, one often considers the minimal situation of a value greater than zero ($V = 0$) as relevant, in which case Eq. (3.2) can be written

$$P(0) = \frac{1}{2} - \frac{1}{\sqrt{2\pi}} \int_0^b \exp\left(-\frac{u^2}{2}\right) du \qquad (3.3)$$

where, now, $b = -E/\sigma = -(v)^{-1}$. Thus a high volatility ($v \gg 1$) makes $|b| \ll 1$, and then, through Taylor series expansion of the integrand, one obtains the following approximation to Eq. (3.3):

$$P(0) \approx \frac{1}{2} - \frac{1}{\sqrt{2\pi}}\, b \qquad (3.4)$$

A low volatility ($v \ll 1$) makes $|b| \gg 1$, and then, through asymptotic expansion of the integral (Bleistein and Handelsman, 1986), App. 3A, a good approximation is

$$P(0) \sim \frac{1}{2}\,[1 + \text{sgn}\,(E)] + (2\pi)^{-1/2} \exp\left(-\frac{b^2}{2}\right) b^{-1} \qquad (3.5)$$

The preceding cumulative probability expressions can be used in two ways: (1) for a given set of parameters of the decision-tree diagram of Fig. 3.1 (or Fig. 3.2), one can evaluate E, σ, and so b and then calculate the corresponding probability $P(0)$ that the contract bid B will exceed or at least cover all possible costs, or (2) one can require that a given probability $P(0)$ be enforced and then determine the b value needed and so determine parameters in the decision-tree diagram in order that one has, say, a 90% chance [$P(0) = 0.9$] of the contract bid covering all costs. The latter situation arises when a corporation mandates a minimum probability of success P_{min} for setting forward a bid for an environmental project. Assuming a profit-bearing situation with at least 50% probability of success requires that the upper limit b in Eq. (3.2) be negative, i.e., that $E \geq 0$. Furthermore, designation of an acceptable probability level P_{min} determines a value of $|b_{min}|$ that, in turn, requires that $E/\sigma \geq |b_{min}|$. Using the symbol $E_2 = E(x^2)$ for the second moment, one obtains the inequality

$$E^2 \geq b_{min}^2 \sigma^2 = b_{min}^2\,(E_2 - E^2) \qquad (3.6)$$

which, with rearrangement of terms, becomes

$$E_2 \geq E^2 \geq \left(\frac{b_{min}^2}{1 + b_{min}^2}\right) E_2 \qquad (3.7)$$

Here $\sigma^2 = E_2 - E^2$ and $\sigma^2 \geq 0$. Inequality (3.7) determines whether the parameters of a problem (B, C, p_s) return an average profit that is in accord with an acceptable corporate level of probability of success. As shown below, both procedures provide insight into the sensitivity of the system to the chosen parameter values.

3.4 ANALYSIS OF ALTERNATE ACTIONS: LIMITED SPILL

Consider first the case where a bid is offered only for the transportation and burial of wastes (action a_1). The expected value is then given by

$$E^{a_1} = B - C_1' - C_3' + p_s(C_1') + p_s(C_3' - C_1 - C_3) \qquad (3.8)$$

The minimum bid that can be offered is $B = C_1 + C_3$, corresponding to zero profit under the optimal conditions. The minimum optimal bid is $B = C_1' + C_3'$, corresponding to a break-even situation under the worst anticipated conditions, whereas the contract under which, on average, one can expect a no-loss situation ($E^{a_1} \geq 0$) is $B = C_1' + C_3' - p_s(C_1' - C_1) - p_s(C_3' - C_3)$. Take the fixed costs of transportation and burial to be $C_1 = 0.2$, $C_3 = 0.5$ (in millions of dollars), respectively, and $C_1' = 2C_1$ and $C_3' = 2C_3$ (cost of transportation in the event of spill increases by a factor of 2, followed by a similar increase in the burial component), the expected value of action a_1 for a minimum bid of $B_{min} = 0.7$ is $E^{a_1}_{min} = -0.07$, $\sigma_{min} = 0.21$, $v = 3$, and $P(0) = 37\%$. The expected value and standard deviation for bids that are multiples of B_{min}, $B_{t=0} = kB_{min}$ with $k \geq 1$, are given by $E' = E^{a_1}_{min} + (k - 1)B_{min}$ and $\sigma' = \sigma_{min}$. Thus an increase of the contract award improves the expected return, but the uncertainty on the outcome remains the same. Depicted in columns 1, 2, 5, and 6 of Table 3.1 are the effects on the expected value E^{a_1}, v, and $P(0)$ of different levels of bids $B_{t=0}$ normalized by the minimum costs $B_{min} = 0.7$.

The second moment $E_2^{a_1}$ is given by $E_2^{a_1} = p_s(B - C_1 - C_3)^2 + (1 - p_s)(B - C_1' - C_3')^2$. Assume that a corporation mandates a profit-bearing situation with a 75% probability of success. Then, from tables of the standardized normal distribution (Benjamin and Cornell, 1970), one obtains the value of $|b_{min}| = 0.675$, and for the fixed costs of the study, one can easily calculate values of E^{a_1} and $E_2^{a_1}$ for various bids. Inequality (3.7) is then satisfied for contract awards exceeding $1.30B_{min}$, as corroborated by Table 3.1. Thus inequality (3.7) provides a quick procedure for estimating the appropriate cutoff level of bids that are within the risk tolerance of a corporation (Clemen, 1996; Lerche and MacKay, 1999).

Correspondingly, one can evaluate the effect of operating the facility and monitoring the wastes for periods of $t = 5$, 10, and 20 years. The expected value for action a_2 is given by

$$E^{a_2} = E^{a_1} + [(M' - M)p_s - M']t \qquad (3.9)$$

TABLE 3.1   Limited Spill: Statistical Measures for Actions a_1 and a_2 ($t = 20$ y)

$B_{t=0}/B_{min}$	E^{a_1}	$B_{t=20}/B_{min}$	E^{a_2}	v	$P(0)$, %
1.00	−0.070	1.29	−0.087	3.00	37
1.02	−0.056	1.32	−0.066	3.70	39
1.04	−0.042	1.34	−0.052	5.00	42
1.06	−0.028	1.37	−0.031	7.70	45
1.08	−0.014	1.39	−0.017	15.0	47
1.10	0.000	1.42	0.000	∞	50
1.15	0.035	1.48	0.045	5.90	57
1.20	0.070	1.55	0.095	3.00	63
1.25	0.105	1.61	0.137	2.00	69
1.30	0.140	1.68	0.186	1.50	75
1.35	0.175	1.74	0.228	1.20	79
1.40	0.210	1.81	0.277	1.00	84
1.45	0.245	1.87	0.319	0.85	88
1.50	0.280	1.94	0.368	0.75	91
1.60	0.350	2.06	0.452	0.60	95
1.70	0.420	2.19	0.543	0.50	98
1.80	0.490	2.32	0.634	0.43	98.9
1.90	0.560	2.45	0.725	0.37	99.6
2.00	0.630	2.58	0.816	0.33	99.9

It is straightforward to show that the second term on the right-hand side of Eq. (3.9) is strictly negative, implying that the expected return including waste monitoring will always be lower than that of constructing the facility and transporting the wastes only. Hence, if the decision maker relies only on comparison of expected returns of alternate actions, then a_1 will always be preferred if the same contract is awarded in both cases.

In addition to the values of the preceding parameters (C_1, C_3, C_1', and C_3'), let the costs of operation and monitoring per year M and M' be 1% and 2% of the contract B, respectively. The minimum possible bid is then given by $0.7 + 0.01t$, the minimum contract under the worst conditions is $1.4 + 0.02t$, and the bid under which there exists at least 50% chance of profit is $0.77 + 0.011t$.

If one considers the minimum bid of $0.7 + 0.01t$, then the expected return would be $E_t = -0.07 - 0.001t$, and the standard error would be $\sigma_t = 0.21 + 0.003t$. Hence both the expected return and its uncertainty vary linearly with the time horizon of the project. The various statistical measures for a time period of 20 years are shown in columns 3, 4, 5, and 6 of Table 3.1. Figure 3.3 plots the probabilities $P(0)$ for actions a_1 and a_2 for various time horizons as a function of the contract bid normalized by $B_{min} = 0.7$.

Table 3.1 and Fig. 3.3 demonstrate that for the same value of the contract award, the expected return of action a_1 is higher, the uncertainty about this average and the volatility are lower, and the probability of success of a_1 is higher. Hence, for the case of the same value of contract, all the statistical criteria indicate the same preferred action a_1; the use of additional criteria to that of the maximum expected utility reinforces the confidence in the selection process.

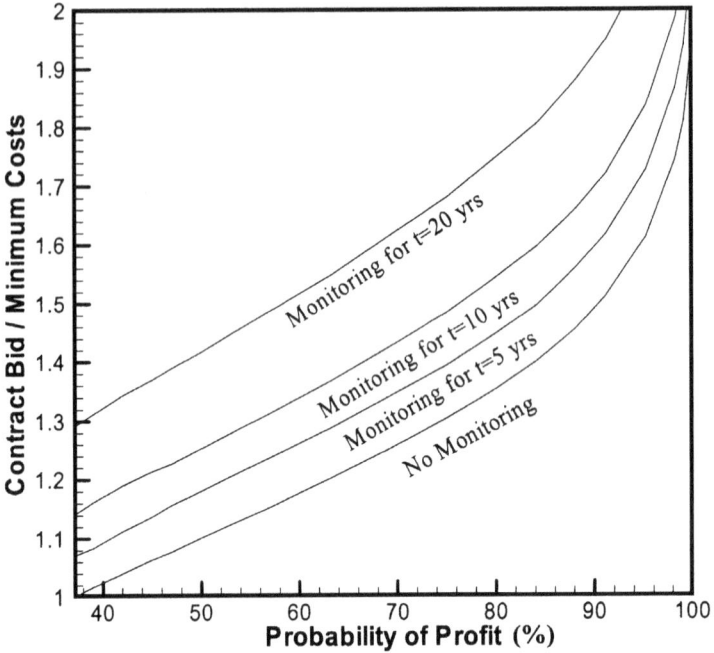

FIGURE 3.3 Probabilities of profits for different actions and contract bids.

Compare now actions a_1 and a_2 with offering bids of 0.805 and 1.035, respectively (row 7 of Table 3.1). The various probability branches and associated costs are depicted in Fig. 3.4. The expected value promised by choosing a particular action is indicated within a box at the left-hand end of each action branch. Figure 3.4 indicates that, according to the MEMV criterion, action a_2 is preferred because the largest expected value is returned. Detailed evaluation reveals that the difference between E^{a_1} and E^{a_2} (in millions of dollars) is \$10,000. This profit is realized by expending an extra capital of \$230,000 with the same high volatility of $v = 5.9$ and $P(0) = 57\%$ as for action a_1 but with a higher standard error of 0.27 instead of 0.21. It is important to realize here that the value that will be received if a_2 is preferred will not be \$0.045 × 10^6 but rather either a gain of \$0.135 × 10^6 or a loss of \$0.765 × 10^6. The potential gain of action a_2 over a_1 is \$30,000, and the loss is \$170,000. Thus, if the action recommended by

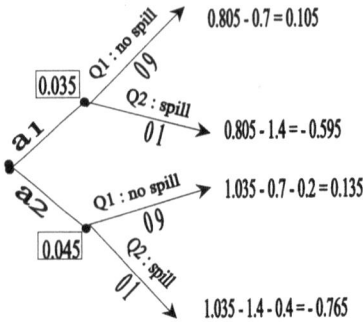

FIGURE 3.4 Comparison between a_1 and a_2 ($t = 20$ y).

the MEMV criterion alone is favored, extra capital in the amount of $230,000 must be expended to realize, at best, a gain of $30,000 over a_1. Such a decision requires a willingness to accept a loss of an extra $170,000 over a_1. In other words, in order to expect a 13% return on the additional investment, one needs to be willing to risk losing 74% of the extra capital required for action a_2. It is clear from this analysis that the exclusive use of the MEMV criterion leads to selection of a high-risk action; the use of additional statistical measures provides a cautionary sign (high volatility and increase of standard error) for a more detailed evaluation that points to the relative merits of each action. Hence, in this case, the MEMV principal provides a misleading "objective" criterion for selection of the optimal action, whereas the use of additional statistical criteria places the selection process within the risk-tolerance framework of the decision maker.

3.5 ANALYSIS OF ALTERNATE ACTIONS: CATASTROPHIC SPILL

In the scientific and economic analyses of environmental projects, catastrophic scenarios often are incorporated in the decision-making process in an effort to quantify the costs of remotely probable future events (Hunter and Mann, 1992; U.S. DOE, 1994; Sandia, 1994). Examples of such cases include possible climate changes that can affect the hydrogeologic conditions around a geologic repository of radioactive wastes, catastrophic well blowouts during oil drilling due to overpressure, spills during the transport of radioactive wastes (or oil) due to extreme events, etc. In this section the impact on the decision-making process of the inclusion of extreme, catastrophic events is addressed. Investigation of such scenarios may alter the outlook of a project, lead a corporation away from a project, or guide a corporation to extremely conservative decisions that undermine the profitability of a project.

Commence with an analysis of action a_1 (see Fig. 3.2). The expected value $E_K^{a_1}$ due to a catastrophic event is now given by

$$E_K^{a_1} = E^{a_1} - p_K[(C_1'' + C_3'') - (C_1' + C_3')] \tag{3.10}$$

where E^{a_1} corresponds to the limited spill and is described by Eq. (3.8). Because the cost of catastrophic events is higher than that of an event of limited extent, the second term on the right-hand side of Eq. (3.10) is strictly negative, and hence $E_K^{a_1} < E^{a_1}$.

The minimum bid that can be offered remains the same as previously, but the minimum optimal bid is now substantially higher, $B = C_1'' + C_3''$. The average break-even bid is now $B = C_1'' + C_3'' - p_s(C_1'' - C_1) - p_s(C_3'' - C_3) - p_f(C_1'' - C_1') - p_f(C_3'' - C_3')$. The fixed costs of transport, burial, and monitoring for no or a limited spill remain the same, with $C_1'' = 100$ and $C_3'' = 1.0$ (in millions of dollars) now reflecting the economic impact of a severe environmental spill. Here the cost of a catastrophic event is borne primarily by the transport component of the project, and such can be the case where, due to regulations, an expensive site-remediation program must be implemented at the location of the spill. However, the analysis applies equally well to leaks at the burial site because the outcome is affected by the overall cost and not by the costs of the individual components of the project.

Again using the minimum bid of $B_{min} = 0.7$, the expected value corresponding to a catastrophic event becomes $E_{K\,min}^{a_1} = -0.17$, but now $\sigma_{K,min} = 3.17$, $v = 18.6$, and $P(0) = 48\%$. Thus, although the average return has not changed significantly, the standard error and the volatility have increased significantly (by factors of 15 and 6, respectively). For bids that are multiples of B_{min}, one has $E_K' = E_{k,min}^{a_1} + (k-1) B_{min}$, $k \geq 1$ and $\sigma_K' = \sigma_{K,min}$. Hence an increase in the contract award improves the expected return but not the uncertainty of this return. Depicted in columns 1, 2, 5, and 7 of Table 3.2 are the effects on the expected value $E_K^{a_1}$, $v_K^{a_1}$, and $P(0)$ of different levels of bids $B_{t=0}^K$ normalized by the minimum costs $B_{min} = 0.7$.

Table 3.2 demonstrates that inclusion of a low-probability, high-cost event has the effects of (1) reversing the expected return from a positive to a negative value in the range of bids of $1.10B_{min}$ to $1.20B_{min}$, (2) increasing the standard error by a factor of 15, (3) significantly increasing the volatility, and (4) substantially reducing the probability of success. Thus, for example, inclusion of a catastrophic event makes a contract bid of $2B_{min}$, which previously almost guaranteed the profitability of a project [$P(0) = 99.9\%$], now appear a risky venture because $\sigma_{K,min} = 3.17$, $\mu = 6.0$, and $P(0) = 57\%$. In the catastrophic case, a 99% probability of success

TABLE 3.2 Catastrophic Spill: Statistical Measures for Actions a_1 and a_2 ($t = 20$ y)

$B_{t=0}^K/B_{min}$	$E_K^{a_1}$	$B_{t=20}^K/B_{min}$	$E_K^{a_2}$	$v_K^{a_1}$	$v_K^{a_2}$	$P(0)$, %
1.00	−0.170	1.29	−0.190	18	17	48
1.02	−0.156	1.32	−0.170	20	18	48
1.04	−0.142	1.34	−0.154	22	20	48
1.06	−0.128	1.37	−0.116	24	27	48
1.08	−0.114	1.39	−0.114	27	28	49
1.10	−0.100	1.42	−0.100	31	32	49
1.15	−0.065	1.48	−0.055	48	58	49
1.20	−0.030	1.55	−0.010	106	317	50
1.25	0.005	1.61	0.035	587	90	50
1.30	0.040	1.68	0.080	79	40	50
1.35	0.075	1.74	0.125	42	25	51
1.40	0.110	1.81	0.170	28	18	51
1.45	0.145	1.87	0.215	21	15	52
1.50	0.180	1.94	0.260	17	12	52
1.60	0.250	2.06	0.350	12	9	53
1.70	0.320	2.19	0.440	9	7	54
1.80	0.390	2.32	0.530	8	6	55
1.90	0.460	2.45	0.620	7	5	56
2.00	0.530	2.58	0.710	6	4	57

is reached only for bids exceeding $12B_{min}$.

Two points are worth emphasizing here. The first point is that unwarranted inclusion of a catastrophic scenario into the decision-making analysis can substantially alter the perspective on the value of a project and guide a corporation away from an investment that, even under a limited-liability case, could be profitable. In a case where a catastrophic event needs to be considered, a corporation may resort to a change of strategy, including insurance coverage for the project, entering into a consortium and thus limiting involvement, or reaching an agreement with the funding agency to limit financial liability in the event of an environmental disaster.

The second point is illuminated by the minor differences in the expected value for the majority of the bids of the limited and catastrophic failure cases (second column, Tables 3.1 and 3.2), whereas the standard error, volatility, and cumulative probabilities are very different. Consider the case where a decision needs to be made among two projects. Both projects consist of burial-site construction and transportation of wastes (but no monitoring). However, although the first project is assessed considering only the prospect of a limited spill, the second project is evaluated under conditions of a catastrophic spill as well. Tables 3.1 and 3.2 demonstrate that reliance on the expected value will fail to differentiate unambiguously the economic consequences of each project. Only the addition of supplementary statistical measures will provide an indication of the viability of each project.

Consider now action a_2 for a monitoring period of 20 years. The expected value is given by

$$E_K^{a_2} = E_K^{a_1} + [(M'' - M)p_s + (M'' - M')p_f - M'']t \qquad (3.11)$$

Equation (3.11) relates the expected values of actions a_1 and a_2 when the same levels of contract bids are considered. The variance of this action is $\sigma^2 = 10.08 + 3.05 \times 10^{-3} \times t + 9 \times 10^{-6} \times t^2$, indicating that the variance changes but little with time. The various statistical measures for a time period of 20 years are shown in columns 3, 4, 6, and 7 of Table 3.2. Comparison of actions a_2 in Tables 3.1 and 3.2 demonstrates that the catastrophic event significantly affects the variance, volatility, and cumulative probability but has little influence on the expected value. Here parameters are as defined previously, with the addition of the value of $M'' = 0.02$ (in millions of dollars).

3.6 PARAMETER UNCERTAINTY EFFECTS

The evaluation of expected value E, variance σ^2, and cumulative probability $P(V)$ has so far been predicated on statistically sharp values for the parameters in the decision tree of Fig. 3.1, with the uncertainty arising from the probabilities along the different possible paths. But there is a second form of uncertainty that arises due to the fact that one does not have statistical sharpness on each and every parameter but rather an estimated range of values within which a parameter can lie.

For instance, one might calculate a success probability p_s of transporting wastes without leakage at, say, $p_s = 0.9$, but different methods of estimation and different underlying assumptions can produce a range of estimates of p_s, say, 0.8 $\leq p_s \leq 0.99$. Similar arguments are relevant for all parameters entering the calculations. Thus project cost estimates are routinely assigned different error ranges at different stages over the life cycle of a project to account for various contingencies (Yeo, 1990).

Two factors are apparent: (1) there is no particularly valid, objective justification for preferring any one value of p_s out of its range over any other value, and (2) one needs to evaluate which ranges of which particular parameters are causing the largest uncertainty in estimates of the expected value E, its variance σ^2, and the cumulative probability $P(V)$. Then one has available the relative importance of each parameter in contributing to system uncertainty, and so one can then determine where to place effort to narrow the ranges of uncertainty of the most important parameters—and with a clear determination of the definition of importance as it relates to the specific system of study.

3.6.1 Means and Variances

Select at random from the range of each variable parameter. Then, for the ith choice of such a set of parameter values, one can calculate an expected value E_i and a variance σ_i^2 together with a cumulative probability $P_i(V)$. Clearly, by performing a series of Monte Carlo computations with N total selections, one produces both a mean expected value

$$<E> = N^{-1} \sum_{i=1}^{N} E_i \tag{3.12}$$

and a direct uncertainty $<\rho^2>$ in $<E>$ due solely to the statistical variations in each parameter given by

$$<\rho^2> = N^{-1} \sum_{i=1}^{N} (E_i - <E>)^2 \tag{3.13}$$

In addition, one also has a mean value for the variance $<\sigma^2>$ from

$$<\sigma^2> = N^{-1} \sum_{i=1}^{N} \sigma_i^2 \tag{3.14}$$

and an uncertainty $<\delta^2>$ in the mean variance calculated from

$$<\delta^2> = N^{-1} \sum_{i=1}^{N} (\sigma_i^2 - <\sigma^2>)^2 \tag{3.15}$$

Because the direct uncertainty $<\rho^2>$ in $<E>$ is due to fluctuations in parameter values, whereas the variance $<\sigma^2>$ is due to the different probabilistic paths, the total uncertainty on $<E>$ is then measured by the variance σ_E^2, where

$$\sigma_E^2 = <\rho^2> + <\sigma^2> \tag{3.16}$$

Equally, for the cumulative probability $P(V)$ there are several measures of worth. Thus one can compute a mean cumulative probability $<P(V)>$ from

$$<P(V)> = N^{-1} \sum_{i=1}^{N} P_i(V) \tag{3.17}$$

and an associated uncertainty $<\delta P(V)^2>$ from

$$<\delta P(V)^2> = N^{-1} \sum_{i=1}^{N} [P_i(V) - <P(V)>]^2 \tag{3.18}$$

One also can compute an approximate mean value $\overline{P}$ using $<E>$ and σ_E in Eq. (3.2). If the difference

$$|<P(V)> - \overline{P}(V)| \le <\delta(V)^2>^{1/2} \tag{3.19}$$

then it is an accurate enough approximation to use $\overline{P}(V)$; otherwise, one must use the ensemble value $<P(V)>$.

3.6.2 Relative Importance

In addition to having the ability to compute ensemble averages and their fluctuations, one also requires knowledge of which uncertainty ranges of which parameters are contributing most to the total system uncertainty. This relative-importance problem can be addressed simply as follows: Each parameter in the decision tree of Fig. 3.1 has a mean value within its range of variation. Let a vector $\mathbf{q}$ denote all the component parameters, with the jth component q_j having a mean value $<q_j>$, with $q_{j,\min} \le q_j \le q_{j,\max}$, where $q_{j,\min}$ and $q_{j,\max}$ are, respectively, the minimum and maximum values of q_j.

Suppose that one were to use only the vector of mean values $<\mathbf{q}>$ to calculate the quantities in the preceding subsection. Then this choice is precisely the same as choosing specific values of the parameters, so $<\rho^2> = 0$, and $<\sigma^2>$ is the same as $\sigma^2(<\mathbf{q}>)$. Now consider the influence of uncertainty in each variable in contributing to variations in σ_E^2, because there is an intrinsic value for σ_E^2 even when the parameters are statistically sharp. The changes in σ_E^2 relative to the value $\sigma^2(<\mathbf{q}>)$ due to variations in each parameter around its mean value are then determined as follows.

Let all except one, say, the jth, of the components of the parameter vector $\mathbf{q}$ be held at their mean values. This jth component of $\mathbf{q}$ is then varied, randomly, around its mean value. The result is obviously a value for $<\rho^2>$ and σ_E^2 that is dependent on $<q_j>$ and also on $q_{j,\min}$ and $q_{j,\max}$. Denote these values as $<\rho(j)^2>$ and $\sigma_E(j)^2$. Repeating the process for all components of **q**, one can then calculate the relative fractional uncertainty contribution RC_j to $<\rho^2>$ from each q_j as

$$RC_j = \frac{<\rho(j)^2>}{\sum\limits_{k=1}^{N} <\rho(k)^2>} \qquad (3.20)$$

Equally, one can compute the relative importance RI_j to σ_E^2 from

$$RC_j = \frac{<\sigma(j)^2>}{\sum\limits_{k=1}^{N} <\sigma_E(k)^2>} \qquad (3.21)$$

Similarly, for the ensemble mean value $<E>$ and the cumulative probability $P(V)$, one can compute relative-importance values from, respectively,

$$RI_j\,(E) = \frac{<E(j)>}{\sum\limits_{k=1}^{N} <E(k)>} \qquad (3.22)$$

and

$$RI_j\,(P) = \frac{P_j(V)}{\sum\limits_{k=1}^{N} P_k(V)} \qquad (3.23)$$

which provide measures of the contributions.

3.7 RELATIVE CONTRIBUTION OF UNCERTAIN PARAMETERS IN SUBSURFACE HYDROLOGY

The heterogeneity of soils and geologic formations is one of the most challenging issues in trying to predict the movement of water and contaminants in the soil. Because of the complexity and heterogeneity of the subsurface environment, the definition of all the properties that describe the dynamics of subsurface physical processes is impossible practically. Additionally, the amount of field data available is always limited due to high cost and technical limitations, and hence a field cannot be characterized completely. This section shows that application of the concepts presented in Sec. 3.6 is not restricted to economic evaluation of the uncertainties of a project but can be applied equally well to quantification of the scientific uncertainties of a project.

3.7.1 Introduction

Unsaturated flow in soil formations is a complex process that is not easily understood, even in relatively homogeneous systems. The *unsaturated zone* is defined as the geologic environment that lies between the land surface and the local water table. The upper part commonly consists of the root zone and weathered soil horizons. Soils and bedrock within this zone are usually unsaturated; i.e., their pores are partially filled with water (Hillel, 1980; Stephens, 1996). One major factor that contributes to this complexity is the spatial variability of soil properties. Field observations have shown that the hydrologic properties of soils vary over several orders of magnitude even in the same formation (Woodbury and Sudicky, 1991; Sandia, 1994). Thus the effect of the spatial variability on predictions of water flow and the transport of contaminants is a major focus of scientific investigations. Studies of unsaturated flow within porous media exhibiting random heterogeneities have employed either analytic approximations through linearizations and perturbation methods or Monte Carlo simulations that have focused on one parameter at a time (Gardner, 1958; Mantoglou and Gelhar, 1987a, 1987b, 1987c; Yeh et al., 1985a, 1985b, 1985c).

This section quantifies the relative contribution of the uncertainty and spatial variability of multiple parameters that enter into unsaturated flow problems. First, an investigation is made of several types of probability distributions that may describe the data, and the degree to which the uncertainty in the distribution type of a physical parameter is contributing to the total system uncertainty is calculated. The assumption of the type of distribution is critical because only a small amount of field data is usually available, and even in relatively homogeneous aquifers, these data can be fitted by more than one type of probability distribution (Yeh, 1989). The significance for site-characterization efforts lies in that if the exact functional form of the probability distribution is required, then a very dense measurement network needs to be implemented to determine unequivocally the statistics of the flow field. Also presented here is the framework that accounts for the uncertainty and the relative contribution and relative importance of various uncertain and/or spatially variable parameters. This aspect has important implications in site-characterization and modeling efforts because it allows the focus of time and resources on those factors of a problem which contribute the most and dominate the total system uncertainty.

3.7.2 Physical Problem

The methodology of Sec. 3.6 is applied to the physical problem of one-dimensional infiltration in unsaturated porous media under constant infiltration rate.

For this problem, field data are used that were collected for the characterization of the flow field at the U.S. Department of Energy site for potential disposition of radioactive wastes, located at Yucca Mountain, Nevada. Figure 3.5 shows the six hydrogeologic units that are considered in the study. These units, from the ground surface to the water table, are the Tiva Canyon welded (TCw), the Paintbrush nonwelded (PTn), the Topopah Spring welded (TSw), the Topopah spring vitrophyre (TSv), the Calico Hills nonwelded-vitric (CHnv), and the Calico Hills nonwelded-zeolitic (CHnz). These units exhibit significant differences in their properties and hydraulic behaviors, as illustrated in Tables 3.3 and 3.4 and discussed in detail below.

The study is concerned with one-dimensional infiltration through the 625-m-thick unsaturated soil formations depicted in Fig. 3.5. If one assumes, as is commonly done (Sandia, 1994), that the infiltration rate is constant at the site and that steady-state conditions have been attained under gravity-capillary equilibrium, then the hydrologic state between the ground surface and the water table (pressure head and saturation against depth) can be described by the flow equation (Hughson and Yeh, 1998)

$$-K(\psi)\left(\frac{d\psi}{dz} + 1\right) = q \qquad (3.24)$$

Here q denotes the constant steady-state infiltration rate (m/s), ψ is the pressure head (m), which is positive when the soil is fully saturated and is negative when the soil is partially saturated, and z is the vertical coordinate, which is positive in the upward direction. To describe the saturation–pressure head relation, Mualem's model is used:

$$S = (1 + |\alpha\psi|^\beta)^{-m} \qquad (3.25)$$

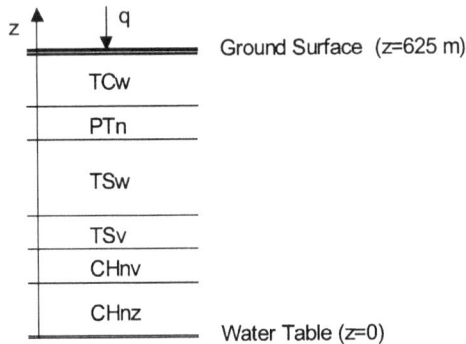

FIGURE 3.5 Schematic of the stratigraphy (not to scale) at the Yucca Mountain site.

TABLE 3.3 Statistics of the Hydrogeologic Parameters for Different Layers (Sandia, 1994)

Rock unit	Thickness, m	log K_s		log α		log β	
		$E(\cdot)$	CV$(\cdot)$	$E(\cdot)$	CV$(\cdot)$	$E(\cdot)$	CV$(\cdot)$
TCw	81	-10.90	0.098	-2.094	0.325	0.206	0.259
PTn	39	-7.960	0.202	-1.134	0.636	0.347	0.647
TSw	299	-10.71	0.084	-1.885	0.265	0.233	0.523
TSv	15	-11	0.062	-2.624	0.167	0.349	0.539
CHnv	64	-8.99	0.115	-1.644	0.302	0.373	0.614
CHnz	127	-10.79	0.093	-2.270	0.275	0.223	0.540

TABLE 3.4 Minima and Maxima of the Parameters for Different Layers (Sandia, 1994)

Rock unit	K_s, m/s		α, 1/m		β	
	Min	Max	Min	Max	Min	Max
TCw	7.00E-13	4.83E-09	0.0003	0.1338	1.349	2.805
PTn	2.86E-12	2.35E-06	0.0104	1.6990	1.187	11.800
TSw	3.05E-13	5.23E-09	0.0021	0.4244	1.155	5.363
TSv	1.52E-12	6.95E-11	0.0002	0.0077	1.377	4.473
CHnv	5.13E-12	2.92E-07	0.0054	0.3752	1.249	9.888
CHnz	2.37E-14	3.14E-09	0.0004	0.2355	1.184	5.914

where S is the saturation of soil, and α, β, and m are empirical parameters with $m = 1 - 1/\beta$. The unsaturated hydraulic conductivity $K(\psi)$ is then given by (van Genuchten, 1980)

$$K(\psi) = K_s \, S^{1/2} \, [1 - (1 - S^{1/m})^m]^2 \qquad (3.26)$$

where K_s is the saturated hydraulic conductivity (m/s). Thus in Eq. (3.24) the parameter of the problem [unsaturated hydraulic conductivity, Eq. (3.26)] depends on the unknown pressure ψ. Through integration of Eq. (3.24), one obtains the expression

$$\int_0^{\psi} \frac{K(\psi)\, d\psi}{K(\psi) + q} = - (z - z_0) \qquad (3.27)$$

where z_0 is the elevation of the water table (pressure equals zero), and z corresponds to a pressure ψ. Table 3.3 shows the statistical properties of the saturated hydraulic conductivity K_s and the van Genuchten α and β parameters as tabulated in the Sandia report (1994). The notation $E(\cdot)$ denotes expected value and CV$(\cdot)$ the coefficient of variation. The minimum and maximum values of these parameters are given in Table 3.4 for each hydrogeologic unit (Sandia, 1994).

Tables 3.3 and 3.4 illustrate the high degree of variability of the parameters within each layer and among the different layers.

3.7.3 Theoretical Framework

Consider the parameters K_s, α, and β as random variables that obey the same probability distribution. The distributions in the analysis are gaussian, log-normal, exponential, uniform, and triangular, with the mean, variance, and range of each parameter defined in Tables 3.3 and 3.4. The first three distributions were chosen because they have been shown to fit data at several sites (Woodbury and Sudicky, 1991; Sandia, 1994), the uniform distribution because it describes the (common) situation where one has knowledge only of the range within which a parameter lies for a geologic formation, and the triangular because it represents the case where, in addition to the minimum and maximum values, one has information about the most commonly occurring value. By selecting a triplet of values from a specific distribution i for the parameter set (K_s, α, β) and by then performing a series of Monte Carlo computations (Deutsch and Journel, 1992) with N total selections, one can create N profiles of ψ with depth z. At each discretization point of the grid one can then average the N equiprobable values of ψ to obtain the mean pressure head and the variance σ^2 of ψ that apply to this point for a specific distribution i:

$$<\psi>_i = \frac{1}{N} \sum_{j=1}^{N} \psi_j \qquad \sigma_i^2 = \frac{1}{N-1} \sum_{j=1}^{N} (\psi_j - <\psi>_i)^2 \qquad (3.28)$$

Now one can calculate, at each point, the global mean $<\psi>_G$, the arithmetic mean of the expected values from the five distributions, and the global variance $<\sigma^2>_G$, the arithmetic mean of the variances obtained from each distribution:

$$<\psi>_G = \frac{1}{5} \sum_{i=1}^{5} <\psi>_i \qquad <\sigma^2>_G = \frac{1}{5} \sum_{i=1}^{5} \sigma_i^2 \qquad (3.29)$$

By calculating the quantity $<\rho^2>_T$,

$$<\rho^2>_T = \frac{1}{5} \sum_{i=1}^{5} (<\psi>_i - <\psi>_G)^2 \qquad (3.30)$$

which is just the divergence of the means of the distributions from the global mean, one can obtain the total uncertainty at each point:

$$\sigma_T^2 = <\rho^2>_T + <\sigma^2>_G \qquad (3.31)$$

Here $<\rho^2>_T$ is a measure of the uncertainty in the mean ψ behavior because of the uncertainty in the type of distribution, and $<\sigma^2>_G$ is the average fluctuation around the mean ψ behavior irrespective of distribution.

Now one can examine, for every point, the relative contribution of each distribution i toward the global mean through the expression

$$RC_p(i) = \frac{(<\psi>_i - <\psi>_G)^2}{\displaystyle\sum_{m=1}^{5} (<\psi>_m - <\psi>_G)^2} \qquad (3.32)$$

and also calculate the relative importance of each distribution i toward the average variance:

$$RC_{\sigma^2}(i) = \frac{\sigma_i^2}{\displaystyle\sum_{m=1}^{5} \sigma_m^2} \qquad (3.33)$$

Finally, by calculating the ratios

$$\frac{<\rho^2>_T}{\sigma_T^2} \qquad \text{and} \qquad \frac{<\sigma^2>_G}{\sigma_T^2} \qquad (3.34)$$

one can evaluate to what degree the total uncertainty is dominated by the lack of knowledge in the type of distribution or by the fluctuations around the mean values. A large value of the first ratio indicates that the choice of the probability distribution model is critical in total uncertainty, and hence more data need to be collected for a clear determination of the shape of the distribution. In contrast, a large value of the second ratio indicates that the fluctuations around the mean ψ behavior are dominating the total system uncertainty, and hence the parameters need to be defined more sharply.

Now hold two of the three parameters K_s, α, and β at their mean values and vary the third according to a distribution i with mean, variance, and range for that parameter obtained from Tables 3.3 and 3.4. By performing Monte Carlo simulations one obtains $<\psi>_{i,k}$, the mean pressure head, and $\sigma_{i,k}^2$, the variance of the pressure head, due to fluctuations in the kth random parameter according to an ith distribution. By repeating the procedure for all parameters, one evaluates the relative importance toward the mean pressure of each parameter k for every distribution i:

$$RI_{j,k}^{<\psi>} = \frac{<\psi>_{i,k}}{\displaystyle\sum_{m=1}^{3} <\psi>_{i,m}} \qquad (3.35)$$

as well as the relative importance toward the variance of each parameter k and distribution i:

$$\mathrm{RI}_{i,k}^{\sigma^2} = \frac{\sigma_{i,k}^2}{\displaystyle\sum_{m=1}^{3} \sigma_{i,m}^2} \tag{3.36}$$

Clearly, this process can be repeated for all distributions ($i = 1,\dots, 5$), and then, for each parameter k, one can calculate the relative importance toward the mean:

$$\mathrm{RI}_{k}^{<\psi>} = \frac{\displaystyle\sum_{i=1}^{5} <\psi>_{i,k}}{\displaystyle\sum_{i=1}^{5}\sum_{m=1}^{3} <\psi>_{i,m}} \tag{3.37}$$

and the variance:

$$\mathrm{RI}_{i,k}^{\sigma^2} = \frac{\displaystyle\sum_{i=1}^{5} \sigma_{i,k}^2}{\displaystyle\sum_{i=1}^{5}\sum_{m=1}^{3} \sigma_{i,m}^2} \tag{3.38}$$

irrespective of distribution. Thus the preceding analysis can provide a ranking of the importance of each parameter in the evaluation of the mean and variance of the pressure for a specific distribution and also irrespective of the choice of distribution.

3.7.4 Results and Discussion

The flow domain was discretized into elements of length of 0.5 m, for a total of 1250 nodes. A constant infiltration rate of 0.1 mm/year was assumed. The upper boundary was considered as a prescribed flux boundary, whereas the lower boundary was treated as a stationary water table. Because the relationship between the three variables K_s, α, and ß cannot be specified from the limited field data available, it was assumed that the variables were random processes, perfectly correlated, with their statistics given by Tables 3.3 and 3.4. The nonlinear equation (3.27) was solved using an iterative scheme, whereas the integration was performed through an adaptive Newton-Cotes nine-point rule (Press et al., 1987). For each node, and for the parameter set (K_s, α, ß), 500 Monte Carlo simulations were performed. Thus, for each choice of distribution, 500 equiprobable ψ profiles were created along the 1250 nodes. This process was repeated for all five types of distributions.

The quantities described in Eqs. (3.28) through (3.34) were then calculated, but in order to simplify the depiction of the results for each distribution, the detailed mean point profiles were averaged over the six formations. This situation represents the case where an evaluation is needed that ranks the different parameters with respect to the total system uncertainty. Of course, one also can use the results of the simulations to compare detailed point profiles, or depth-average over each layer only, and rank the parameters at each point of the flow field or for each individual layer.

Figure 3.6 plots the contribution of each type of distribution toward the global mean, as described in Eq. (3.32). Figure 3.7 plots the relative importance of each distribution model toward the average variance, as depicted in Eq. (3.33). These two figures indicate that a gaussian model assumption for the parameters will produce a mean ψ profile, as well as fluctuations around this profile, that are significantly larger than those for any other statistical model. Hence these results demonstrate that the assumption of a particular distribution is critical in the prediction of the mean hydraulic behavior and the fluctuations about this mean in an unsaturated zone field. This conclusion is supported by the values of the ratios in Eq. (3.34), where the first term, which measures the importance of parameter assumptions to total uncertainty, is 75.3%, whereas the second term, which measures the contribution of fluctuations, is 24.7%.

3.8 SUMMARY

The following conclusions can be drawn from this chapter:

1. The criterion of the maximum expected monetary value (MEMV) is widely used in decision-making analyses of environmental and gas and oil projects to

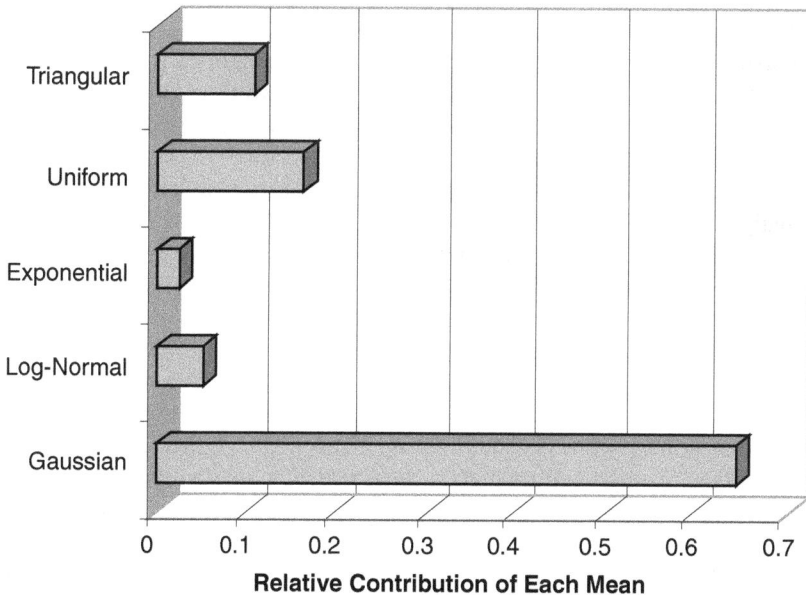

FIGURE 3.6   Relative contribution toward the global mean.

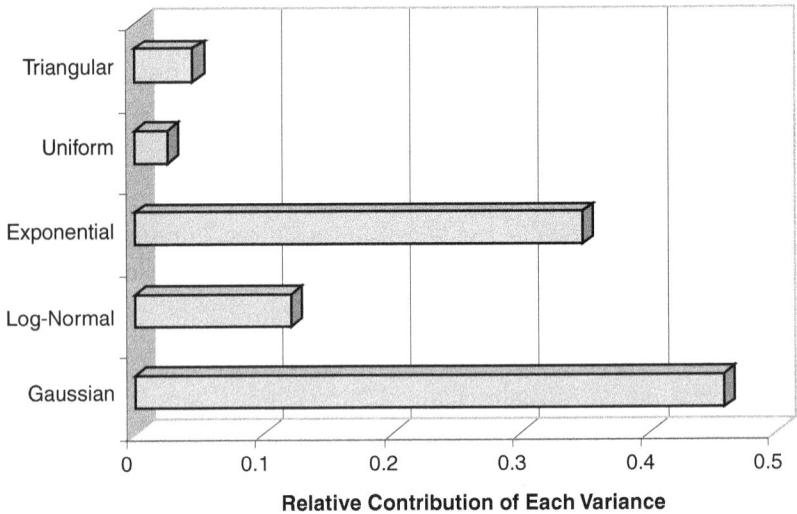

FIGURE 3.7   Relative contribution toward the average variance.

select a preferred action as the one returning the maximum expected return. Sole use of this criterion may lead to erroneous decisions in the presence of uncertainty.

2. The MEMV fails to differentiate the economic consequences of limited and catastrophic failures in a project. The use of additional statistical measures, such as standard error, volatility, and cumulative probability, provides insight into the selection process, illuminating the implications of each case. Thus Eqs. (3.1) and (3.3) that provide expressions for the volatility and the probability of profit, respectively, should be used in conjunction with the MEMV criterion to analyze a project.

3. Consideration of a low-probability, high-cost (catastrophic) event in a decision analysis can have the effects of
   a. Reversing the expected return from a positive to a negative value for a range of contract awards of a project.
   b. Significantly increasing the standard error and volatility.
   c. Substantially reducing the probability of success.

4. Consideration of catastrophic scenarios in risk analyses of environmental projects, as illustrated in Sec. 3.5 of this chapter, should almost always never be performed in a routine decision-making manner. Unjustified inclusion of catastrophic scenarios into such an analysis can alter substantially the perspective of a project and guide a corporation away from an investment that, even under a limited-liability case, could be profitable. Thus significant effort should be expended by risk-management groups to properly account for the

inclusion of catastrophic scenarios in their analyses. Use of Eqs. (3.1) and (3.3) in conjunction with the MEMV is a good starting point for illumination of the consequences of catastrophic events on the profitability of projects. This should be followed by a more detailed evaluation of strategy alternatives and insurance options that allow a corporation to be involved in a project while safeguarding against catastrophic events. Subsequent chapters provide a detailed exposition of such alternatives.

5. When parameters of a decision tree take on values lying within estimated ranges, a series of Monte Carlo simulations provides the framework for relevant decision making. Section 3.6 and Eqs. (3.12) to (3.19) provide the background for such analyses.

6. Use of the concept of relative uncertainty can guide selection of those parameters which dominantly influence the total system uncertainty, thus allowing one to concentrate resources on efforts to minimize the range in such dominant parameters. Equations (3.20) to (3.23) may be used to illuminate the critical components of a project.

APPENDIX 3A: APPROXIMATIONS FOR CUMULATIVE PROBABILITY

Equation (3.3) gives the probability of having a value greater than zero as

$$P(0) = \frac{1}{2} - \frac{1}{\sqrt{2\pi}} \int_0^b \exp\left(-\frac{u^2}{2}\right) du \qquad (3.39)$$

where $b = -E/\sigma = -(v)^{-1}$.

For a low volatility $v \ll 1$ ($|b| \gg 1$) combined with a positive expected value ($E > 0$), the upper limit of integration in Eq. (3.39) tends to $-\infty$, and hence $P(0)$ approaches the value of 1. When $v \ll 1$ but the expected value is negative ($E < 0$), the upper limit of integration in Eq. (3.34) approaches $+\infty$, and the value of $P(0)$ becomes approximately zero. One can recover the preceding limits through the asymptotic expression

$$P(0) \approx \frac{1}{2}[1 + \mathrm{sgn}(E)] \qquad \text{when } v \ll 1 \qquad (3.40)$$

which is exact when $v \to 0$.

To provide a correction to Eq. (3.40) for the cases where $v \ll 1$ (but not equal to zero), consider the integral

$$\frac{1}{\sqrt{2\pi}} \int_0^b \exp\left(-\frac{u^2}{2}\right) du = \frac{1}{2} - \frac{1}{\sqrt{2\pi}} \int_b^{+\infty} \exp\left(-\frac{u^2}{2}\right) du$$

$$= \frac{1}{2} - \frac{b}{\sqrt{2\pi}} \int_1^{+\infty} \exp\left(-\frac{b^2 s^2}{2}\right) ds$$

$$= \frac{1}{2} - \frac{b}{\sqrt{2\pi}} \int_0^{+\infty} \exp\left[-\frac{b^2}{2}(1+x)^2\right] dx$$

$$= \frac{1}{2} - \frac{b}{\sqrt{2\pi}} e^{-b^2/2} \int_0^{+\infty} \exp\left[-\frac{b^2}{2}(2x+x^2)\right]$$

dx

$$\approx \frac{1}{2} - \frac{b^{-1}}{\sqrt{2\pi}} e^{-b^2/2} \tag{3.41}$$

where, in the second and third equalities, the transformations $u = bs$ and $s = 1 + x$, respectively, have been employed, and the final result was obtained by taking into account that the exponential integrand, for large b ($|b| \gg 1$), decays rapidly away from the neighborhood of zero, thus rendering the value of the integral (approximately) equal to $1/b^2$.

Combining Eqs. (3.41) and (3.40) with (3.39), one obtains

$$P(0) \approx \frac{1}{2} [1 + \text{sgn}(E)] + (2\pi)^{-1/2} \exp\left(-\frac{b^2}{2}\right) b^{-1} \tag{3.42}$$

CHAPTER 4
UTILITY THEORY AND WORKING-INTEREST OPTIMIZATION IN A HAZARDOUS WASTE TRANSPORT AND BURIAL OPPORTUNITY

4.1 INTRODUCTION

In considering environmental projects, decision makers are often faced with the prospect of low-probability but high-cost uncertain events that can significantly affect corporate fiscal integrity (Lerche and Mackay, 1999). This prospect is accentuated by the potential of long-term liabilities arising from unpredictable spills during transportation of hazardous material and leakage at the burial site that can lead to subsurface plumes that are difficult to characterize and remediate (Paleologos and Fletcher, 1999). On the other hand, Chap. 3 made the point that inclusion of such considerations (which, after all, have a low probability of occurring) in decision-making situations should be done cautiously because such analyses can guide management teams away from projects that under normal conditions would be profitable. This chapter discusses an alternative that corporations have in bidding on environmental projects that allows them to limit financial exposure to extreme events.

We expound on the familiar example of transport and long-term storage of hazardous (or radioactive) waste analyzed in preceding chapters by distinguishing four dominant possible scenarios: (1) successful transport without spillage from the location of origin and storage without leakage, (2) transport without accidental spillage but leakage at the burial site, (3) spillage during conveyance of the waste but no further contamination problems at the storage location, and (4) transport with spillage and unpredicted leakage at the repository site. The decision-tree diagram of Fig. 4.1 depicts these four events together with the associated probabilities and total anticipated costs C_1, C_2, C_3, and C_4, respectively. Thus C_1 is the estimated expenditure under optimal conditions; C_2 may, additionally, include the costs of excavation of leaking containers, repackaging, remediation, and increased monitoring at the operating facility; C_3 may, in addition to C_1, consist of the costs of collecting and transporting spilled waste as well as of any other contaminated material from a spillage location; and C_4 may reach a level where it encompasses all the previous financial obligations (Paleologos and Lerche, 1999). On the other side of an environmental opportunity is the funding agency for hazardous (or radioactive) waste control, which offers a contract at a fixed price G for transport, burial, and monitoring for a given time. Usually the contract price G will be less than the costs for the worst-case scenario C_4 and may be less than C_3 and C_2 as well. The values of each scenario (contract price minus costs) are shown in Fig. 4.1.

Under such conditions, a decision maker wishing to determine whether to assume the project has two major concerns: (1) Can a company make a profit if it undertakes the entire project, or should it enter into a consortium and accept only a fractional working interest W in this venture, thus limiting future gains but also mitigating the effect of potential catastrophic losses? (2) Given that each corporation has an internal tolerance to risk, what is the working interest that maximizes the return to a company under the preceding uncertain conditions and difficult-to-estimate costs? The objective of this chapter is to provide the tools that allow a corporation to consider involvement in a project (and to maximize such involvement) that under worst-case-scenario analyses would have been deemed nonprofitable.

4.2 WORKING-INTEREST EVALUATION

Consider a two-branch decision situation (or lottery L) that can lead either to a monetary outcome A or, alternatively, to a monetary outcome B with equal probabilities $p = 0.5$. The expected value of this gamble, or the value that will be returned if one is involved in a large number of these situations, is $E_1 = (A + B)/2$. An amount CE is termed the *certainty equivalent of L*, or *risk-adjusted value* (RAV) in oil-industry terminology (Lerche and MacKay, 1999), if an individual is willing to exchange the rights to the preceding lottery for the certain amount CE.

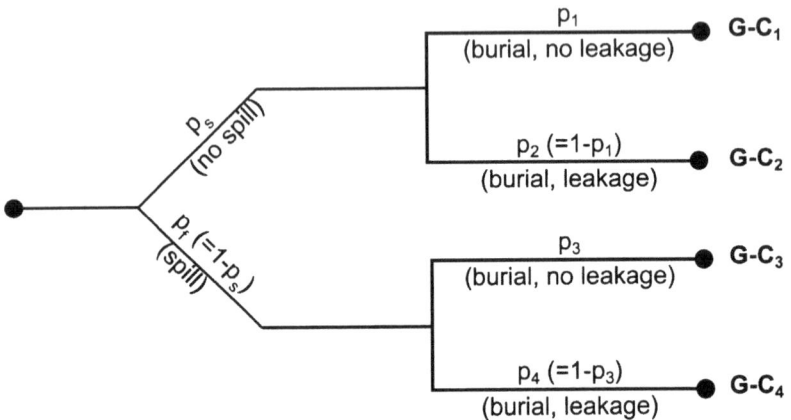

FIGURE 4.1 Decision tree for transport and burial of hazardous wastes.

This individual is characterized as *risk-averse* if and only if CE $< E_1$, *risk-seeking* if CE $> E_1$, and *risk-neutral* if CE $= E_1$, and the relations hold consistently for all similar decision situations (Krzysztofowicz, 1983a, 1986; Raiffa, 1997).

A utility function U is a way to standardize monetary outcomes and consequences of a decision-making situation that are not measurable in monetary terms into the same units, and it encodes the strength of an individual's preference for specific outcomes and risk attitudes toward the uncertainty of such outcomes. A decision maker's attitude toward risk is approximated, for a particular situation, by the shape of the utility function. Thus U is concave in the range of dollar amounts where risk-averse behavior is expected, convex where risk-seeking behavior is sought, and linear where the risk aspects of the various alternatives are ignored (risk-neutral attitude). Exponential utility models such as

$$U(x) = 1 - \exp\left(-\frac{x}{\text{RT}}\right) \tag{4.1}$$

have been employed in several decision-making situations (Cozzolino, 1977a; Krzysztofowicz, 1986; Keeney and Raiffa, 1993; MacKay and Lerche, 1996) to represent risk-averse preferences. Here x stands for the dollar value of an opportunity, and RT represents the maximum amount a corporation can lose without endangering its cash-flow situation or total assets. As the dollar amount x becomes large, $U(x)$ tends to 1, $U(0)$ equals zero, and for negative x (losses), the utility function turns negative. The parameter RT in Eq. (4.1) is denoted as *risk tolerance* and determines the risk-averse behavior of a utility function. Larger values of RT make $U(x)$ flatter (indicating a more risk-tolerant behavior), whereas smaller values make the function more concave (representing a more risk-averse attitude).

Consider now the decision-making situation of Fig. 4.1. The expected value E of the four-branch system is given by

$$E = G - p_s(p_1 C_1 + p_2 C_2) - p_f(p_3 C_3 + p_4 C_4) \tag{4.2}$$

and the variance can be easily calculated to yield

$$\begin{aligned}
\sigma^2 = E_2 - E^2 &= p_s p_1(1 - p_s p_1)C_1^2 + p_s p_2(1 - p_s p_2)\,C_2^2 \\
&\quad + p_f p_3(1 - p_f p_3)C_3^2 + p_f p_4(1 - p_f p_4)C_4^2 \\
&\quad + 2[\,p_s^2 p_1 p_2 C_1 C_2 + p_f^2 p_3 p_4 C_3 C_4 + p_s p_f(p_1 C_1 + p_2 C_2)\,(p_3 C_3 + p_4 C_4)\,]
\end{aligned} \tag{4.3}$$

where E_2 is the second statistical moment. A measure of risk is often assigned by the volatility v, defined in Eq. (3.1), an estimate of the stability of the mean value relative to fluctuations about the mean. A small value of volatility ($v \ll 1$) implies that there is little uncertainty in the expected value, whereas a large volatility ($v \gg 1$) implies considerable uncertainty.

While the expected value of a project may be high and the volatility small, nevertheless it can be the case that if failure does occur, the total project costs may be so large as to bankrupt, or cause serious financial damage to, a corporation (Yeo, 1990; Paleologos and Lerche, 1999). Under such conditions, it makes sense to take less than 100% working interest in a project. A smaller fraction of a project will reduce potential gains but also will mitigate the effects of catastrophic losses. Thus, with a working-interest fraction W, the expected value of the project depicted in Fig. 4.1 becomes

$$E^W = p_s[p_1 W(G - C_1) + p_2 W(G - C_2)] + p_f[p_3 W(G - C_3) + p_4 W(G - C_4)]$$

$$= W[G - p_s(p_1 C_1 + p_2 C_2) - p_f(p_3 C_3 + p_4 C_4)] = WE \tag{4.4}$$

on the assumption that fractional working interest does not affect the probabilities of success or failure of the project.

4.2.1 Exponential Utility Function

Using the exponential model in Eq. (4.1) for the utility function, the expected utility EU for a working fraction W of the project in Fig. 4.1 is given by

$$EU = 1 - p_s\{p_1 \exp[-W(G - C_1)/RT] + p_2 \exp[-W(G - C_2)/RT]\}$$
$$- p_f\{p_3 \exp[-W(G - C_3)/RT] + p_4 \exp[-W(G - C_4)/RT]\} \tag{4.5}$$

where the relations $p_s + p_f = 1$, $p_1 + p_2 = 1$, and $p_3 + p_4 = 1$ have been used. From the definitions of CE and the utility function, it follows that the expected utility equals the utility value of the certainty equivalent (Clemen, 1996; Raiffa, 1997), which, for this case, translates into

$$U(CE) = 1 - \exp(-CE/RT) = EU \tag{4.6}$$

Cozzolino (1977b, 1978), and Lerche and MacKay (1999) developed similar relations to Eqs. (4.5) and (4.6) based on energy-related analogues for gas and oil exploration projects. Substitution of Eq. (4.5) into Eq. (4.6) allows evaluation of the certainty-equivalent amount, in other words, the value of the project after adjustment for the risks of spillage and leakage and the risk attitude of a corporation:

$$CE = -RT \ln(p_s\{p_1 \exp[-W(G - C_1)/RT] + p_2 \exp[-W(G - C_2)/RT]\}$$
$$+ p_f\{p_3 \exp[-W(G - C_3)/RT] + p_4 \exp[-W(G - C_4)/RT]\}) \tag{4.7}$$

Equation (4.7) can be simplified to

$$CE = WG - RT \ln\{p_s[p_1 \exp(WC_1/RT) + p_2 \exp(WC_2/RT)]$$
$$+ p_f[p_3 \exp(WC_3/RT) + p_4 \exp(WC_4/RT)]\} \tag{4.8}$$

If the decision maker's risk tolerance RT is extremely high compared with the costs of the various alternatives, then as RT $\to \infty$, CE reduces to (App. 4A)

$$CE \to W[G - p_s(p_1C_1 + p_2C_2) - p_f(p_3C_3 + p_4C_4)] = WE \qquad (4.9)$$

which one recognizes immediately as the expected value of a project where a fractional working interest W has been obtained.

If the risk tolerance is much smaller than any of the costs of the alternatives, then since $C_4 \gg C_i(i = 1, 2, 3)$, it follows that as RT $\to 0$, then (App. 4A)

$$CE \to W(G - C_4) \qquad (4.10)$$

In this case, because a corporation has set zero risk tolerance (utility function is extremely curved, representing an excessively risk-averse attitude), the project will be considered only if the contract price G exceeds the highest anticipated costs C_4 of the spill-plus-leakage scenario. Equation (4.10) may be used when the probability and magnitude of future liabilities are deemed to be high (Paleologos and Fletcher, 1999) or when an agency (or society as a whole) has an extremely low tolerance, as can be the case with spills or leakage during the transport and burial of radioactive wastes (Paleologos and Lerche, 1999). Thus, depending on the value of the risk tolerance, the assessed costs, and their corresponding probabilities, there will, in general, be a working interest W_{max} that maximizes CE_{max}. Then, if the value of the project (including considerations related to potential risks and risk attitudes) is greater than or equal to zero ($CE_{max} \geq 0$), it is worthwhile to become involved in the project at any fractional working interest that returns CE ≥ 0. If $CE_{max} \leq 0$, then there is no range of working interest that will return a positive risk-adjusted value to a corporation.

When $W = 0$, then Eq. (4.8) yields CE $= 0$, whereas at 100% working interest ($W = 1$), CE ≥ 0 only when

$$G \geq RT \ln\{p_s[p_1 \exp(C_1/RT) + p_2 \exp(C_2/RT)]$$
$$+ p_f[p_3 \exp(C_3/RT) + p_4 \exp(C_4/RT)]\} = G_1 (RT) \qquad (4.11)$$

Thus, for a contract of fixed price G, assessed costs for each scenario must be kept low, as must also catastrophic probability chances, for a given risk tolerance if there is to be an involvement at 100% of the project.

In general, CE regarded as a function of W has a maximum when $\partial CE/\partial W = 0$, which occurs at a value of W satisfying the equation

$$(G - C_1)p_sp_1 \exp(WC_1/RT) + (G - C_2)p_sp_2 \exp(WC_2/RT)$$
$$+ (G - C_3)p_fp_3 \exp(WC_3/RT) + (G - C_4)p_fp_4 \exp(WC_4/RT) = 0 \qquad (4.12)$$

If $G > C_4 \geq C_i$ (for $i = 1, 2, 3$), then Eq. (4.12) has no solution for W. The implication is that CE is a monotonically increasing (or decreasing) function of W in $0 \leq W \leq 1$. Because CE $= 0$ when $W = 0$ and CE ≥ 0 when inequality (4.11) is obeyed, it follows that when $G \geq G_1(\text{RT})$, one takes 100% working interest in the project; otherwise, when $G \leq G_1(\text{RT})$, then CE ≤ 0 everywhere, and one takes zero working interest; i.e., one does not become involved in the project.

The more usual situation, however, is when the contract price G is less than C_4 (and occasionally less than C_3 and/or C_2) but is never less than C_1 or else one does not even consider involvement in the project. In the situation where $C_4 > G > C_1$, C_2, C_3, dividing Eq. (4.12) by $\exp(C_3W/\text{RT})$, separating the term containing cost C_4 from the rest, and taking the logarithm of the resulting expressions yield

$$
\begin{aligned}
W_{\max} = &[\text{RT}/(C_4 - C_3)] \ln([p_f p_4 (C_4 - G)]^{-1} \times \{(G - C_1)p_s p_1 \\
&\times \exp[-W_{\max}(C_3 - C_1)/\text{RT}] + (G - C_2)p_s p_2 \\
&\times \exp[-W_{\max}(C_3 - C_2)/\text{RT}] + (G - C_3)p_f p_3\})
\end{aligned}
\tag{4.13}
$$

provided that $0 \leq W_{\max} \leq 1$.

For instance, if $W_{\max}(C_3 - C_1) \ll \text{RT}$, then also $W_{\max}(C_3 - C_2) \ll \text{RT}$ because $C_3 \geq C_2 \geq C_1$. In such a case, the exponential terms in Eq. (4.13) may be approximated by unity, and then a solution to Eq. (4.13) is

$$
W_{max} = \frac{\text{RT}}{(C_4 - C_3)} \ln \left[\frac{G(p_s + p_f p_3) - p_s(p_1 C_1 + p_2 C_2) - p_f p_3 C_3}{p_f p_4 (C_4 - G)} \right]
\tag{4.14}
$$

Then one has only to check whether $W_{\max}(C_3 - C_1)/\text{RT} \ll 1$ and whether $0 \leq W_{\max} \leq 1$ for Eq. (4.14) to be an accurate approximation.

In general, one solves Eq. (4.13) numerically for the particular parameters (probabilities, value of contract, and costs) of the problem to obtain a curve of $W_{\max}$ versus RT. Then, for each value of RT, by inserting the corresponding $W_{\max}$ (if $0 \leq W_{\max} \leq 1$) into Eq. (4.8), one obtains the maximum certainty equivalent $CE_{\max}$ as a function of the risk tolerance. Equally, one then checks if $CE_{\max} \geq 0$. If so, then, by both increasing and decreasing W away from $W_{\max}$, one can obtain the two values of W at which CE [Eq. (4.8)] crosses zero. These two values then provide the range of W inside of which CE will exceed zero; i.e., the project is profitable for all other parameters held fixed. Denote these values of W by W_- and W_+ such that $0 \leq W_- \leq W_{\max} \leq \max(1, W_+)$. Then, by setting Eq. (4.8) equal to zero, one can obtain the two values of the working interest, that is,

$$
\begin{aligned}
W_\pm = \text{RT}/(G - C_4) \ \ln(p_s \{ &p_1 \exp[W_\pm(C_1 - C_4)/\text{RT}] \\
&+ p_2 \exp[W_\pm(C_2 - C_4/\text{RT}]\} \\
&+ p_f \{p_3 \exp[W_\pm(C_3 - C_4)/\text{RT}]\} + p_4)
\end{aligned}
\tag{4.16}
$$

that make the risk-adjusted value of the project equal to zero. One should notice now that $W_- = 0$ [because $\ln(1) = 0$, rendering CE in Eq. (4.8) equal to zero] but that W_+ is positive. If $W_+ < 1$, then there is a finite range of working interests that permits a profit, whereas if $W_+ > 1$, then any working interest, up to full involvement ($W = 1$) in the project, will return a profit, with the highest profit attained at a working interest W_{max}. Analytically, it is difficult to take these transcendental equations much further. The next subsection illustrates the dependencies between the parameters when a parabolic model is used for the utility function.

4.2.2 Parabolic Utility Function

While the exponential model for the utility function is a popular choice, it is difficult to handle analytically. Appendix 4B provides a quadratic approximation to the exponential form that is easier to analyze, thereby providing insight into the response of CE to changes in the values of the parameters. The parabolic approximation for CE is more risk averse at high working interest values than is the exponential CE formula [Eq. (4.7) or (4.8)] but is almost identical at low working interests. This property of the parabolic CE formula parallels decisions in the oil industry, where managers are often willing to accept an opportunity to around 50% involvement, even when the risk tolerance is high, but are less eager to accept smaller working interests. If one attempts to model this diversification preference by lowering the risk tolerance in the exponential model, then the entire range of working interest involvement is affected. However, if the parabolic form is used at the same risk tolerance, only the high end of the range of working interests is reduced, thus providing a correction to this managerial preference.

Consider an N-component decision tree with probability p_i along the ith branch and of value ε_i. For a working interest W and a risk tolerance RT, the exponential CE formula [see Eq. (4.7)] is

$$CE = -RT \ln \left[\sum_{i=1}^{N} p_i \exp - \left(\frac{W\varepsilon_i}{RT} \right) \right] \qquad (4.16)$$

The corresponding quadratic approximation is given by (App. 4B)

$$CE = WE \left(1 - \frac{Wv^2E}{2RT} \right) \qquad (4.17a)$$

or, alternatively, in the standard quadratic form $f(x) = ax^2 + bx + c$ by

$$CE(W) = aW^2 + bW \qquad a = \frac{-v^2E^2}{2RT} \qquad b = E \qquad (4.17b)$$

The expected value E in Eq. (4.17a) or (4.17b) is given by

$$E = \sum_{i=1}^{N} p_i \varepsilon_i \qquad (4.18a)$$

the square of the volatility by

$$v^2 = \frac{\sigma^2}{E^2} \tag{4.18b}$$

and the variance by

$$\sigma^2 = E_2 - E^2 = \sum_{i=1}^{N} p_i \varepsilon_i^2 - E^2 \tag{4.18c}$$

Thus, for the decision tree of Fig. 4.1, we have four association pairs (p_i, ε_i), given by $(p_s p_1, G - C_1)$, $(p_s p_2, G - C_2)$, $(p_f p_3, G - C_3)$, and $(p_f p_4, G - C_4)$, so that one obtains Eq. (4.2) for the expected value E and Eq. (4.3) for the variance σ^2. On substitution of Eqs. (4.2) and (4.3) into Eq. (4.17b), the parabolic approximation is obtained for the certainty equivalent amount of the decision tree of Fig. 4.1.

From the properties of parabolic functions (see any standard text on mathematical analysis, e.g., Zevas, 1973), when $E > 0$, one has that

$$W_{max} = -\frac{b}{2a} = \frac{RT}{v^2 E} \tag{4.19}$$

and on $W = W_{max}$, the maximum CE is attained at

$$CE_{max} = -\frac{b^2}{4a} = \frac{RT}{2v^2} \tag{4.20}$$

The two roots of CE (CE = 0), defining the range of working interests within which the certainty equivalent amount is positive (and hence the project is profitable), are given by

$$W_- = 0 \qquad W_+ = \frac{2RT}{v^2 E} \tag{4.21}$$

If $E < 0$, then CE is intrinsically negative [parameters a and b in Eq. (4.17b) are both negative] and so one should take no working interest in the project ($W = 0$).

4.3 APPLICATIONS

A numerical illustration is now given using the parabolic utility function for ease of presentation. Consider a hypothetical situation where estimates of spillage during transportation are about 1% so that, referring to Fig. 4.1, $p_f = 0.01$ and $p_s = 0.99$. Also assume that the probabilities of leakage (or not) during the burial phase of the project are not affected by what has occurred during transportation of the waste; that is, $p_1 = p_3$ and $p_2 = p_4$. For illustration, take the leakage probability to be 10% of the spillage probability; that is, $p_2 = p_4 = 0.001$ (and then $p_1 = p_3 = 0.999$). In the event of leakage at the burial site (but no spill during transportation), remediation costs (which may include excavation of leaking containers, repackaging and reburial of material, as well as addressing pollution problems at the site) are estimated to be 10 times higher than normal operating costs; that is, $C_2 = 10C_1$. If there is spillage, then not only must the spilled material be collected and repackaged but so too must any contaminated material. Set the cost of such an event at twice the basic no spill/no leakage estimated cost; that is, $C_3 = 2C_1$. The highest costs occur in the spill/leakage scenario when collection of contaminated material and disinterment, repackaging, reburial, and remediation costs have to be borne. Set these costs at 20 times those of the basic scenario; that is, $C_4 = 20C_1$.

The decision maker's concern is to estimate whether a corporation should be involved and at what level in this project depending on the contract price offered G and for given but variable amounts of risk tolerance RT. Thus one first estimates the statistical moments E, σ^2, and v^2 [Eqs. (4.2), (4.3), and (4.18b), respectively], which depend on costs and probabilities of each scenario and the contract price G. For the numerical values given above, setting $m = G/C_1$, one obtains

$$E = C_1(m - 1.0191) \tag{4.22a}$$

$$\sigma^2 = 0.0934C_1^2 \tag{4.22b}$$

$$v^2 = \frac{0.0934}{(m - 1.0191)^2} \tag{4.22c}$$

The certainty equivalent amount is then given by Eq. (4.17b), where the parameters a and b of the quadratic formula are

$$a = \frac{-0.0467C_1^2}{RT} \quad \text{and} \quad b = C_1(m - 1.0191) \tag{4.23}$$

Parameter a is always negative, and $b = E = C_1(m - 1.0191)$ needs to be positive for a corporation to even consider a contract offered at a price G. When $b \leq 0$, then CE in Eq. (4.17b) is intrinsically negative, and one should take no working interest in the project, leading to the condition

$$m = \frac{G}{C_1} > 1.0191 = m_* \tag{4.24}$$

i.e., the contract price needs to be about 2% higher than the basic no spill/no leakage estimated costs for the project to even be considered.

The range of working interests within which CE is positive is defined by the two end values [Eq. (4.21)]

$$W_- = 0 \qquad W_+ = 21.4(m - m_*)\frac{RT}{C_1} \qquad\qquad (4.25)$$

provided that $m > m_*$. The maximum working interest is attained at the value

$$W_{max} = 10.7(m - m_*)\frac{RT}{C_1} \qquad\qquad (4.26)$$

and at this value the maximum certainty equivalent amount becomes [Eq. (4.20)]

$$CE_{max} = 5.35RT(m - m_*)^2 \qquad\qquad (4.27)$$

Consider now that the contract price G is not particularly high but that G satisfies Eq. (4.24) so that involvement in the project can be undertaken. Then the requirement for W_+ to equal unity, and hence of the project returning a positive risk-adjusted value irrespective of the fractional involvement, is given by [Eq. (4.25)]

$$m = m_* + 0.0467\frac{C_1}{RT} = m_1 \qquad\qquad (4.28)$$

From Eq. (4.26), the requirement for complete involvement in a project ($W = 1$) under conditions of optimal risk-adjusted return is

$$m = m_* + 0.0933\frac{C_1}{RT} = m_2 > m_1 \qquad\qquad (4.29)$$

If G/C_1 is less than m_1, then both W_+ and W_{max} will be less than unity; i.e., only a partial involvement in the project will return a positive risk-adjusted value, whereas if G/C_1 is greater than m_1 but less than m_2, then the whole range of working interest is available to a decision maker, but the maximum return is attained at less than full involvement in the project.

Figure 4.2 plots W_{max} versus (RT/C_1) for various values of m, whereas Fig. 4.3 provides curves of the maximum CE standardized by the basic cost C_1 for increasing risk tolerance as a function of m. The complementary curves of W_{max} and CE_{max}/C_1 versus m, for various values of RT/C_1, are given in Figs. 4.4 and 4.5, respectively. Several factors are apparent from these graphs. First, as the tolerance to risk increases (RT/C_1 increasing), the maximum working interest a corporation can take increases; the higher the contract price G relative to the costs of the basic no spill/no leakage scenario C_1, the faster the optimal working interest reaches 100% for a fixed corporate risk tolerance (see Fig. 4.2). The maximum CE increases linearly with RT/C_1 until $W_{max} = 1$ when CE_{max}/C_1 reaches the value of $CE_{max}/C_1 = 0.5(m - m_*)$, because, from Eq. (4.29), one obtains (RT/C_1)($m - m_*$) $= 0.0933$, and then CE_{max}/C_1 becomes independent of RT/C_1 and is directly proportional to m. The critical curve at which this crossover occurs is shown as a dashed line in Fig. 4.3. The complementary Figs. 4.4 and 4.5 show how the maximum working interest rises as the contract price G/C_1 increases, with a more rapid growth of W_{max} toward unity as the risk tolerance increases (see Fig. 4.4), and how CE_{max}/C_1 first increases quadratically with increasing G/C_1 (see Fig. 4.5), but once the curve $(RT/C_1)(m - m_*) = 0.0933$ is crossed (dashed line), where $W_{max} = 1$, then CE_{max}/C_1 increases only linearly with increasing contract price G/C_1.

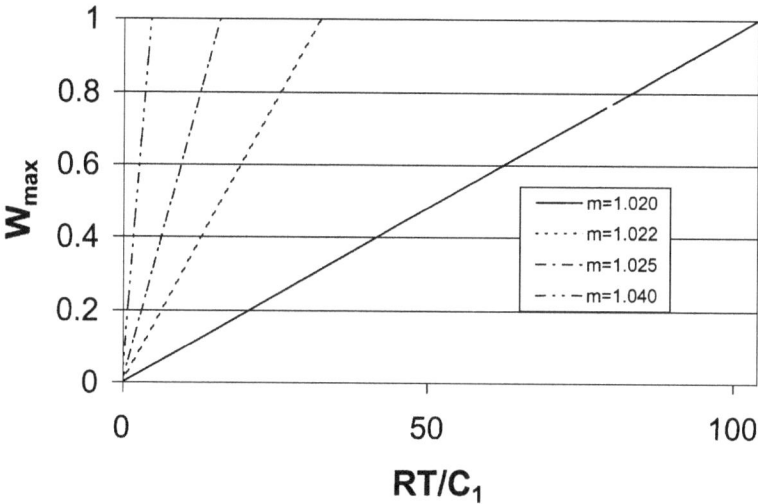

FIGURE 4.2   Maximum working interest versus risk tolerance for various contract awards.

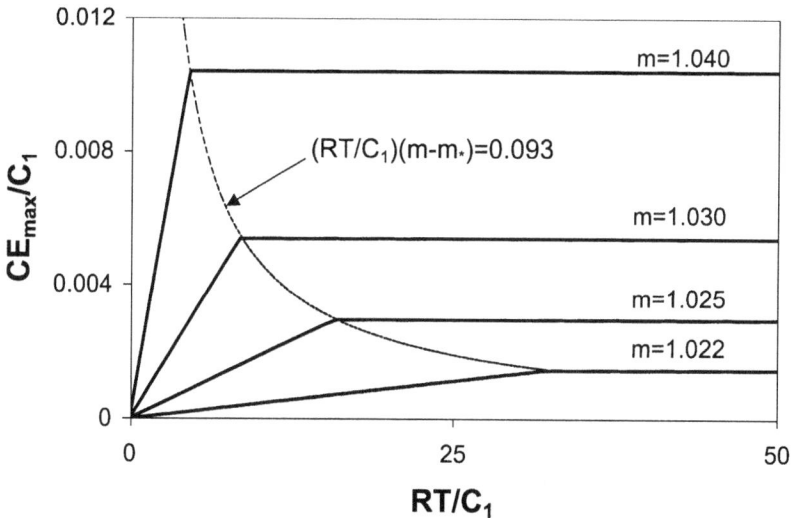

FIGURE 4.3   Maximum certainty equivalent amount versus risk tolerance for various contract awards.

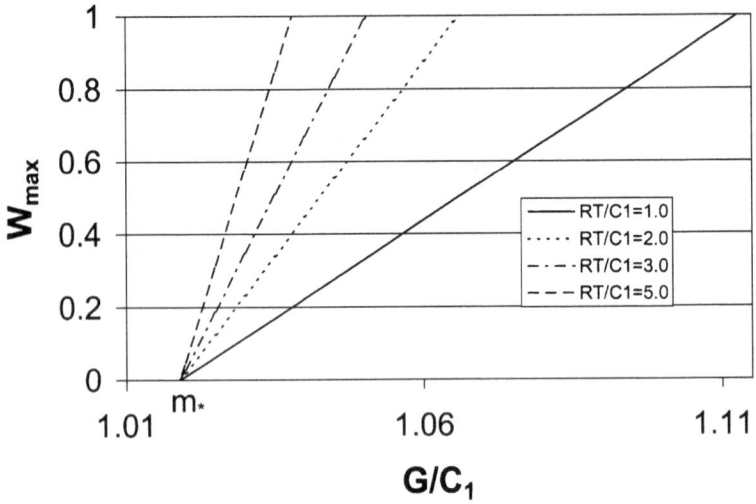

FIGURE 4.4 Maximum working interest versus contract award for various risk tolerances.

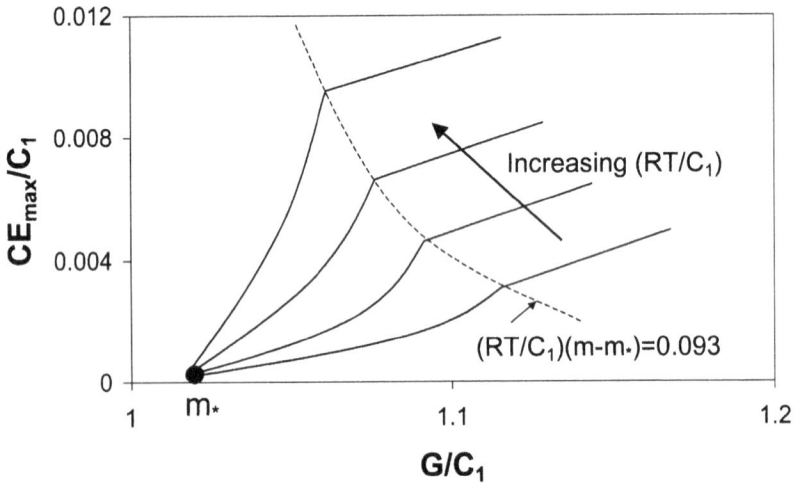

FIGURE 4.5 Maximum certainty equivalent amount versus contract award for various risk tolerances.

4.4 SUMMARY

One of the major concerns decision makers face in deciding whether to accept a contract for transport and burial of hazardous wastes is whether there will be spillage during transportation and/or leakage after burial, both of which could significantly affect corporate profitability should either occur. To mitigate the possible negative impact of such events, two avenues of potential damage control usually are invoked. First, each corporation has some predetermined risk tolerance that will not be exceeded without extremely good reason. Second, each corporation will limit potential exposure to the negative aspects (spillage and/or leakage) of a project by taking less than 100% working interest in a contract. While such fractional involvement lowers potential gains, it also mitigates potential liabilities, thus preserving corporate fiscal integrity.

Indeed, there exist situations where if the contract award does not exceed a certain amount, only partial involvement in the project may guarantee a positive certainty equivalent amount (the worth of the project after adjustment for various potential risks and for the risk attitude of a decision maker), and even then, the maximum return may be attained at less than full involvement in a project. This chapter has illustrated how such decisions can be made depending on a level of tolerance to risk and subject to the estimated probabilities of success and failure (spillage and/or leakage), together with the associated costs of various probable events. Exponential and parabolic utility models were employed to demonstrate how decision makers can evaluate a project and assess, for a given contract price, whether one wishes to become involved in the project, what fraction of the contract one is prepared to accept, and how this acceptance could influence corporate profits.

In particular, if a corporation's attitude toward risk can be approximated by an exponential utility function, then Eqs. (4.8) to (4.10) can be used to analyze the risk-adjusted value of a project, and Eqs. (4.12) to (4.15) can be employed to maximize the return from partial involvement in such project. Equivalently, if a corporation's risk attitude can be described by a parabolic model, then Eq. (4.17b) provides the risk-adjusted value of a transport and burial waste project, and Eqs. (4.19) to (4.21) can be used to estimate the maximum return from such a project.

APPENDIX 4A: LIMITING EXPRESSIONS FOR CERTAINTY EQUIVALENT

The certainty equivalent for an exponential utility is given in Eq. (4.8) as

$$CE = WG - RT \ln \Phi \, (RT) \tag{4.30}$$

where the function $\Phi(RT)$ corresponds to

$$\Phi(RT) = p_s \left[p_1 \exp\left(\frac{WC_1}{RT} \right) + p_2 \exp\left(\frac{WC_2}{RT} \right) \right]$$

$$+ p_f \left[p_3 \exp\left(\frac{WC_3}{RT} \right) + p_4 \exp\left(\frac{WC_4}{RT} \right) \right] \qquad (4.31)$$

When $RT \to \infty$, then the exponential functions in Eq. (4.31) tend to one, that is, $\Phi(RT) \to 1$, and because $\ln(1) = 0$ in the second term on the right-hand side of Eq. (4.30), one has an indeterminate limit of infinity multiplied by zero. Designating

$$H(RT) = RT \ln \Phi(RT) = \frac{\ln \Phi(RT)}{1/RT} \qquad (4.32)$$

allows one, as $RT \to \infty$, to obtain in Eq. (4.32) the indeterminate form of zero divided by zero; then one can use l'Hôpital rule to find the limit of $H(RT)$ as $RT \to \infty$ by equating it to the limit of the fraction of the derivatives of nominator and denominator in Eq. (4.32).

Now, one has that

$$\Pi(RT) = \frac{[\ln \Phi(RT)]'}{(1/RT)'}$$

$$= \left(-\frac{W}{RT^2} \right) \frac{\begin{array}{c} p_s[p_1 C_1 \exp(WC_1/RT) + p_2 C_2 \exp(WC_2/RT)] \\ + p_f[p_3 C_3 \exp(WC_3/RT) + p_4 C_4 \exp(WC_4/RT)] \end{array}}{-\Phi(RT)/RT^2} \qquad (4.33)$$

where the prime at the left-hand side of Eq. (4.33) stands for differentiation with respect to RT. The limit of Eq. (4.33), as $RT \to \infty$, is easily seen to be

$$\lambda = W[p_s(p_1 C_1 + p_2 C_2) + p_f(p_3 C_3 + p_4 C_4)] \qquad (4.34)$$

because $\Phi(RT)$ in the denominator tends to one, the powers of RT cancel out, and the exponential terms tend to one. Combining Eqs. (4.30) and (4.34) yields the limit of the certainty equivalent as $RT \to \infty$:

$$CE \to W[G - p_s(p_1 C_1 + p_2 C_2) - p_f(p_3 C_3 + p_4 C_4)] = WE \qquad (4.35)$$

When $RT \to 0$, it is not fruitful to differentiate Eq. (4.32) because all the derivatives of $H(RT)$ will contain, in the numerator, exponential terms that tend to infinity. For this case, make use of the theorem that states that if three functions are ordered as $f(x) \leq \phi(x) \leq g(x)$ and $\lim f(x) = \lim g(x) = 1$ as $x \to \xi$, then $\lim \phi(x) = 1$ for $x \to \xi$. Consider now the expression $\Pi(RT)/(WC_4)$. From Eq. (4.33), one has that

$$\frac{[\ln \Phi(RT)]'}{(1/RT)'} < WC_4 \frac{\begin{array}{c} p_s[p_1 \exp(WC_1/RT) + p_2 \exp(WC_2/RT)] \\ + p_f[p_3 \exp(WC_3/RT) + p_4 \exp(WC_4/RT)] \end{array}}{\Phi(RT)} = WC_4$$

(4.36)

because in the numerator of Eq. (4.33), all C_i, for $i = 1, 2, 3$, have been replaced with the largest cost ($C_4 > C_i$). Based on Eq. (4.36), one has that

$$\frac{\Pi(RT)}{WC_4} < 1 < 1 + RT \qquad (4.37)$$

Hence the function $\Pi(RT)/(WC_4)$ is bounded above by the function $1 + RT$, and furthermore, $\lim(1 + RT) = 1$ as $RT \to 0$.

From Eq. (4.33), it follows that

$$\frac{\Pi(RT)}{WC_4} > Wp_f p_4 C_4 \frac{\exp(WC_4/RT)}{WC_4 \Phi(RT)} \qquad (4.38)$$

because in Eq. (4.33) the numerator (which consists of additions of positive numbers) has been decreased. Dividing the numerator and denominator of the right-hand side of Eq. (4.38) by $\exp(WC_4/RT)$ yields

$$\frac{\Pi(RT)}{WC_4} > \frac{p_f p_4}{\begin{array}{c} p_s\{p_1 \exp[W(C_1-C_4)/RT] + p_2 \exp[W(C_2-C_4)/RT]\} \\ + p_f\{p_3 \exp[W(C_3-C_4)/RT] + p_4\} \end{array}} = Q(RT)$$

(4.39)

Notice that in Eq. (4.39) all the arguments of the exponential terms are negative (because $C_4 > C_i$, $i = 1, 2, 3$); hence all these terms tend to zero when $RT \to 0$, which in turn makes $Q(RT) \to 1$ as $RT \to 0$. Thus the function $\Pi(RT)/(WC_4)$ is bounded below by the function $Q(RT)$, and furthermore, $\lim Q(RT) = 1$ as $RT \to 0$.

Combining Eqs. (4.37) and (4.39) yields

$$Q(RT) < \frac{\Pi(RT)}{WC_4} < 1 + RT \qquad (4.40)$$

i.e., function $\Pi(RT)/(WC_4)$ is bounded below and above by two functions that both tend to unity as $RT \to 0$, leading to

$$\Pi(RT) \to WC_4 \qquad \text{when } RT \to 0 \qquad (4.41)$$

Using l'Hôpital's rule yields

$$RT \ln \Phi(RT) \to WC_4 \qquad \text{as } RT \to 0 \qquad (4.42)$$

which, combined with Eq. (4.30), gives

$$CE \to W(G-C_4) \qquad \text{when } RT \to 0 \qquad (4.43)$$

APPENDIX 4B: PARABOLIC APPROXIMATION TO EXPONENTIAL UTILITY

Here the parabolic approximation to the exponential form of the utility function is developed. Consider, for simplicity, a two-branch decision tree with probabilities p_1 and p_2 $(p_2 = 1 - p_1)$ and branch values ε_1 and ε_2, respectively. For example, such can be the case when one is faced with the random outcome of a venture that may either succeed with probability p_1 resulting in a value ε_1 or fail with probability p_2 $(p_2 = 1 - p_1)$ returning a value ε_2 (which can be negative). The expected utility for a working fraction W of this decision tree is given by

$$EU = p_1 \left[1 - \exp\left(\frac{-W\varepsilon_1}{RT} \right) \right] + p_2 \left[1 - \exp\left(\frac{-W\varepsilon_2}{RT} \right) \right] \qquad (4.44)$$

Using exponential expansions truncated at order 2, that is, $\exp(-x) = 1 - x + (x^2/2) + \cdots$, Eq. (4.44) yields

$$EU = \frac{WE}{RT} - \frac{W^2 E_2}{2RT^2} \qquad (4.45)$$

where the relations $E = p_1\varepsilon_1 + p_2\varepsilon_2$ and $E_2 = p_1\varepsilon_1^2 + p_2\varepsilon_2^2$ have been used. The second moment is given through

$$v^2 = \frac{E_2 - E^2}{E^2} \Rightarrow E_2 = v^2 E^2 + E^2 \qquad (4.46)$$

Substituting Eq. (4.46) in Eq. (4.45) gives

$$EU = \frac{WE}{RT} - \frac{W^2 v^2 E^2}{2RT^2} - \frac{W^2 E^2}{2RT^2} \qquad (4.47)$$

which, after rearrangement, becomes

$$EU = \frac{1}{RT} \left[WE \left(1 - \frac{Wv^2E}{2RT} \right) \right] - \frac{W^2E^2}{2RT^2} \tag{4.48}$$

Consider now the exponential utility of the certainty equivalent

$$U(CE) = 1 - \exp\left(-\frac{CE}{RT} \right) \approx \frac{CE}{RT} - \frac{CE^2}{2RT^2} \tag{4.49}$$

where, again, the exponential term is truncated to second order. Setting

$$CE = WE \left(1 - \frac{Wv^2E}{2RT} \right) \tag{4.50}$$

and raising Eq. (4.50) to the second power give

$$CE^2 \approx W^2E^2 \tag{4.51}$$

because the remaining terms are of order higher than 2 and so are not retained. Substituting Eqs. (4.50) and (4.51) into Eq. (4.49) yields

$$U(CE) = \frac{1}{RT} \left[WE \left(1 - \frac{Wv^2E}{2RT} \right) \right] - \frac{W^2E^2}{2RT^2} \tag{4.52}$$

i.e., the utility of the certainty equivalent is equal to the expected utility [Eq. (4.48)] of the decision tree. Hence, Eq. (4.50) provides the parabolic approximation to the exponential utility of the certainty equivalent amount. Following the same procedure, it is simple to show that Eq. (4.45) holds for a decision tree with an arbitrary number N of branches, Eqs. (4.46) to (4.52) remaining the same as above. Thus Eq. (4.17), with Eqs. (4.18a) to (4.18c) providing the statistical moments, gives the parabolic approximation to the exponential utility of CE of an N-component decision tree.

CHAPTER 5

CATASTROPHIC EVENTS, INSURANCE, AND UNILATERAL REGULATORY CHANGES

5.1 INTRODUCTION

The evaluation and management of risks in business projects are of increasing importance to corporations because of the magnitude of liability costs (Drewnowski, 1996) and of cost overruns (Yeo, 1990). Environmental projects in particular are burdened by the consequences of extreme events for which very little historical data exist and which, as illustrated in previous chapters, can alter the outcome of a project significantly. This lack of information severely limits the capabilities to assess the probabilities and consequences of distinct alternatives, a situation that is further accentuated by the complex interaction of physical, chemical, geographic, socioeconomic, and other factors that enter into decision-tree analyses of environmental projects (Bagneschi, 1998).

Environmental risks are inherent in most business operations and can have substantial ramifications on balance sheets and, consequently, on shareholder equity if they are not properly addressed (Voorhees and Woellner, 1997). Most corporations involved in activities that can affect the environment face the threat of legal and financial liabilities that may result from poor environmental management practices, improper waste-disposal methods, changes in environmental regulations, and inadequate risk management (coupled with the potential loss of reputation). Thus, for example, in 1997 the U.S. Environmental Protection Agency (EPA) levied fines of about $169 million against corporations for violations of environmental laws and, furthermore, referred 278 criminal and 426 civil cases (involving corporate compliance assurance programs) to the U.S. Department of Justice (Telego, 1998). During the same period, corporate liability in the United States had reached $250 billion (Merkl and Robinson, 1997), with accrued liability for environmental risks related to real estate property estimated at $2 trillion. This amount corresponds to around 16 to 20% of the total value of all property in the United States (Freeman and Kunreuther, 1997).

A significant factor that contributes to the preceding problems is poor understanding of the complex interactions of the scientific controls that influence the outcome of environmental solutions. Consequently, it is difficult to quantify scientifically short- and long-term liabilities, as well as the potential for limited and catastrophic losses (Paleologos et al., 1998). Thus liability risks can span the whole range of environmental activities: from phase I and II property assessments, to site investigations and sampling, to feasibility and remediation studies, to permit authorization, and to facility compliance audits and industrial hygiene surveys (Dixon, 1996). A survey of 33 environmental engineering firms found that 21% of phase I reports submitted in 1996 either had misestimated uncertainties related to technical aspects of the project or had neglected to document adverse environmental conditions from past use of a property. Accordingly, a realistic estimation of potential liabilities was impossible (Dunn, 1997). Internationally, the broad legal implications and economic ramifications of risks from operations in developing countries and the potential of liability transfer within the legal framework of the United States are being increasingly recognized by project finance professionals (Drewnowski, 1996; de Souza Porto and de Freitas, 1996).

In addition to environmental liabilities, cost overruns present another form of risk to clients and firms. Thus, in 1990, the average cost for private-sector environmental remediation projects in the United States was 25 to 50% over the initial budget (Al-Bahar and Crandell, 1990; Diekmann and Featherman, 1998). Correspondingly, the cost overruns related to major military, energy, and information-technology projects in the United States, the United Kingdom, and a number of developing countries were between 40 and 500% (Morris and Hough, 1987). Cost overruns arise primarily from (1) external risks due to modifications in the scope of a project and changes in the legal, economic, and technologic environments, (2) technical complexity of a project due to size, duration, or technical difficulty, (3) inadequate project management manifested in the control of internal resources, poor labor relations, and low productivity, and (4) unrealistic cost estimates because of improper quantification of the uncertainties involved (Yeo, 1990; Minato and Ashley, 1998). Thus Moorehouse and Millet (1994) found that poor assessment of uncertainties and their corresponding consequences, together with inadequate staff training, comprised the two primary causes of financial failure in environmental projects. Jeljeli and Russell (1995) found that underestimation of a project budget, improper insurance coverage, and lack of technical expertise constituted the major factors in liability risks.

This chapter discusses several insurance alternatives that can be used by environmental corporations to provide protection against the effect of extreme events, liabilities, and cost overruns. Detailed analyses and examples are provided for projects in the oil and environmental industries that illustrate the influence of insurance coverage on the profitability of projects. The chapter concludes with a presentation of two alternatives, the first consisting of the use of optioning against regulatory changes and the second dealing with the inclusion into a project's budget of the cost of a risk reserve fund that a corporation creates in order to meet potential liability claims.

Two conditions can occur in environmental waste-disposal problems that require that a corporation consider some form of insurance. First is the possibility of a catastrophic event occurring during transport and/or after burial of the waste, e.g., a truck overturning and spilling waste, thereby involving massive and expensive cleanup costs. A corporation would surely like to take out insurance against this possibility. Second is the possibility that the regulatory agencies may consider a unilateral change in the environmental stringency conditions for transport and/or burial of a material *after* a project is under way. In this case, the corporation could be involved in further costs, thereby lowering potential profitability of a preexisting contract. The corporation surely would like to option against the possibility of such an event occurring prior to the chance that the contract terms could be changed. These two forms of insurance are not equivalent. In the catastrophic loss event situation, one would like to pay an insurance premium to cover the unknown catastrophic costs should they occur. In the regulatory stringency conditions situation, one usually knows ahead of time precisely how such more stringent conditions, if enacted, would influence the corporate profitability, and one would like to have a contingency option operating that would be activated if and only if the regulatory agency does indeed enact the new more stringent regulations.

We consider first the catastrophic loss problem. A prevalent concern that is becoming manifest in decision-tree analyses for hydrocarbon exploration projects has to do with environmental cleanup if something goes wrong. For instance, if a project is saddled with the small chance of a catastrophic blowout while drilling and the associated high cleanup costs, then it is all too easy to take what could have been a very worthwhile exploration project and downgrade it to less than worthwhile. Equally, even if a blowout scenario is not included while drilling, a similar catastrophic loss condition is often envisioned by management: economic oil (or gas) is found, and once into production, there is the chance of an oil spill with, again, enormous environmental cleanup costs. Accordingly, management often uses the catastrophic loss excuse to get out of (or never get into) good exploration projects. Thus management bias can lead to a reevaluation that takes, *on average,* a very good prospect and makes it seem very poor.

The point about all these management biases, however, is that if only the expected value (average value) is used to quantify the risk, then one is completely overlooking the fact that there is also an uncertainty on the expected value that ameliorates the "catastrophic loss" effect. The purpose of this chapter is to show how this amelioration operates in practice. Two case histories will be outlined: catastrophic failure during (1) the exploration assessment and (2) the development assessment.

5.2   CATASTROPHIC LOSS IN EXPLORATION ASSESSMENTS

A blowout (or other catastrophic failure) while exploring, due to a natural cause, will lead to loss of equipment and massive environmental cleanup costs. The classic way to assess how catastrophic failure chance operates is as follows: First, an evaluation of an opportunity is made, yielding an estimate of costs C, potential gains G, and chance p_s of successfully finding hydrocarbons. If this were all that were to be done, then one can use the decision-tree diagram of Fig. 5.1a to estimate an expected value of

$$E = p_s G - C \qquad (5.1)$$

and a variance σ^2 on the expected value of

$$\sigma^2 = p_s p_f G^2 \qquad (5.2)$$

where p_f is the failure probability, $p_f \equiv 1 - p_s$. Conventionally, one would then estimate the probability P_+ of obtaining an investment return greater than or equal to zero from

$$P_+ = \sigma^{-1}(2\pi)^{-1/2} \int_0^\infty \exp - \left(\frac{(x - E)^2}{2\sigma^2} \right) dx \equiv \pi^{-1/2} \int_{-d}^\infty \exp(-u^2)\, du \quad (5.3a)$$

where

$$d = -\frac{E}{2^{1/2}\sigma} \qquad (5.3b)$$

However, the difficulty arises when a mandated catastrophic failure is required to be included with vanishing small probability p_c.

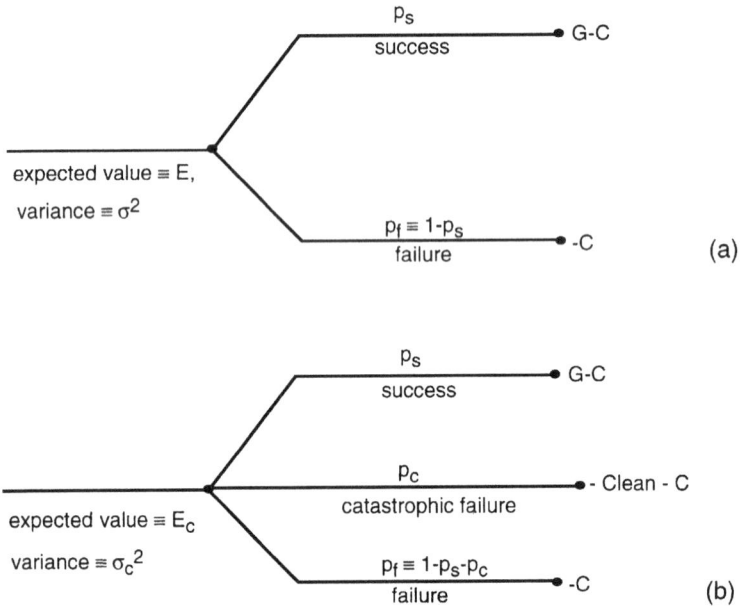

FIGURE 5.1 Decision-tree diagram illustrating (a) a conventional evaluation scheme for an exploration opportunity and (b) changes brought about by inclusion of a "catastrophic failure" possibility.

A depiction of inclusion of such a failure in a classic decision tree is given in Fig. 5.1b, where "Clean" is the amount one would have to pay for cleanup and equipment loss if the catastrophic failure occurs. And "Clean" is usually very large compared with normal costs C of the project and often even compared with anticipated gains G if the project were to succeed. In this case, the expected value E and the variance σ^2 in the *absence* of the catastrophic loss are as given by Eqs. (5.1) and (5.2), respectively, with the probability P_+ of a positive return again given by Eq. (5.3b).

Inclusion of the catastrophic loss changes both the expected value E_c and variance σ_c^2, respectively, to

$$E_c = p_s G - C - p_c \text{Clean} \equiv E - p_c \text{Clean} \qquad (5.4a)$$

and

$$\sigma_c^2 = p_s G^2 + p_c (\text{Clean})^2 - (p_s G - p_c \text{Clean})^2 \qquad (5.4b)$$

The probability P_c of making a return on investment greater than zero is now given by

$$P_C = \sigma_C^{-1} (2\pi)^{-1/2} \int_0^\infty \exp\left(\frac{-(x - E_C)^2}{2\sigma_C^2} \right) dx \equiv \pi^{-1/2} \int_{-D_c}^\infty \exp(-u^2)\, du \quad (5.5a)$$

where

$$D_c = \frac{-E_c}{2^{1/2}\sigma_C} \qquad (5.5b)$$

The point here is that *both* the expected value *and* its uncertainty (as measured through the variance) are used to assess the worth of the project. And the relevant comparison to make is between the probability of obtaining a positive return from the project both in the absence and presence of the catastrophic failure zone. Consider then, for illustrative purposes, the numerical example sketched in Fig. 5.2, both in the absence (Fig. 5.2a) and presence (Fig. 5.2b) of the catastrophic loss scenario.

For the assessment in the *absence* of the catastrophic failure scenario, one has, from Fig. 5.2a, that

$$E = 60 \qquad \sigma = 0.55 \times 10^3 \qquad \frac{E}{2^{1/2}\sigma} = 7.7 \times 10^{-2} \qquad (5.6a)$$

while for the assessment *including* the catastrophic loss scenario, one has, from Fig. 5.2b, that

CHAPTER FIVE

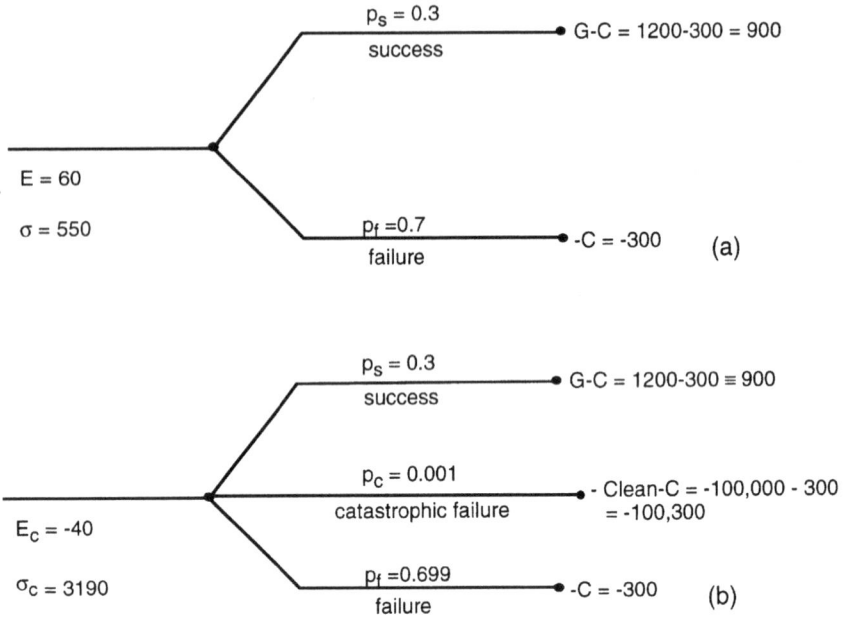

FIGURE 5.2 Decision-tree diagram illustrating (*a*) numerical values for Fig. 5.1a and (*b*) numerical values for Fig. 5.1b.

$$E_c = -40 \qquad \sigma_C = 3.19 \times 10^3 \qquad \frac{E_c}{2^{1/2}\sigma_C} = -0.89 \times 10^{-2} \qquad (5.6b)$$

Thus, in this case, while the expected value is shifted from positive to negative by the inclusion of the catastrophic option, the standard error in the mean is increased by a factor of 6, suggesting less accuracy in the mean.

Indeed, a calculation of the probability of obtaining a positive return in the absence of the catastrophic option loss yields

$$P_+ = 54.3\% \qquad (5.7a)$$

whereas including the catastrophic loss yields the positive return probability of

$$P_c = 49.6\% \qquad (5.7b)$$

Thus the catastrophic loss scenario is ameliorated correctly when the inclusion of the variance on the mean is given. In the specific example given, just over a 4% drop occurs in the chance of being profitable, so the small probability of a catastrophic loss is being included correctly—something that is not possible to have or even visualize if one operates with only the expected value as a measure of worth of the opportunity.

5.3 CATASTROPHIC LOSS AFTER OIL IS FOUND

A different type of management concern is to argue that even if an opportunity is to be drilled and if it were to find oil, then there is the probability of a catastrophic oil spill with attendant massive cleanup costs. Such arguments are again often used to downgrade good prospects. In this case, the decision-tree diagram is different from that of the prior case. As sketched in Fig. 5.3a, the concern here is with catastrophic failure *only* if oil is found.

In this case, with p_n as the probability of no oil spill and with "Clean" the cost of cleanup if an oil spill does occur, from Fig. 5.3a one can write the expected value E_c and its variance in the presence of a catastrophic oil spill as

$$E_c = p_s G - C - p_s (1 - p_n)\text{Clean} \tag{5.8a}$$

and

$$\sigma_C^2 = p_s \{(1 - p_s)G^2 - 2(1 - p_s)(1 - p_n)\,G\text{Clean}$$
$$+ (1 - p_n)[1 - p_s(1 - p_n)]\text{Clean}^2\} \tag{5.8b}$$

Consider, for illustration, the numerical situation depicted in Fig. 5.3b. Then

$$E_c = -240 \quad \text{and} \quad \sigma_C = 3.12 \times 10^3 \tag{5.9a}$$

so that

$$\frac{E_c}{2^{1/2}\sigma_C} = -5.44 \times 10^{-2} \tag{5.9b}$$

For comparison, note that if the catastrophic oil spill scenario is omitted, then the corresponding values are

$$E = +60 \quad \text{and} \quad \sigma = 0.55 \times 10^3 \tag{5.10a}$$

so that

$$\frac{E}{2^{1/2}\sigma} = 7.64 \times 10^{-2} \tag{5.10b}$$

In this case, there is a reversal of the expected value (from profitable to nonprofitable *on average*), but there is also an increase in the uncertainty (standard error) on the expected value by an order of magnitude.

The corresponding probabilities of obtaining a positive return are (1) in the *absence* of the catastrophic scenario, one has

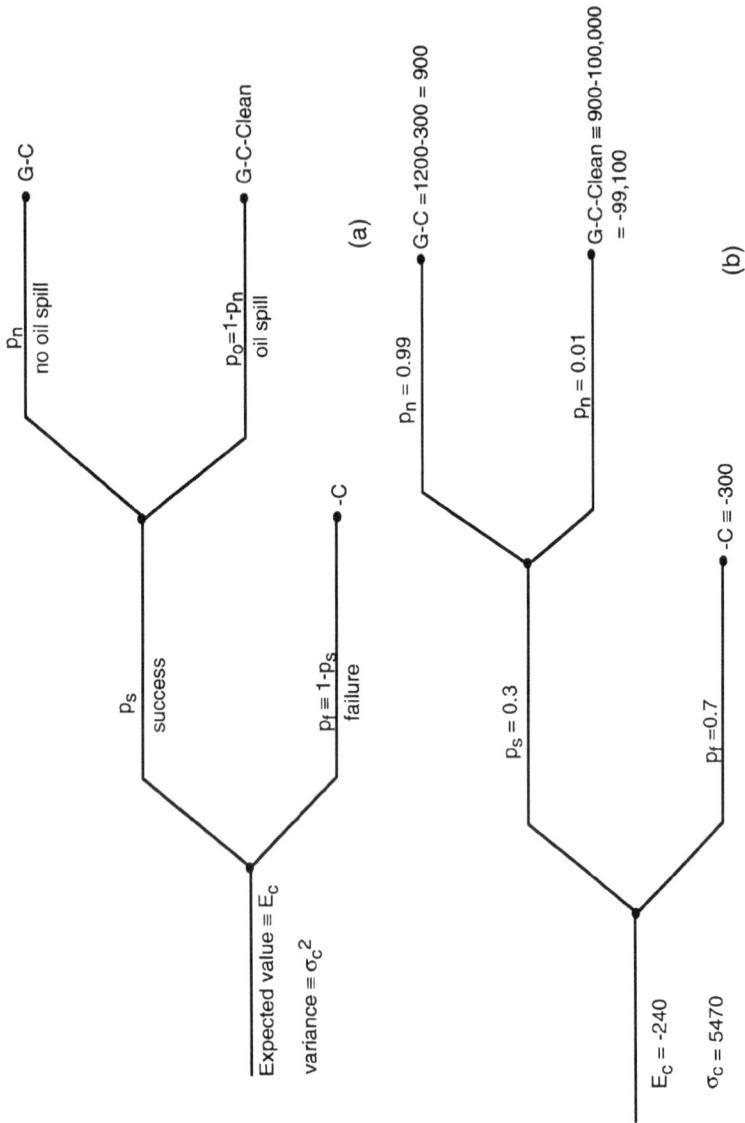

no oil spill p_n •G-C

oil spill $p_o = 1 - p_n$ •G-C-Clean

success p_s

failure $p_f = 1 - p_s$ •-C

Expected value ≡ E_c

variance ≡ σ_c^2

(a)

$p_n = 0.99$ •G-C = 1200-300 = 900

$p_n = 0.01$ •G-C-Clean = 900-100,000 = -99,100

$p_s = 0.3$

$p_f = 0.7$ •-C = -300

$E_c = -240$

$\sigma_c = 5470$

(b)

FIGURE 5.3 Decision-tree diagram illustrating (*a*) how a catastrophic oil spill after finding oil is included and (*b*) numerical values for the depiction of Fig. 5.3*a*.

$$P_+ = 54.3\% \tag{5.11a}$$

and (2) in the presence of the catastrophic scenario, one has

$$P_c = 48.3\% \tag{5.11b}$$

Thus inclusion of the variance (into the calculation of the influence of a catastrophic oil spill) lowers the chance of making a profit by only 6%. Thus amelioration is again properly taken into account.

5.4 INSURANCE FOR HYDROCARBON EXPLORATION AND DEVELOPMENT RISKS

Insurance is one of the costs that is becoming of increasing concern to oil companies as they strive to limit corporate exposure to loss under either exploration or development conditions. On the exploration side, one of the hazards that insurance is designed to mitigate is the loss of an exploration rig due to adverse weather conditions (a hurricane), fire, blowout, or other factors of low probability but high cost to the corporation if they occur. Thus an exploration project has associated with it not only the usual estimates of gains G, normal exploration costs C, and the success probability p_s of encountering commercial hydrocarbons but also the chance p_{fc} of a catastrophic failure with costs C_1 that can be many times the "normal" exploration costs. This scenario is depicted in Fig. 5.4a. To mitigate against having to absorb the high costs C_1 should this low-probability event be realized, a company will insure a fraction f of such potential catastrophic losses at an insurance premium J. Then, should such a catastrophic event occur, corporate liability is limited to $C_1(1 - f)$ plus the insurance cost J. This situation is sketched in Fig. 5.4b. The question is: What is the maximum insurance cost the company should pay?

Equally, in a hydrocarbon development project, there is the chance p_s that the wells involved in the project could yield a net gain G, but there is also the chance that there is a net loss C, perhaps due to too many wells being involved or too high a lease cost being paid or too low an oil selling price. In any event, however, part of the risk occurs *after* oil is found and development undertaken. There is always the chance p_{fc} of an oil spill or a blowout with some attendant cleanup costs C_1 that are usually sufficiently high that they can convert a likely profitable development project to a losing proposition. This situation is sketched in Fig. 5.5a. To mitigate against this potential for disaster, a corporation usually will take out insurance at a premium cost J *after* oil has been found. The insurance company will then pick up the fraction f of the catastrophic costs C_1 so that corporate liability is limited to $(1 - f)C_1$ if the catastrophe occurs. This insured situation is sketched in Fig. 5.5b. Once again, the question is: How much should the corporation pay for insurance?

No insurance

(a)

Insurance

(b)

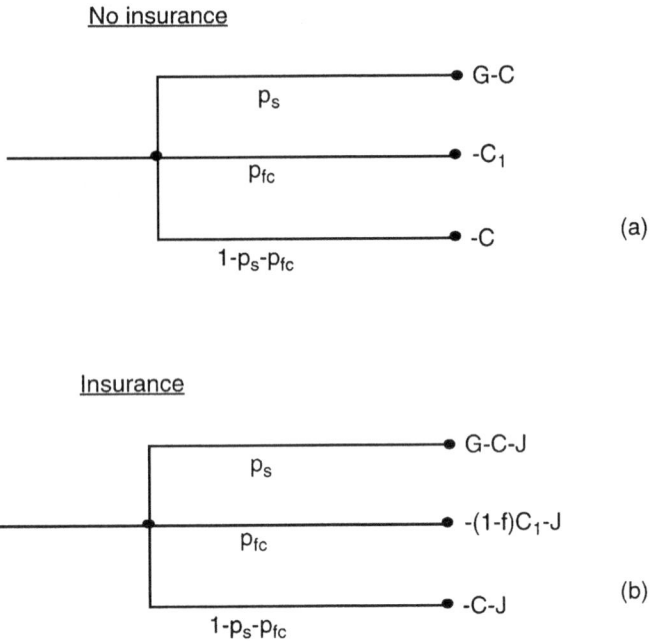

FIGURE 5.4 Decision-tree diagrams showing the effects of (*a*) excluding insurance premium and (*b*) including insurance premium for an exploration project.

The two situations just described and the decision-tree depictions of Figs. 5.4 and 5.5 are not identical. Neither, therefore, are their insurance costs. Indeed, a very large number of different types of decision trees can be constructed to incorporate facets of insurance costs for different components of the various branches of the trees. The two examples just described are designed to show the differences that can arise in formulating insurance needs and costs.

We consider each of the examples in turn to develop the typical logic lines for handling such problems and also give numerical illustrations to show practically the influence of insurance costs on the worth of a project and the amelioration of massive potential costs of catastrophic events.

5.5 GENERAL CATASTROPHIC LOSS CONDITIONS

5.5.1 Mathematical Considerations

The first illustration to be considered is the catastrophic loss situation depicted in Fig. 5.4*a* (without insurance) and *b* (with insurance). The main effect of paying

No insurance

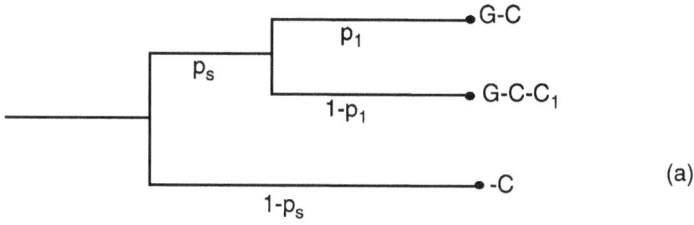

(a)

Insurance

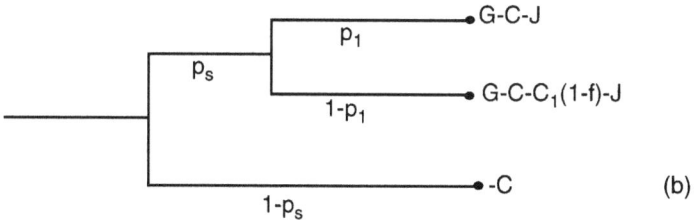

(b)

FIGURE 5.5   As for Fig. 5.4 but only after oil is found so that an oil spill catastrophe is a real possibility.

the insurance cost J is to decrease the catastrophic loss costs the corporation would have to bear (if the catastrophic loss scenario occurred) from C_1 to $(1 - f)C_1$. Thus insurance would pick up the cost fC_1 of the total. The question to be addressed is the price to be paid J for insurance versus fractional coverage f of the catastrophic condition.

In the *absence* of insurance cover, as depicted in Fig. 5.4a, the expected value E_0 of the project is

$$E_0 = p_s G - C(1 - p_{fc}) - p_{fc} C_1 \qquad (5.12)$$

and the variance σ_0^2 on this expected value is

$$\sigma_0^2 = p_s(1 - p_s)G^2 + 2p_s p_{fc} G(C_1 - C) + p_{fc}(1 - p_{fc})(C_1 - C)^2 \quad (5.13)$$

While the probability p_{fc} of catastrophic failure is usually estimated to be extremely small, the problem is that the costs C_1, if such a failure does occur, are so large that the product $p_{fc} C_1$ usually will dominate the expected value E_0. Thus an otherwise highly profitable venture ($E_0 \gg 0$) in the absence ($p_{fc} = 0$) of a catastrophic chance can be converted to an expected major cash drain on the corporation ($E_0 \ll 0$) if $p_{fc} > 0$—even for very small values of p_{fc}.

If an insurance premium J is paid so that the fraction f of the catastrophic costs would be paid by the insurance company, then the corporate risk is reduced to $(1 - f)C_1$. Accordingly, the corresponding expected value E_1 is now

$$E_1 = E_0 + p_{fc}fC_1 - J \qquad (5.14)$$

with variance σ_1^2 around this expected value of

$$\sigma_1^2 = p_s(1 - p_s)G^2 + 2p_s p_{fc}G[C_1(1 - f) - C] \\ + p_{fc}(1 - p_{fc})[C_1(1 - f) - C]^2 \qquad (5.15)$$

which, of course, is smaller than σ_0^2 because of the lowering of the catastrophic costs charged to the corporation. In addition, the new expected value E_1 is larger than the "old" expected value E_0 by the amount $p_f fC_1 - J$, representing the pickup of expected catastrophic costs by the insurance company rather than by the corporation.

As a first requirement, then, it would seem that one should pay an insurance premium J only when the fractional coverage f of catastrophic losses satisfies the inequality

$$J < p_{fc}fC_1 \qquad (5.16)$$

because otherwise E_1 will be smaller than E_0, and so the expected value to the corporation would be reduced. In addition, because $\sigma_0^2 < \sigma_0^2$, the uncertainty on the E_1 value is less than that on the E_0 value; i.e., E_1 is statistically sharper than E_0.

A further measure of worth of insurance is guided by the probability of making a profit, *including* the variance on the expected value. Thus let P denote the probability of making a profit greater than or equal to zero. Then

$$P = \sigma^{-1}(2\pi)^{-1/2} \int_0^\infty \exp\left(\frac{-(x - E)^2}{2\sigma^2}\right) dx \qquad (5.17)$$

which can be rewritten in the form

$$P = \pi^{-1/2} \int_{-\alpha}^\infty \exp(-y^2)\ dy \qquad (5.18)$$

where $\alpha = E/(2^{1/2}\sigma)$.

Note that even if the expected value is negative, there is still a chance of turning a profit ($P > 0$), but this chance declines as E becomes more negative at a fixed variance.

Thus, in the *absence* of taking out insurance, the probability P_0 of making a profit is

$$P_0 = \pi^{-1/2} \int_{-a}^{\infty} \exp(-y^2)\, dy \tag{5.19}$$

where $a = E_0/(2^{1/2}\sigma_0)$, whereas when insurance *is* taken out, then the probability P_1 of making profit is

$$P_1 = \pi^{-1/2} \int_{-b}^{\infty} \exp(-y^2)\, dy \tag{5.20}$$

where $b = E_1/(2^{1/2}\sigma_1)$.

If one is to pay insurance, then, at the very least, one would like the probability of a profit to increase; i.e., one requires at a minimum that $E_1/\sigma_1 > E_0/\sigma_0$—which incorporates inequality (5.16) on the amount of insurance to pay versus coverage obtained.

However, a more stringent condition is usually in force. A corporation often will insist that a project be undertaken only if a given probability exists of the project being profit-making. Thus, if P is some fixed value, then the question reduces to finding the values of fractional coverage f and insurance premium J so that Eq. (5.20) will yield a *specific,* predetermined value for P.

A remarkably good practical approximation for Eq. (5.20) in $0.1 \le P \le 0.9$ is that

$$P \cong \frac{1}{2} + \frac{b}{\pi^{1/2}} \tag{5.21a}$$

or, conversely, if P is given, then one has

$$b \approx \pi^{1/2}(P - \tfrac{1}{2}) \tag{5.21b}$$

Equation (5.21b) is useful for a prescribed corporate mandate on P, the probability of making a profit, because one then requires

$$E_1 \approx (2\pi)^{1/2}\sigma_1(P - 0.5) \tag{5.22}$$

Thus the approximate Eq. (5.22) provides a constraint on insurance premium J and coverage fraction f of the catastrophic loss costs and, provided $P \ge 50\%$, automatically requires $E_1 > 0$—a "standard" constraint often invoked for decision-making investments. Because both E_1 and σ_1 involve the coverage fraction f, but because only E_1 involves the insurance premium J, it is easiest to rearrange Eq. (5.22) to provide the expression

$$J \approx p_s G - C(1 - p_{fc}) - p_{fc}(1 - f)C_1 - (2\pi)^{1/2}\sigma_1(P - 0.5) \tag{5.23}$$

Thus one can insert into Eq. (5.23) the fractional coverage required and the corporate-mandated probability of a profit and so figure the appropriate premium to pay, subject to the constraint $0 \le J \le p_{fc}fC_1$, of course.

5.5.2 Numerical Illustration

Consider a typical set of parameters for an exploration project. Let the success probability $p_s = 30\%$, with normal costs C about a quarter of expected gains (that is, $G = 4C$). The catastrophic probability usually is estimated to be very small, typically 1% or less. We carry through the numerical illustration with $p_{fc} = 10^{-2}$ for definiteness.

Estimating catastrophic costs C_1 is another matter entirely. We perform the calculations with $C_1 = 100G = 400C$ so that there would be no chance of a profit in the expected value E_0 without insurance because then

$$E_0 = 0.3 \times G - C(1 - 0.3 - 10^{-2}) - 10^{-2} \times C_1$$

$$= C(0.3 \times 4 - 1 + 0.3 + 10^{-2} - 10^{-2} \times 400)$$

$$= -3.49C < 0 \tag{5.24}$$

Thus, without some form of insurance, the high cost of catastrophe, if such occurs even at the 1% chance, makes the expected value of the project very negative.

For the same numerical values one has the variance σ_1^2 given through

$$\frac{\sigma_1^2}{C^2} = 3.36 + 9.6(0.9975 - f) + 1584(0.9975 - f)^2 \tag{5.25}$$

The connection between insurance premium J, the mandated corporate success probability P, and the fractional coverage f is, from Eq. (5.23), given by about

$$\frac{J}{C} = -2.79 + 4f - (2\pi)^{1/2}(P - 0.5)\,[3.36 + 9.6\,(0.9975 - f)$$

$$+ 1584\,(0.9975 - f)^2]^{1/2} \tag{5.26a}$$

provided that

$$0 \le \frac{J}{C} \le 4f \tag{5.26b}$$

and inequality (5.26b) is automatically satisfied for a corporate-mandated profit probability greater than 50%. At 100% coverage ($f = 1$) of the catastrophic losses, one has the insurance premium of

$$\frac{J}{C} \approx 1.21 - 4.64(P - 0.5) \tag{5.27}$$

Thus, at 50% probability of a profit, one should pay 121% of normal costs C to cover the likelihood of catastrophic costs (i.e., 0.33% of catastrophic costs).

5.6 *INSURANCE COVERAGE AFTER OIL IS FOUND*

5.6.1 General Considerations

If oil were to be found, then the insurance coverage one would take is predicated on the probability of oil spillage, as depicted in Fig. 5.5. The point here is that one must fold this potential cost of insuring against spillage into the assessment of the worth of the project ahead of drilling. In this case, two factors are operative: (1) the fact that until (or unless) one finds commercial amounts of oil, one does not take out insurance, and (2) the fact that the corporation must make a decision to undertake the exploration project *ahead* of knowing whether oil is present or not. Thus the corporation must assess potential gains G and potential cleanup costs C_1 if an oil spill were to occur, as well as reduction in potential gains due to the insurance premium J the corporation would pay if oil were to be found. And these assessments must all be done at the stage of deciding whether to undertake the exploration project or not. Thus the decision-tree diagram *excluding* insurance coverage but *including* the potential for an oil spill is given in Fig. 5.5a, whereas including insurance cover to a fraction f of potential oil-spill costs C_1 at an insurance premium J is shown in Fig. 5.5b.

The expected value of the opportunity in the *absence* of insurance cover is

$$E_0 = p_s[G - C_1(1 - p_1)] - C \tag{5.28}$$

whereas the variance σ_0^2 on this mean value is just

$$\sigma_0^2 = p_s\{(1 - p_s)[G - C_1(1 - p_1)]^2 + p_1(1 - p_1)C_1^2\} \tag{5.29}$$

Note that if $p_1 = 1$ (no catastrophic oil spill), then σ_0^2 reduces correctly to $p_s(1 - p_s)G^2$, which is the usual variance limit.

If an insurance premium J is paid, with a corresponding catastrophic insurance coverage of fC_1 paid by the insurance company, then the corresponding mean expected value is E_1, where

$$E_1 = p_s[G - J - (1 - f)(1 - p_1)C_1] - C \tag{5.30}$$

and the relevant variance on this expected value is

$$\sigma_1^2 = p_s\{(1 - p_s)[G - J - (1 - f)(1 - p_1)C_1]^2 + p_1(1 - p_1)(1 - f)^2C_1^2\} \tag{5.31}$$

The probability P_0 of a profitable operation in the *absence* of insurance cover is

$$P_0 = \pi^{-1/2}\int_{-b_0}^{\infty} \exp(-u^2)\, du \tag{5.32}$$

where $b_0 = E_0/(2^{1/2}\sigma_0)$.

In the presence of insurance cover in the amount fC_1 at a premium J, the corresponding probability P_1 of a profitable income is

$$P_1 = \pi^{-1/2} \int_{-b_1}^{\infty} \exp\left(-u^2\right) du \tag{5.33}$$

where $b_1 = E_1/(2^{1/2}\sigma_1)$.

For a *fixed* corporate probability on profit, it follows from Eq. (5.33) that b_1 must then be a fixed dimensionless number n, say, where

$$P_1 = \pi^{-1/2} \int_{-n}^{\infty} \exp\left(-u^2\right) du \tag{5.34}$$

with

$$E_1 = 2^{1/2}\sigma_1 n \tag{5.35}$$

If a probability P_1 *greater* than 50% is required of operating at a profit, then n must be positive. In this case, because σ_1 is intrinsically positive, it follows that $E_1 \geq 0$; otherwise, there is no chance of satisfying a corporate mandate of greater than 50% chance of profit.

The requirement $E_1 \geq 0$ can be written

$$G \geq \frac{C}{p_s} + J + (1 - f)(1 - p_1)C_1 \tag{5.36}$$

Failure to satisfy inequality (5.36) means that one *cannot* have a greater than 50:50 chance of making a profit.

Once inequality (5.36) *is* satisfied, then the probability of making a profit greater than or equal to that n value given by Eq. (5.35) can be written

$$E_1^2 = \frac{2n^2}{(1 + 2n^2)}\left(E_1^2 + \sigma_1^2\right) \tag{5.37}$$

Rearranging factors allows Eq. (5.37) to be written as the quadratic form

$$a^2 p_s(\lambda - p_s) + 2p_s aC + \lambda p_1(1 - p_1)(1 - f)^2 C_1^2 - C^2 = 0 \tag{5.38}$$

where

$$a = G - J - (1 - f)C_1(1 - p_1) \tag{5.39a}$$

and

$$\lambda = \frac{2n^2}{1 + 2n^2} \leq 1 \tag{5.39b}$$

The relevant solution of the quadratic Eq. (5.38) for the insurance premium J is

$$J = G - (1 - f)(1 - p_1)C_1 + C[p_s(\lambda - p_s)]^{-1}$$
$$\times \{p_s\lambda - (\lambda p_s)^{1/2}[1 - (\lambda - p_s)p_1(1 - p_1)(1 - f)^2 C_1^2/C^2]^{1/2}\} \qquad (5.40)$$

Note that if the insurance cover is 100% ($f = 1$), then Eq. (5.40) provides the maximum insurance premium one should pay of

$$J = G + C[p_s(\lambda - p_s)]^{-1}(p_s\lambda)^{1/2}[1 - (p_s\lambda)^{1/2}] \qquad (5.41)$$

at a given probability of success.

If the insurance cover is zero ($f = 0$), then Eq. (5.38) provides

$$J = G - (1 - p_1)C_1 - C[p_s(\lambda - p_s)]^{-1}(p_s\lambda)^{1/2}$$
$$\times \{[1 - (\lambda - p_s)p_1(1 - p_1)C_1^2/C^2]^{1/2} - (p_s\lambda)^{1/2}\} \qquad (5.42a)$$

However, for a greater than 50% probability of success, one has the requirement (on $f = 0$) of

$$J \leq G - \frac{C}{p_s} - (1 - p_1)C_1 \qquad (5.42b)$$

And both Eqs. (5.42a) and (5.42b) *cannot* be satisfied simultaneously, so one should not then pay any premium.

5.6.2 Numerical Illustration

Consider that a success probability of $p_s = 1/3$ obtains, with potential gains G of 10 times normal costs C, that is, $G = 10C$. Then, in the absence of any potential for catastrophic loss ($p_1 = 1$), the expected value of the opportunity is

$$E_0 = p_s G - C = 7/3 C > 0 \qquad (5.43a)$$

and the variance is

$$\sigma_0^2 = 2/9 \times 10^2 C^2 \qquad (5.43b)$$

so that $E_0/(2^{1/2}\sigma_0) = 0.35$ with the associated probability of making a profit at $P = 0.7$.

Including a 1% chance ($p_1 = 0.99$) of a major oil spill after production commences, with an associated cleanup cost of $C_1 = 1000C$, then yields an expected project value of

$$E_1 = p_s[G - C_1(1 - p_1)] - C = -C < 0 \qquad (5.44a)$$

and the corresponding variance on E_1 is

$$\sigma_1^2 = 3301C^2 \tag{5.44b}$$

Hence $E_1/(2^{1/2}\sigma_1^2) = -0.0123$. In this case, the probability of making a profit is reduced to $P = 0.48$.

If insurance is taken out in the amount $J = jC$, and if the insurance cover is 80% of the catastrophic losses should they occur (that is, $f = 0.8$), then the expected value E_2 of the project is now given by

$$\frac{E_2}{C} = \frac{5 - j}{3} \tag{5.45a}$$

suggesting that the insurance premium should be less than $5C$ if the expected value is to be positive.

The variance σ_2^2 on E_2 is then given by

$$\sigma_2^2 = {}^2\!/_9 j^2 - 8j + 209.4 = {}^2\!/_9 (j - 6)^2 + 201 \tag{5.45b}$$

In this case, one has the ratio

$$\frac{E_2}{2^{1/2}\sigma_2^2} = 0.5 \times \frac{5 - j}{3} \times 2^{1/2} \times [{}^2\!/_9 (j - 6)^2 + 201]^{-1/2} \tag{5.46}$$

The largest value of this ratio (and so the greatest probability of making a profit) is when $j = 0$, of course, i.e., if one can get 80% catastrophic insurance cover for nothing! As the premium j of insurance cover increases, the probability of making a profit systematically decreases until, at $j = 5$, there is only a 50% chance of a profit with 80% catastrophic coverage.

Thus one should not pay an insurance premium of more than 5 times the normal operating costs C for 80% catastrophic coverage if one wishes to retain a greater than 50% chance of making a profit. Accordingly, 0.5% of the total catastrophic costs (or 0.625% of the 80% covered costs) is an adequate insurance premium for the corporate-mandated constraint of greater than 50% chance of making a profit.

5.7 INSURING AGAINST CATASTROPHIC LOSS FOR ENVIRONMENTAL PROJECTS

Imagine that a government contract has been let for a waste project, and a corporation has undertaken the transport, storage, and continued monitoring after burial. The corporation anticipates a positive value V_1 to the project, provided that its preproject estimates of spillage during transport and/or leakage after storage are pretty good estimates of the probability of occurrence of such events.

However, the corporation is also concerned that if a catastrophic spill or post-storage leak does occur, then not only will the corporation not make any profit on the contract but it also could incur such massive costs that it would find itself bankrupt. The corporation wishes to take out insurance against this potential liability and needs to estimate how much it should pay for such insurance.

Let the probability of a catastrophic spill be p_S, and let the probability of a catastrophic leak after storage be p_L, with such large costs that the value of the project to the corporation would be reduced from V_1 to $-S$ (in the case of spillage) and $-L$ (in the case of leakage after storage). The probability of the project operating at the value V_1 is then $1 - p_L - p_S$. The corporation considers the cost of insurance to be I. The insurance will do two things. First, if no catastrophic spillage or leakage occurs during the life of the project, then the value of the project is reduced to $V_1 - I$ from V_1. Second, if catastrophic spillage and/or leakage does occur, then the insurance company will be responsible for a fraction f of the costs in either case, with the corporation being responsible for $1 - f$ of the costs. Thus, as far as the corporation is concerned, the spillage loss now is $(1 - f)S - I$ and the leakage loss is $L(1 - f) - I$, because no matter which eventuality obtains, the corporation must still pay the insurance premium I. A decision-tree diagram illustrating this scenario is given in Fig. 5.6.

The expected value E_0 of the project in the *absence* of taking out insurance is

$$E_0 = (1 - p_S - p_L)V_1 - p_S S - p_L L \tag{5.47}$$

Without Insurance

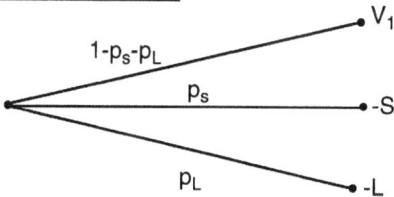

With Insurance

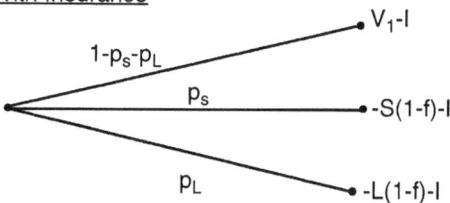

FIGURE 5.6 Decision-tree diagram showing the differences with and without insurance coverage for a general environmental project with government funding of the contract.

and the variance σ_0^2 on E_0 is

$$\sigma_0^2 = (1 - p_S - p_L)V_1^2 + p_S S^2 + p_L L^2 - E_0^2 \tag{5.48}$$

In the case of an insurance premium being paid, the expected value E_1 of the project to the corporation is

$$E_1 = (1 - p_S - p_L)V_1 - (1 - f)(p_S S + p_L L) - I \tag{5.49a}$$

whereas the variance σ_1^2 on E_1 is

$$\sigma_1^2 = [(1 - p_S - p_L)V_1 - I]^2 + p_S[S(1 - f) + I]^2 + p_L[L(1 - f) + I]^2 - E_1^2 \tag{5.49b}$$

The corporation can decide, based on several different criteria, how much it wants to pay for insurance.

The simplest criterion is that the expected value E_1 should be positive. In this case, the relationship between insurance premium I and fraction f of catastrophic losses to be covered is

$$I \leq (1 - p_S - p_L)V_1 - (1 - f)(p_S S + p_L L) \tag{5.50a}$$

and if E_1 is to be positive, one also has the further constraint in force

$$1 \geq f \geq 1 - \frac{(1 - p_S - p_L)V_1}{p_S S + p_L L} \tag{5.50b}$$

A second criterion is that one assesses the difference between E_0 and E_1 *on the assumption both are positive and that* $E_0 > E_1$ and so pays no more than the difference as insurance premium, that is,

$$2I \leq f(p_S S + p_L L) \tag{5.51}$$

A third criterion for assessing insurance worth incorporates the variance around each expected value in both the absence and the presence of insurance as follows: In the absence of insurance, the cumulative probability P_0 that the corporation will make a profit is

$$P_0 = \pi^{-1/2} \int_a^\infty \exp(-y^2)\, dy \tag{5.52a}$$

where $a = -E_0/(2^{1/2}\sigma_0)$. In the presence of the insurance premium, the probability P_1 that the corporation will make a profit is

$$P_1 = \pi^{-1/2} \int_{-b}^\infty \exp(-y^2)\, dy \tag{5.52b}$$

where $b = -E_1/(2^{1/2}\sigma_1)$.

One would like to ensure that if the insurance premium is paid, then the cumulative probability of making a profit is greater than or equal to the cumulative probability in the absence of insurance; i.e., one would arrange that I and f be such that

$$\int_b^\infty \exp(-y^2)\, dy \geq \int_a^\infty \exp(-y^2)\, dy \qquad (5.53a)$$

which can be reduced to the simpler requirement $b \leq a$, which, in turn, requires that

$$\frac{E_1}{\sigma_1} \geq \frac{E_0}{\sigma_0} \qquad (5.53b)$$

for E_1 and E_0 both positive.

Thus in this case one plots the variation of I versus f to obtain the regimes where the constraint of inequality (5.53b) is satisfied. In this way the fractional amount of insurance coverage f versus insurance premium I to ensure a profit can be obtained.

5.8   OPTIONING AGAINST POTENTIAL REGULATORY CHANGES

The problem here is that one is fairly sure of the changes in value of the project likely to be brought about if the regulatory agencies do go ahead with the enactment of stricter environmental standards, but one is not sure that there will be a change, only a chance p_c of a change. Therefore, what the corporation would like to do is to organize a put option with its partners or with an insurance company. If no change occurs, then the put is not in place. If there is a change, however, then the partners (or insurance coverage) are required to contribute so that the worth of the project to the corporation is at least the discounted expected value, i.e., the value $E_0 = (1 - p_c)V_1 + p_cV_2$, where V_2 is the reduced value of the project if a change should occur. In this way, in the event of a change, the corporation only absorbs the difference $V_1 - E_0 = p_c(V_1 - V_2)$, with partners contributing the value $E_0 - V_2 = (1 - p_c)(V_1 - V_2)$.

In this case the decision-tree diagram from the corporate perspective is modified as shown in Fig. 5.7. The corporation has to decide on the value of the put.

Again, the argument is relatively simple. The value of the average put is just

$$\text{put} = E_1 - E_0 = p_c(1 - p_c)(V_0 - V_1) \qquad (5.54)$$

<u>Without PUT option</u>

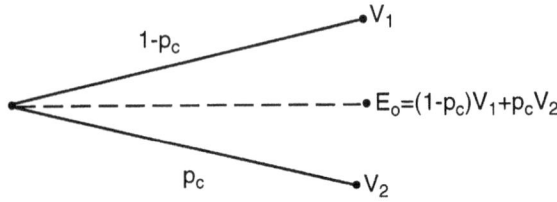

<u>With PUT option</u>

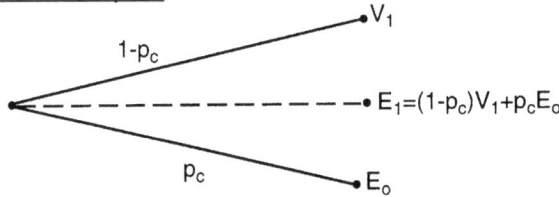

FIGURE 5.7   Decision-tree diagram showing the effect of a put option designed to provide limitations on financial risk to a corporation in the event of unilateral regulatory changes in environment cleanup rules.

However, because E_1 and E_0 have variances $\sigma_1^2 = p_c \sigma_0^2$ and $\sigma_2^2 = p_c(1 - p_c)(V_0 - V_1)^2$, respectively, so too the put has a variance σ_p^2 given by

$$\sigma_p^2 = (1 + p_c)\sigma_0^2 \qquad (5.55)$$

Hence the probability P_0 of a put value exceeding zero is given by

$$P_0 = \pi^{-1/2} \int_c^\infty \exp(-y^2)\, dy \qquad (5.56)$$

where $c = -\mathrm{put}/(2^{1/2}\sigma_p)$, with put given by Eq. (5.54).

5.9   RISK RETENTION

Historically, the primary risk-management technique for business managers has been to transfer financial losses to insurance schemes. This practice often has promoted complacency in the development of preventive procedures to reduce losses (Howard, 1998). Furthermore, liability insurance may be unavailable or may contain provisions that exclude coverage for potential release of hazardous wastes (Ness, 1992). Even when commercial insurance is purchased, the risk is not completely eliminated because of high deductibles, relatively low individual and aggregate coverage limits, and short time limits on claims-made policies (Frano, 1991; Paek, 1996). An innovative approach at self-insurance is the risk reserve fund (Architects and Engineers Insurance Company) of 66 major engineering, architectural, and environmental firms, which aims to cover professional liabilities. The majority of environmental claims against this fund has been related to permits, personal injuries, fee disputes, reporting, and asbestos identification (Janney et al., 1996). Despite such creative approaches to risk limitation, the environmental industry continues to be characterized by (1) high scientific uncertainty and complexity of problems (e.g., the movement of a contaminant plume in the subsurface depends on the geologic, hydrologic, geochemical, and biologic characteristics of a site, for which very little information is usually available), (2) hazardous work conditions and the potential for a major health impact on operators and the general population, and (3) strict regulations that usually do not allow for indemnification clauses and which do not rely on standards of liability based on fault and negligence. Thus, unless the cost of liability is addressed by appropriate management techniques, such as risk retention, the danger exists that the environmental field will be occupied by unqualified contractors willing to take high risks (Paek, 1996).

Risk retention is based on the argument that potential liability costs of environmental projects can be treated as legitimate business revenues and may be priced into the projects. A corporation involved in multiple environmental projects will then attribute a weighted risk-based premium to each project according to its likelihood of producing a liability. The funds gathered in this manner can be managed as a risk reserve fund with the purpose of meeting potential future claims (Frano, 1991). The costs of administering the fund also may be priced into the projects (Tietenberg, 1998). Such an approach encourages experienced contractors to invest in environmental projects and leads to procedures for rational allocation of risks (according to the characteristics of the different project types) within the resources of a corporation without resorting to truncation of the effects of claims or penalties through bankruptcy (Larson, 1996; Beard, 1990).

The rest of this section is organized as follows: First, a procedure is developed to incorporate the cost of a risk reserve fund in a project budget proposal in order to meet potential liability claims. Expressions are then provided that allow this fund to be used for a number of liability situations, including cases of distributing the cost to projects of similar monetary value awarded in the same fiscal year or in variable time periods, as well as of proposals of different magnitudes initiated in distinctly different years. Two simple numerical illustrations of environmental remediation are then presented and conclusions given.

5.9.1 Evaluation of Risk Reserve Fund

The cash involvement E of an environmental project usually consists of direct costs D, overhead H, and profit P such that

$$E = D + H + P \tag{5.57}$$

Let H be a multiple of the direct costs, $H = Dh$, where h is an overhead markup (%), and let P be a multiple of direct costs and overhead, $P = (D + H)p$, where p is a profit markup (%). Equation (5.57) then can be rewritten as

$$E = D(1 + h)(1 + p) \tag{5.58}$$

Equation (5.57) can be modified to include that portion of the reserve fund RF that is contributed by a single project:

$$E_R = D + H + \text{RF} + P \tag{5.59}$$

If RF is considered as a multiple of direct costs and overhead only, such that

$$\text{RF} = (D + H)r = D(1 + h)r \tag{5.60}$$

where r is a risk markup (%), then, substituting Eqs. (5.58) and (5.60) into Eq. (5.59), one obtains

$$E_R = D(1 + h)(1 + r + p) \tag{5.61}$$

Equation (5.61) gives the total project cash involvement, including cost appropriated for future potential liabilities, and differs from other expressions in the literature (Frano, 1991; Paek, 1996), where the risk markup also was applied to the profit P. From Eq. (5.60), the appropriate risk markup r can be written as

$$r = \frac{\text{RF}}{D(1 + h)} \tag{5.62}$$

which, when applied to Eq. (5.61), provides the modified project cash involvement that is retained for future claims. Expressions for RF for different liability situations are now considered.

5.9.2 Projects Awarded at One Time Exposed to Liability L

Based on published data of liability claims, for each specific environmental sector, assume that one can define an average level of monetary claim L, an average time t, and a frequency of occurrence q, respectively, of such a claim. Thus, for each particular category of project exposed to a claim L, one can establish the number N of similar projects that will lead to one liability claim. Consider that the

N projects are initiated at the same time $t = 0$. The portion of the reserve fund that corresponds to project m, that is, RF_m ($m = 1, 2,..., N$), consists of a part R_m that grows (in t years) to cover that portion of the liability L corresponding to project m and a part A_m that covers the expenses of managing the fund over t years (Fig. 5.8) for project m. Thus

$$RF_m = R_m + A_m \tag{5.63}$$

Using simple annual compounding fixed at a fractional rate i per year (Garrison, 1991), the risk part R_m of the reserve fund, booked on a project basis at the time of the contract award ($t = 0$), after t years has a value of

$$R_m(1 + i)^t = qL \tag{5.64}$$

and is used to cover the portion of the liability qL where $q = 1/N$. The underlying assumption here is that projects of similar monetary value and of similar liability risk characteristics should contribute equally toward a future claim L.

Summing Eq. (5.64) over the total number of projects N that are used to meet the future liability L yields

$$(1 + i)^t \sum_{m=1}^{N} R_m = L \sum_{m=1}^{N} \frac{1}{N} = L \tag{5.65}$$

Because all projects share similar risk characteristics, the risk booked to the contract of each project is the same, that is, $R_1 = R_2 = \cdots = R_m = \cdots = R_N = R$. Equation (5.65) then can be written as

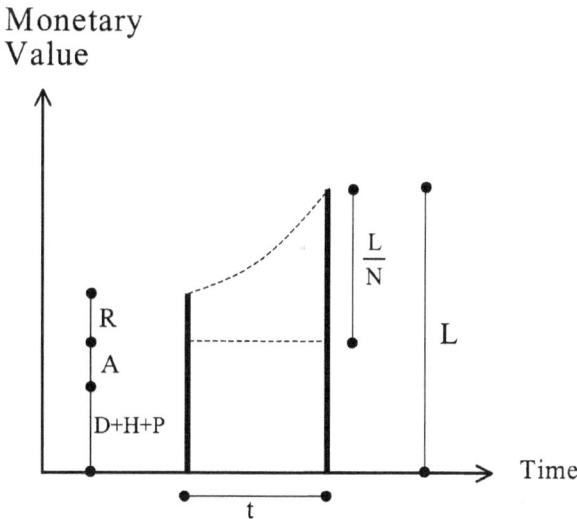

FIGURE 5.8 Contribution of risk price (similar projects awarded at the same time).

$$NR(1 + i)^t = L \tag{5.66}$$

Thus the risk booked and collected (NR) from all projects will grow after t years to meet a liability L. If a corporation can establish both an expected fiscal level and an expected time of liability L and t, respectively, then if the corporation wishes to be involved in N projects, Eq. (5.66) supplies the value of R; otherwise, the corporation can decide on the risk premium R and determine the number of projects N with which it can become involved. Figure 5.8 illustrates the contributions of similar projects to the potential liability.

In addition to meeting future liabilities, the reserve fund RF also must generate cash flow to manage the fund over t years. The value after t years of this management account A is given by (Garrison, 1991)

$$A = \frac{C}{N}\left(\frac{1}{i}\right)\left[1 - \frac{1}{(1 + i)^t}\right] \tag{5.67}$$

where C/N is the annual cost of managing the total reserve fund divided by the number of projects contributing to the fund. Combining Eqs. (5.63), (5.66), and (5.67) gives the total cost of a project's reserve fund, that is,

$$RF = \frac{L}{N(1 + i)^t} + \frac{C}{N}\left(\frac{1}{i}\right)\left[1 - \frac{1}{(1 + i)^t}\right] \tag{5.68}$$

which, when substituted into Eq. (5.62), provides the risk markup, that is,

$$r = \frac{L}{ND(1 + h)(1 + i)^t} + \frac{C}{ND(1 + h)}\left(\frac{1}{i}\right)\left[1 - \frac{1}{(1 + i)^t}\right] \tag{5.69}$$

The first term on the right-hand side of Eq. (5.69) provides coverage to potential liability claims of a project, whereas the second part supplies the costs for managing the reserve fund. Finally, using Eqs. (5.61) and (5.69), the modified total cash involvement for a project can be calculated.

5.9.3 Projects Awarded at Different Times Exposed to Liability L

Consider now an environmental company that plans to allocate risk premiums to N proposals of similar monetary value (and risk characteristics) because of their exposure to an expected liability L. In contrast to the preceding case, these projects have been (or will be) initiated at various times, and hence the company wishes to account for the different time lags t_m from the expected occurrence of a liability. Equation (5.64) is then modified as

$$R_m(1 + i)^{t_m} = q_m L \qquad m = 1, 2, \ldots, N \tag{5.70}$$

where $q_m L$ is a proportional contribution of project m to L. Index m in Eq. (5.70) indicates an increasing sequence of time lags such that project 1 is the closest in time to the occurrence of the claim and project N is the farthest. To solve Eq. (5.70), again use historical data to define an expected magnitude and time of claim for a specific environmental sector. A decision now needs to be made on the level of the risk price R_m and the liability ratio of each project q_m, respectively. Note that q_m cannot equal $1/N$ because $q_m = 1/N$ implies that $R_N < \cdots < R_2 < R_1$ (for a shorter time period one needs to invest a larger amount in order that it grows to a fixed amount), which would mean that recent projects would carry a disproportionate load of the costs of a potential claim. On the other hand, because the proposals are of similar monetary value (and have the same probability of facing a claim), it is reasonable to apply the same risk premium $R_m = R$ to all projects. Then, because of the variable time lags, these equal risk premiums will grow to different amounts, and earlier projects will contribute more toward L than more recent projects. Figure 5.9 illustrates these concepts for three projects of the same monetary value initiated at different times and also shows how the risk premium of each project contributes according to its initiation history. Equation (5.70) is then simplified to

$$R(1 + i)^{t_m} = q_m L \qquad m = 1, 2, ..., N \qquad (5.71)$$

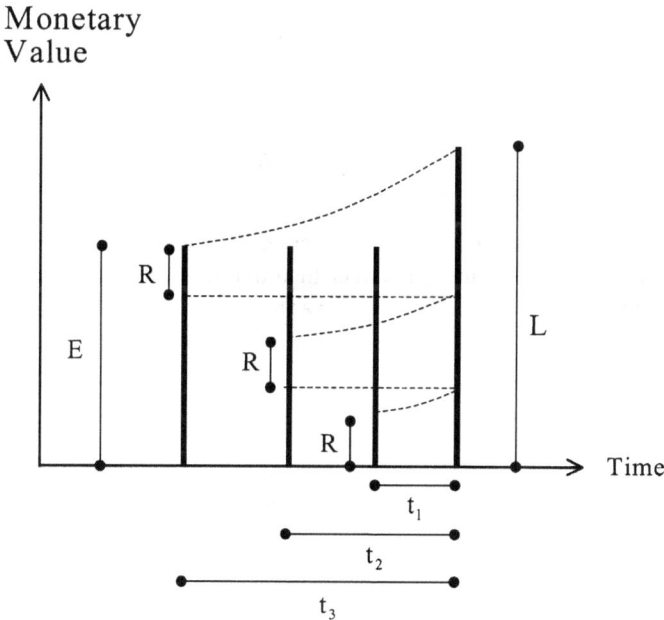

FIGURE 5.9   Variable contribution of risk price (similar projects awarded at different times).

Summing Eq. (5.71) over all projects gives

$$R \sum_{m=1}^{N} (1 + i)^{t_m} = L \sum_{m=1}^{N} q_m = L \tag{5.72}$$

because the sum of all contributions must be L. Rearranging Eq. (5.72) yields

$$\sum_{m=1}^{N} (1 + i)^{t_m} = \frac{L}{R} \tag{5.73}$$

which, when combined with Eq. (5.71), provides the proportion of each project's contribution toward L as

$$q_m = \frac{(1 + i)^{t_m}}{\displaystyle\sum_{m=1}^{N} (1 + i)^{t_m}} \tag{5.74}$$

Rearranging Eq. (5.73) yields the expression for the risk premium as

$$R = \frac{L}{\displaystyle\sum_{m=1}^{N} (1 + i)^{t_m}} \tag{5.75}$$

which, for projects initiated at the same time ($t_m = t$), reduces to Eq. (5.66), that is,

$$R = \frac{L}{\displaystyle\sum_{m=1}^{N} (1 + i)^{t}} = \frac{L}{N(1 + i)^{t}} \tag{5.76}$$

An earlier project now generates a longer series of cash flows, and hence the present-day value of its annuity is larger than that of a recent project. Equation (5.67) is then modified to account for the variable time lags as

$$A_m = \frac{C}{N} \left(\frac{1}{i}\right) \left[1 - \frac{1}{(1 + i)^{t_m}}\right] \tag{5.77}$$

where, again, C/N is the constant annual cost of managing the total reserve fund divided by the number of projects. The total value of a project's reserve fund is now given by

$$RF_m = \frac{L}{\displaystyle\sum_{m=1}^{N} (1 + i)^{t_m}} + \frac{C}{N} \left(\frac{1}{i}\right) \left[1 - \frac{1}{(1 + i)^{t_m}}\right] \tag{5.78}$$

Finally, the risk markup that needs to be applied to each project is

$$r_m = \frac{L}{D(1 + h) \sum\limits_{m=1}^{N} (1 + i)^{lm}} + \frac{C}{ND(1 + h)} \left(\frac{1}{i}\right) \left[1 - \frac{1}{(1 + i)^{lm}}\right] \qquad (5.79)$$

If one wishes to account consistently for the different time lags in the calculation of the administrative expenses, then instead of the proportionality constant $1/N$ in Eqs. (5.77) through (5.79), one needs to use Eq. (5.74), which was derived for risk prices, and then Eq. (5.77) is written as

$$A_m = \frac{C (1 + i)^{lm}}{\sum\limits_{m=1}^{N} (1 + i)^{lm}} \left(\frac{1}{i}\right) \left[1 - \frac{1}{(1 + i)^{lm}}\right] \qquad (5.80)$$

which reduces to Eq. (5.67) for similar time lags. Equations (5.78) and (5.79) should then be modified accordingly.

5.9.4 Projects of Variable Magnitude Awarded at Different Times and Exposed to Liability L

Expressions are now developed for risk prices and administrative costs for the general case where a corporation undertakes projects of variable cash involvement initiated at different times and is faced with a potential future liability L (Fig. 5.10). Again based on historical data of liability claims, the corporation establishes a magnitude L for all projects and determines the time lags of each project. The equation for the risk price is now given by

$$R_m(1 + i)^{lm} = q_m L \qquad m = 1, 2, ..., N \qquad (5.81)$$

However, note that the number of unknowns is $2N$ (N unknown risk prices R_m and N unknown liability ratios q_m), whereas the number of available equations is $N + 1$ [N expressions from Eq. (5.81) and the equation $\Sigma_{m=1}^{N} q_m = 1$]. Thus, in general, one cannot solve for both R_m and q_m independently. It is necessary to define, in advance, $N - 1$ of the variables.
Define the liability ratios as

$$q_m = \frac{E_m t_m}{\sum\limits_{m=1}^{N} E_m t_m} \qquad \text{for } m = 1, 2, ..., N \qquad (5.82)$$

Here, $E_m = D + H + P$ is the original project cash involvement (without incorporation of the reserve fund), which can be estimated in a straightforward manner for each project. Note that Eq. (5.82) allocates a different proportion of the liability to each project and that each proposal's contribution takes into account both the relative magnitudes of the original cash involvement estimates and the time lags from the expected occurrence of a claim. Indeed, Eq. (5.82) does not render small projects

Monetary
Value

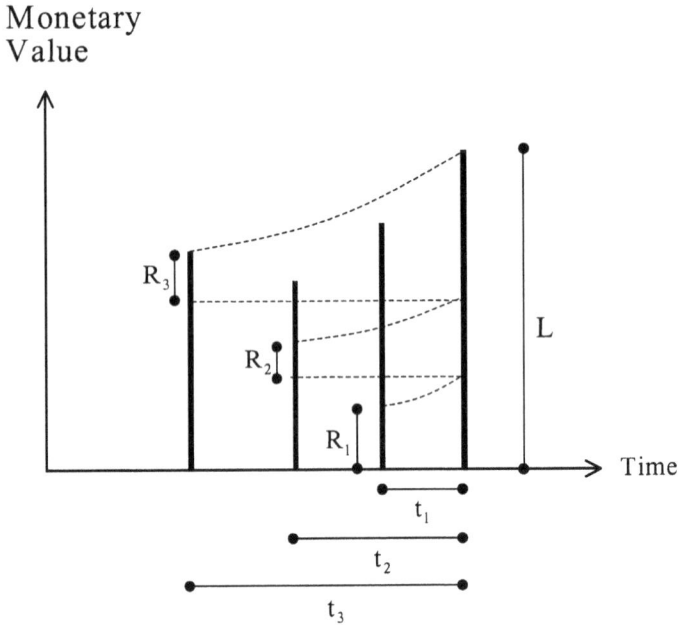

FIGURE 5.10 Variable contributions of risk price (projects of variable monetary value awarded at different times).

noncompetitive, nor does it unfairly burden recent contracts. Furthermore, summing Eq. (5.82) over all N projects yields unity, which is the correct behavior for fractional liability contributions.

Substituting Eq. (5.82) into Eq. (5.81) yields

$$R_m(1 + i)^{t_m} = \frac{E_m t_m}{\sum_{m=1}^{N} E_m t_m} L \tag{5.83}$$

which, upon summation over all N projects, gives

$$\sum_{m=1}^{N} R_m (1 + i)^{t_m} = \frac{L}{\sum_{m=1}^{N} E_m t_m} \sum_{m=1}^{N} E_m t_m = L \tag{5.84}$$

This result is the same obtained by summing Eq. (5.81). Simple rearrangement of Eq. (5.81) provides the risk price as

$$R_m = \frac{L}{(1 + i)^{t_m}} \frac{E_m t_m}{\sum_{m=1}^{N} E_m t_m} \tag{5.85}$$

Similarly, the administrative expenses for each project can be weighted by q_m to yield

$$A_m = \frac{C_m E_m t_m}{\sum\limits_{m=1}^{N} E_m t_m} \left(\frac{1}{i}\right)\left[1 - \frac{1}{(1+i)^{tm}}\right] \tag{5.86}$$

which reduces to Eq. (5.67) if all cash involvements are the same and if all projects are initiated at the same time. Finally, the values of the reserve fund and risk markup are given, respectively, by

$$RF_m = \frac{L}{(1+i)^{tm}} \frac{E_m t_m}{\sum\limits_{m=1}^{N} E_m t_m} + \frac{C_m E_m t_m}{\sum\limits_{m=1}^{N} E_m t_m} \left(\frac{1}{i}\right)\left[1 - \frac{1}{(1+i)^{tm}}\right] \tag{5.87}$$

and

$$r_m = \frac{1}{D(1+h)}\left\{\frac{L}{(1+i)^{tm}} \frac{E_m t_m}{\sum\limits_{m=1}^{N} E_m t_m} + \frac{C_m E_m t_m}{\sum\limits_{m=1}^{N} E_m t_m}\left(\frac{1}{i}\right)\left[1 - \frac{1}{(1+i)^{tm}}\right]\right\} \tag{5.88}$$

5.9.5 Numerical Illustrations

Liability Risk and Reserve Fund. The following illustration applies the procedures to estimate the price charged for liability risk on an environmental contract. An environmental company is preparing a lump-sum proposal for remediation work of soil and groundwater contamination at an industrial site and would like to account for a potential liability claim in the lump-sum proposal. Based on a site visit, input from contractors, and company proposal development meetings, the following parameters are used by the company: The direct costs for the project are estimated to be $700,000. The company anticipates partitioning the cost of a $3 million potential liability claim over 10 projects within the same fiscal year. The claim is expected to arise in 10 years. The annual cost to administer the reserve fund is $50,000; the interest rate the fund will be earning is approximately 10%; the company overhead markup rate is 20%; and the company profit markup is 15%. The parameters are $D = \$700,000$, $N = 10$, $L = \$3$ million, $t = 10$, $C = \$50,000$, $i = 0.10$, $h = 0.20$, and $p = 0.15$.

The proportion of a single project toward the liability is $q = 1/N = 0.10$. The risk price for a project [Eq. (5.66) or Eq. (5.76)] is

$$R = \frac{3,000,000}{10(1+0.10)^{10}} = \$115,663 \tag{5.89}$$

The administrative expenses of a single project [Eq. (5.67)] are

$$A = \frac{50,000}{10 \times 0.10} \left[1 - \frac{1}{(1 + 0.1)^{10}} \right] = \$30,723 \tag{5.90}$$

The reserve fund then [Eq. (5.63) or Eq. (5.68)] is

$$RF = 115,663 + 30,723 = \$146,486 \tag{5.91}$$

and the risk markup that needs to be applied to a project budget is

$$r = \frac{146,386}{700,000 \times 1.2} = 17.4\% \tag{5.92}$$

The proposal that would have been submitted without accounting for liability risk [Eq. (5.58)] amounts to a cash involvement of

$$E = 700,000(1.2)(1.15) = \$966,000 \tag{5.93}$$

and with the risk markup, the corresponding proposal cash involvement is

$$E_R = 700,000(1.2)(1.324) = \$1,112,160 \tag{5.94}$$

The difference $(E_R - E)$ gives the value of \$146,386.

Partitioning Liability Costs.   Consider now the case where a company wants to distribute the same liability cost among 10 projects, with 4 projects of type A with $D_A = \$700,000$, 3 projects of type B with $D_B = \$300,000$, and 3 projects of type C with $D_C = \$1,500,000$. All other variables remain as in the preceding example for all types of projects. Now the liability ratio for type A projects [Eq. (5.82) with $t_m = t$] becomes $q_A = 8.5\%$, the risk price [Eq. (5.85)] is $R_A = \$98,313$, and the administrative expenses [Eq. (5.86)] are $A_A = \$26,114$. The portion of the reserve fund contributed by a single type A project is $RF_A = \$124,427$, and the risk markup [Eq. (5.62) or Eq. (5.88)] is $r_A = 15\%$. The proposal should then have the value of $E_{R,A} = \$1,092,000$, which is slightly lower than that calculated previously.

Now assess the contributions to the reserve fund for the smaller type B projects and the larger type C projects, respectively. The corresponding liability ratios are $q_B = 3.7\%$ and $q_C = 18.3\%$, indicating that the four type A projects contribute 34%, the three type B projects contribute 11%, and the three type C projects contribute 55% to the total reserve fund. The risk price of type B projects is $R_B = \$42,795$ and of type C projects is $R_C = \$211,663$. Administrative expenses are $A_B = \$11,367$ and $A_C = \$56,223$, and the portions of the reserve fund are $RF_B = \$54,162$ and $RF_C = \$267,886$, respectively. Finally, the risk markup for either type B or C projects is 15%, providing the modified cash involvements $E_{R,B} = \$468,000$ and $E_{R,C} = \$2,340,000$, respectively.

5.10 SUMMARY

Environmental companies face the risk of liability claims arising from unpredictable extreme events that can significantly affect their resources and financial future. Property assessments, compliance audits, site investigations, feasibility studies, environmental permits, implementation of remediation systems, construction activities, and land and company acquisitions all pose risks. However, because of the great potential losses, inclusion of extreme events in economical risk models can swamp completely, in an abusive manner, the characteristics of the opportunity that would exist in the absence of the catastrophic loss conditions if attention is restricted solely to the expected value of the opportunity.

Nevertheless, such inclusions also should allow for the uncertainty (measured here by the variance or standard error) in the expected value. When such considerations are allowed for, then the abuse of the expected value of a project by such inclusions is not only properly ameliorated but also, as shown here, can be included appropriately in a balanced, rational, objective, reproducible assessment of project worth. The purpose of the case histories given here and of their numerical illustrations was to show how such allowances can be made by specific example.

It is strongly recommended that one should avoid using just the expected value as an absolute measure of opportunity worth. Instead, one should, more correctly, include the standard error, as well as the probability of obtaining a positive return from an opportunity. In this way, low-risk but high potential cleanup cost scenarios are correctly included. And the cases used here have been tailored to maximize the illustration of the way such factors are to be handled.

Two major hydrocarbon exploration and development scenarios were used to show how evaluation of insurance premium versus insurance coverage can be assessed. The first situation considered insurance against catastrophic loss due to a major adverse event while drilling, irrespective of whether commercially attractive hydrocarbon reserves are found or not; the second situation considered insurance against a hydrocarbon spill *after* hydrocarbons are found. Numerical examples indicated how the insurance premium to be paid is related to the insurance cover provided for a fixed corporate requirement on the probability to make a profit. The examples illustrated how considerations of mean expected value and variance around the mean can both be incorporated in a quantitative measure of worth of insurance cover versus premium paid. Providing financial risk control against unilateral changes in government conditions was addressed through the use of a put option, which also involved both the mean expected value and its uncertainty.

Finally, this chapter provided a procedure that addresses limited-liability claims for a number of situations. The approach is useful for small and medium-sized environmental companies that are unable to afford sophisticated risk-management techniques but need rather a simple procedure to estimate liability risks. This goal is achieved by incorporating the cost of future claims within the cash involvement from projects and by treating liability costs as legitimate business expenses. Part of a premium charged on a project basis is used to meet future potential claims, and part is used to administer the funds collected.

Projects of similar magnitude awarded during the same fiscal year share the same risk price, administrative expenses, and risk markup. For projects of similar magnitude that are (or will be) initiated at various years, these variables are weighted according to the time lag of a project from an expected occurrence of a liability claim. For proposals of variable revenue, awarded at different times, liability costs are distributed proportionally to both a project budget and time lag from an expected claim occurrence. The procedure presented does not render small or recent projects noncompetitive and provides an easy-to-implement risk-management tool for environmental companies.

Depending on an environmental firm's risk tolerance, companies may choose to use the risk reserve fund to fully or partially cover a liability and/or retain, at the same time, insurance coverage. In this case, the risk reserve fund may pay directly either for losses or for the cost of insurance policies. A reserve fund cannot address all situations, such as claims arising from catastrophic losses or from a small number of projects because an incorporation of such a cost could render a proposal noncompetitive. For such cases, other techniques, such as that of working-interest optimization (presented in Chap. 4), allow a company to decide on bids on parts of a project only.

CHAPTER 6
LIMITING RISK USING FRACTIONAL WORKING INTEREST

6.1 INTRODUCTION

Chapter 4 presented a procedure for assessing the potential involvement of a corporation in an environmental project when the corporation has a limited tolerance to risk. This procedure was based on two major assumptions: (1) that a model for assessing risk was in place (the exponential model was taken as the preferred risk model of choice by the corporation, although the parabolic risk model also was introduced to show how different models provide different responses) and (2) that all gain and cost parameters of a project were available and, in addition, that the level of corporate risk tolerance was known. Methods of estimating some of the relevant parameters, not only for direct project costs but also for indirect costs such as insurance, were given in Chap. 5. In this way, one can be confident that each corporation has available what it estimates is the sum total of all costs relevant for a project. Thus all the pieces are now on hand to provide a better determination of when a corporation should become involved in an environmental project and, more important, to allow an assessment of the best fractional participation that the corporation should take in a project.

In the case of environmental projects, individual corporations tend to estimate the expected value of a project and then multiply the result by a risk factor to assess the prioritized worth versus risk to the corporation.

Nearly 20 years ago Cozzolino (1977a, 1978) showed how to allow for risk tolerance factors in estimating a risk-adjusted value to a project. Cozzolino (1977b, 1978) used an exponential model of fractional working interest in a project that should be followed in order to ascertain the maximum likely return on a project subject to the constraints of risk tolerance avoidance.

The requirements of such a Cozzolino-type model force an exponential risk-aversion factor to be employed by any corporation wishing to use the procedure. However, not all corporations choose exponential risk aversion; some have a hyperbolic tangent type of risk weighting, and others have empirical models based on prior evaluations of projects and the anticipated value to the corporate assets, and so on.

It would be useful if a method could be designed along the lines of Cozzolino's basic precepts but that allowed arbitrary functional forms (as employed by different corporations) to be used to assess relevant working interest W and risk-adjusted value RAV of a project. In this way one could then determine quickly the sensitivity, precision, uniqueness, and resolution of W and RAV not only to variations in expected gains and/or losses in relation to chances of success/failure and to variations in risk tolerance RT but also to the functional form of risk-aversion formula being used to make the estimate. This ability would then go a long way toward allowing evaluation of projects under a variety of settings.

The purposes of this chapter are to show precisely how such a general methodology can be set up and to illustrate the practical use of such a framework by direct example.

6.2 GENERAL METHODS

Cozzolino (1977a, 1978) constructed his RAV formula based on utility theory applied to a chance-node decision-tree diagram such as that given in Fig. 6.1a. The notation of Fig. 6.1a is V = value (positive), p_s = probability of success, C = expected cost of nonsuccess (positive), and $p_f (\equiv 1 - p_s)$ = probability of the project not succeeding. At the chance-node point of Fig. 6.1a, the expected value, or weighted average of the two possible outcomes of the project, is

$$E_1 = p_s V - p_f C \qquad (6.1)$$

which is positive provided $p_s V > p_f C$. The second moment of the project value is $E_2 = p_s V^2 + p_f C^2$, so a measure of the uncertainty in the outcome is provided by the variance

$$\sigma^2 = E_2 - E_1^2 \equiv (V + C)^2 (p_s p_f) \qquad (6.2)$$

A measure of risk is often assigned by the volatility v, defined by

$$v = \frac{\sigma}{E_1} \equiv \frac{(V + C)(p_s p_f)^{1/2}}{p_s V - p_f C} \qquad (6.3)$$

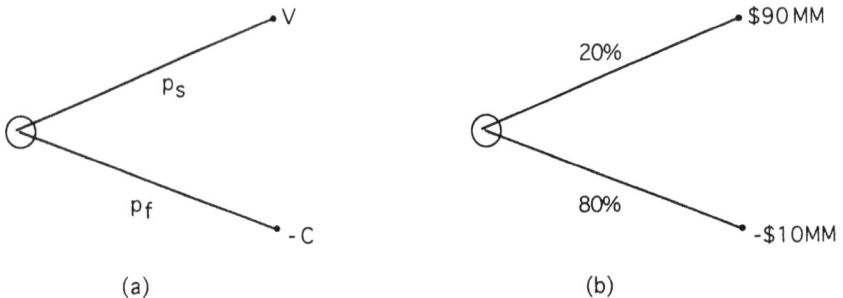

(a)   (b)

FIGURE 6.1 (a) Sketch of the chance-node diagram indicating a value of V with a probability of p_s together with a loss of $-C$ at a probability of $p_f (\equiv 1 - p_s)$. (b) As for part (a) but with V = \$90 million, p_s = 20%, p_f = 80%, and C = \$10 million, as used in the numerical illustration in text. (\$1MM = \$1 million.)

which evaluates the stability of the estimated mean value E_1 relative to the fluctuations about the mean. A small volatility ($v \ll 1$) implies that there is little uncertainty in the expected value, whereas a large volatility ($v \gg 1$) implies a considerable uncertainty in the expected value.

While the expected value E_1 of a project may be high and the volatility small, nevertheless, it can be the case that if failure does occur, then the total project costs C may be so large as to bankrupt or cause serious financial damage to the corporation. Under such conditions, it makes corporate sense to take less than 100% working interest in the project. A smaller fraction of the project will cut potential gains but also will cut catastrophic potential losses. Thus, with a working-interest fraction W, the expected value to the corporation at the chance node is

$$E_1(W) = p_s(WV) - p_f(WC) \tag{6.4}$$

on the assumption that fractional working interest does not change the probabilities of success or failure of the project.

The effective corporate value is reduced from V to WV, whereas the potential losses are reduced from C to WC.

6.2.1 Cozzolino's Formula

Cozzolino's (1977b, 1978) determination of risk-adjusted value RAV in relation to the risk tolerance ($\equiv$ risk threshold) RT of a corporation can be given simply using an analogy from geochemistry. Imagine that two energy states exist with activation energies E_1 ($\equiv WV$) and E_2 ($\equiv -WC$). At a given temperature T, the rate at which a compound can be lost by decay to state E_1 is given by the Arrhenius formula:

$$p_s \exp\left(\frac{-E_1}{\mathrm{RT}} \right) \tag{6.5a}$$

where R is the gas constant, and p_s is the probability of the reaction pathway being along the path determined by energy state E_1. Decay along the path of E_2 is then proportional to

$$p_f \exp\left(\frac{-E_2}{\mathrm{RT}} \right) \tag{6.5b}$$

and because only two paths exist, $p_s + p_f = 1$.

Thus the total decay is then proportional to

$$p_s \exp\left(\frac{-E_1}{\mathrm{RT}} \right) + p_f \exp\left(\frac{-E_2}{\mathrm{RT}} \right) \tag{6.5c}$$

If one were to represent the total decay of Eq. (6.5c) by an equivalent activation energy RAV through a single equivalent pathway, then the equivalence demands of an Arrhenius formula are

$$\exp\left(\frac{-RAV}{RT}\right) = p_s \exp\left(\frac{-E_1}{RT}\right) + p_f \exp\left(\frac{-E_2}{RT}\right) \tag{6.6}$$

Hence

$$RAV = -RT \ln\left(p_s \exp\left(\frac{-WV}{RT}\right) + p_f \exp\left(\frac{WC}{RT}\right)\right) \tag{6.7}$$

The intrinsic assumption here is that the activation energy rates of conversion are controlled by the exponential Arrhenius formula. With RT understood as risk tolerance, Eq. (6.7) is Cozzolino's formula relating estimated risk-aversion value RAV to risk tolerance RT, working interest W, and the value V, cost C, and probabilities of success/failure, with $p_s + p_f = 1$.

Maximum Working Interest. Note that for a given value of RT, Eq. (6.7) has a maximum value of RAV with respect to working interest W when W takes on the value

$$W_{max} = \frac{RT}{C + V} \ln\left(\frac{p_s V}{p_f C}\right) > 0 \qquad \text{if } E_1 > 0 \tag{6.8}$$

and at this value of W_{max}, the maximum value of RAV is

$$RAV(max) = -RT \ln\left[p_s \left(\frac{p_f C}{p_s V}\right)^{V/(V + C)} + p_f \left(\frac{p_s V}{p_f C}\right)^{C/(C + V)}\right]$$

$$\equiv VW_{max} - RT \ln\left[p_s \left(1 + \frac{V}{C}\right)\right] \tag{6.9}$$

Note that the requirement $0 \le W \le 1$ (no less than 0% or more than 100% working interest can be taken in a project) then implies, from Eq. (6.8), that

$$p_s C < p_s V < p_f C \exp\left(\frac{C + V}{RT}\right) \tag{6.10}$$

If Eq. (6.10) is *not* satisfied, then RAV does not have a maximum in the range $0 \le W \le 1$, so RAV is then either monotonically increasing or decreasing as W increases from zero to unity.

As W tends to zero, then, from Eq. (6.7), one has RAV $(W = 0) = 0$ and

$$\left.\frac{dRAV}{dW}\right|_{W = 0} = p_s V - p_f C > 0 \tag{6.11}$$

Thus RAV is positive increasing at small W provided $p_s V > p_f C$, i.e., the expected value at the chance node of Fig. 6.1a is positive. In such a case, because there is no maximum in the range $0 \le W \le 1$, the largest positive value of RAV occurs at the maximum $W = 1$, indicating that 100% interest in the project should be taken. Equally, if $p_s V < p_f C$, then RAV is negative, decreasing throughout $0 \le W \le 1$, indicating that the project should not be invested in at all.

When Eq. (6.10) is satisfied, then there is a range of values of W around W_{max} where RAV can be positive, so some positive risk-adjusted return is likely even if a working interest is taken other than that which maximizes the RAV.

Apparent Risk Tolerance. By rearrangement, Eq. (6.8) can be used in a different manner in the form

$$\text{RT} = W_{max} \frac{C + V}{\ln(p_s V/p_f C)} \tag{6.12}$$

If an arbitrary working-interest value W is used in Eq. (6.12) to replace the optimal value W_{max}, the apparent risk tolerance RT_A is then given by the left side of Eq. (6.12). This apparent risk tolerance expresses the ability to see to what extent a corporate mandate of risk tolerance has overrisked or underrisked a particular project or the extent to which a particular working-interest choice permits the apparent risk tolerance to be in reasonable accord with the corporate-mandated value. It is particularly useful in determining what the corporate attitude toward risk tolerance is, based on prior working-interest decisions.

"Break-Even" Working Interest. The "break-even" value of RAV is conventionally set to zero, which, for the Cozzolino formula [Eq. (6.7)], occurs at a working interest W_0, determined from

$$p_f \exp\left(\frac{W_0(C + V)}{\text{RT}}\right) + p_s - \exp\left(\frac{W_0 V}{\text{RT}}\right) = 0 \tag{6.13a}$$

provided $0 \le W_0 \le 1$.

For $V \gg C$, Eq. (6.13a) has the approximate solution

$$W_0 \cong 2\text{RT}V^{-1} \ln\left(1 + \frac{3Cp_f}{Vp_s}\right) \tag{6.13b}$$

so that $W_0 < 1$ in RT $\le \frac{1}{2}V\{\ln[1 + (3Cp_f/Vp_s)]\}^{-1}$.

Maximum Risk Tolerance. Occasionally, a corporation requests that a particular fixed working interest be taken. The question then is: How does the RAV relate to the risk tolerance? In this situation, RAV has a maximum, RAV(max), with respect to RT at a value of RT_m given through

$$\text{RT}_m = \frac{W[-p_s V \exp(-WV/\text{RT}_m) + p_f C \exp(WC/\text{RT}_m)]}{\ln[p_s \exp(-WV/\text{RT}_m) + p_f \exp(WC/\text{RT}_m)]} \tag{6.14}$$

with

$$\text{RAV(max)} = W\left[p_s V \exp\left(-\frac{WV}{RT_m}\right) - p_f C \exp\left(\frac{WC}{RT_m}\right)\right] \qquad (6.15)$$

The nonlinearity of Eq. (6.14) with respect to RT_m precludes an analytical expression being available expressing RT_m in terms of W, V, C, p_s, and p_f, but simple hand-calculator values for RAV versus RT (for fixed values of the remaining parameters) can be used to estimate quickly whether the risk tolerance for a required working interest is less than the corporation limit. An example of this point will be given later.

6.2.2 Hyperbolic Risk Aversion

Not all corporations model risk attitudes with a risk aversion that is exponentially weighted; an exponential weight corresponds to a risk-aversion factor RAF(V; RT) proportional to $\exp(-E_1/RT)$. Alternative formulas for risk aversion are as many and varied as the individual corporations involved. Here, the hyperbolic rule is investigated corresponding to a risk-aversion factor proportional to $1 - \tanh(E_1/RT)$. The idea behind a hyperbolic rule is that there is greater stability in the management of high-loss scenarios than there is with the exponential rule.

In this case, the equivalent to the Cozzolino formula is to suppose that each state is weighted with the hyperbolic tangent so that one would write a total weighting as

$$p_s[1 - \tanh(WV/RT)] + p_f[1 - \tanh(-WC/RT)]$$

$$\equiv p_s + p_f + p_f \tanh(WC/RT) - p_s \tanh(WV/RT)$$

$$= 1 + p_f \tanh(WC/RT) - p_s \tanh(WV/RT) \qquad (6.16)$$

If the total weighting [Eq. (6.16)] is again taken to correspond to an exponential equivalent in the form

$$\exp\left(-\frac{RAV}{RT}\right) \equiv 1 + p_f \tanh\left(\frac{WC}{RT}\right) - p_s \tanh\left(\frac{WV}{RT}\right)$$

then the risk-adjusted value RAV is given through

$$\text{RAV} = -RT \ln\left[1 + p_f \tanh\left(\frac{WC}{RT}\right) - p_s \tanh\left(\frac{WV}{RT}\right)\right] \qquad (6.17)$$

which is equivalent to Cozzolino's formula, but for hyperbolic tangent weighting of risk aversion rather than exponential.

For any functional form it is then clear how to proceed. For instance, with a weighting proportional to $1 - G(E_1/RT)$, where G is an arbitrary but specified function, then one weights each state in the same manner to obtain the general equivalent formula

$$\text{RAV} = -\text{RT} \ln\left[1 - p_f G\left(-\frac{WC}{\text{RT}} \right) - p_s G\left(\frac{WV}{\text{RT}} \right) \right] \tag{6.18}$$

Thus all risk-aversion factors can be brought, by a general reduction, to corresponding risk-aversion values related to the risk tolerance RT and the working interest W.

For the hyperbolic RAV formula [Eq. (6.17)], a maximum of RAV, RAV(max), with respect to working interest exists at $W = W_m$, where W_m is given through

$$\cosh\left(\frac{W_m C}{\text{RT}} \right) = \left(\frac{p_f C}{p_s V} \right)^{1/2} \cosh\left(\frac{W_m V}{\text{RT}} \right) \tag{6.19a}$$

with approximate solution

$$W_m \approx \frac{1}{2} \frac{\text{RT}}{V} \ln\left(\frac{4 p_s V}{p_f C} \right) \tag{6.19b}$$

provided $0 \le W_m \le 1$, while the break-even value, RAV = 0, occurs at $W = W_0$ given by

$$p_f \tanh\left(\frac{W_0 C}{\text{RT}} \right) = p_s \tanh\left(\frac{W_0 V}{\text{RT}} \right) \tag{6.20a}$$

with approximate solution

$$W_0 \cong \frac{\text{RT}}{C} \ln\left(\frac{p_s}{p_f} \right) \tag{6.20b}$$

provided $0 \le W_0 \le 1$, i.e., provided $\text{RT} \le C \ln (p_f/p_s)$

Similar arguments for fixed working interest but variable values of risk tolerance can be gone through as for the Cozzolino formula, but it is easier to visualize patterns of behavior with a numerical example.

6.3 NUMERICAL ILLUSTRATION

To illustrate the similarities and differences between the Cozzolino formula, [Eq. (6.7)] and the hyperbolic formula [Eq. (6.17)] for risk-aversion values, consider the situation of Fig. 6.1b, in which the probability of success is 20% ($p_s = 0.2$) and so the probability of failure is 80%, in which the value of the success path is

$V = \$90$ million, whereas the cost of failure is $C = \$10$ million. It is requested that the best working interest in the project be evaluated for risk tolerances ranging from RT = \$10 million through RT = \$100 million.

Note from the information supplied that $E_1 = p_s V - p_f C = \$10$ million and the volatility is $v = 4$, so there is a large degree of uncertainty in the expected value of the project at the chance node.

6.3.1 Results from Cozzolino's Formula

For the parameters just given, the Cozzolino formula can be written in the parametric form

$$\text{RAV} = -10W\{x^{-1}\ln[0.2\exp(-9x) + 0.8\exp(x)]\} \qquad (6.21a)$$

with $10W/\text{RT} = x$. For a fixed value of RT, it follows that

$$W = \frac{\text{RT}x}{10} \qquad (6.21b)$$

so W increases as x increases, whereas for a fixed working interest, one has

$$\text{RT} = \frac{10W}{x} \qquad (6.21c)$$

so RT increases as x decreases.

Thus, as x is varied in the range $0 \le x \le \infty$, all possibilities are covered; RAV can then be plotted either as a function of W for fixed RT or as a function of RT for fixed W.

Fixed Risk Tolerance. Plotted in Fig. 6.2 are the estimated RAV values (in \$ million) as the working interest W varies from 0 to 100% for different risk tolerance values from RT = \$10 million to \$100 million. To be seen from Fig. 6.2 is that as RT increases, so that a greater degree of risk is considered acceptable to the corporation, then the range of working interest broadens over which a positive RAV is obtained. This broadening represents the point that as the risk tolerance increases, then the RAV tends asymptotically to $W(p_s V - p_f C) > 0$. Thus the range of working interest over which a positive value for RAV can be found increases with increasing RT.

The break-even value (RAV = 0) then provides a range of working interest W versus risk tolerance where RAV > 0. Plotted in Fig. 6.3 are the ranges of values of W versus RT at which RAV = 0. The implication from Fig. 6.3 is that the larger the risk tolerance value, the greater is the working interest that should be taken in the project.

Fixed Working Interest. Plotted in Fig. 6.4 are the curves of RAV versus RT for different working interests. To be noted from Fig. 6.4 is that as the risk tolerance

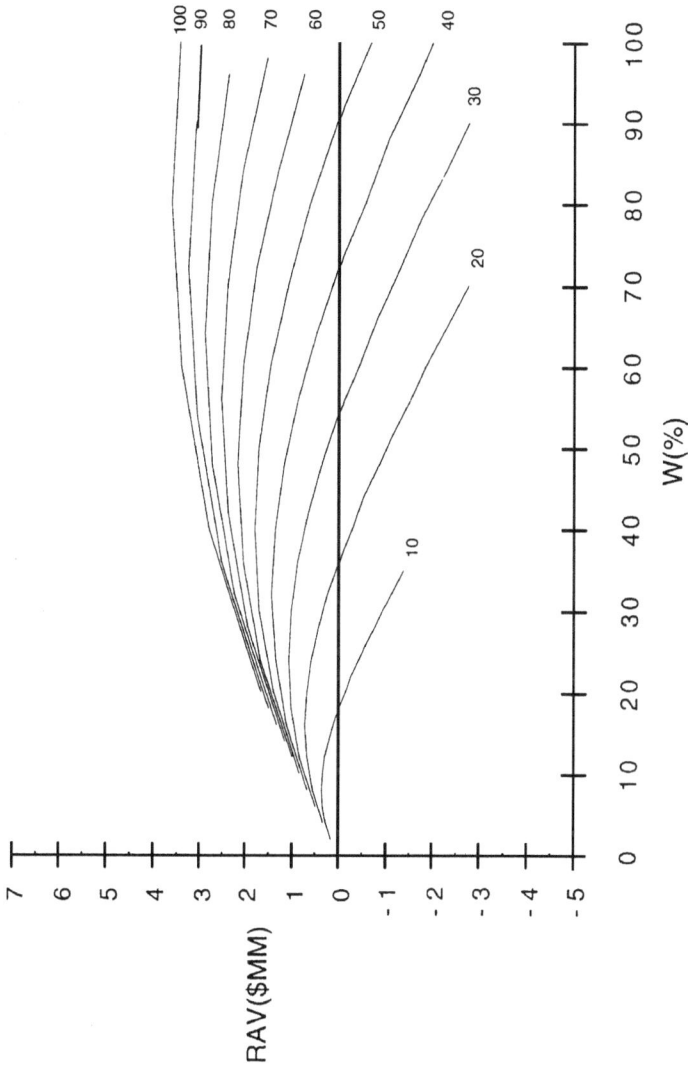

FIGURE 6.2 Risk-adjusted value RAV (in $ million) versus working interest W (%) for various risk tolerance values (labeled on each curve in $ million). Note that the RAV has positive and negative values until RT crosses about $55 million, when any working interest up to 100% will be profitable. The Cozzolino risk-adjustment formula was used with the data of Fig. 6.1b.

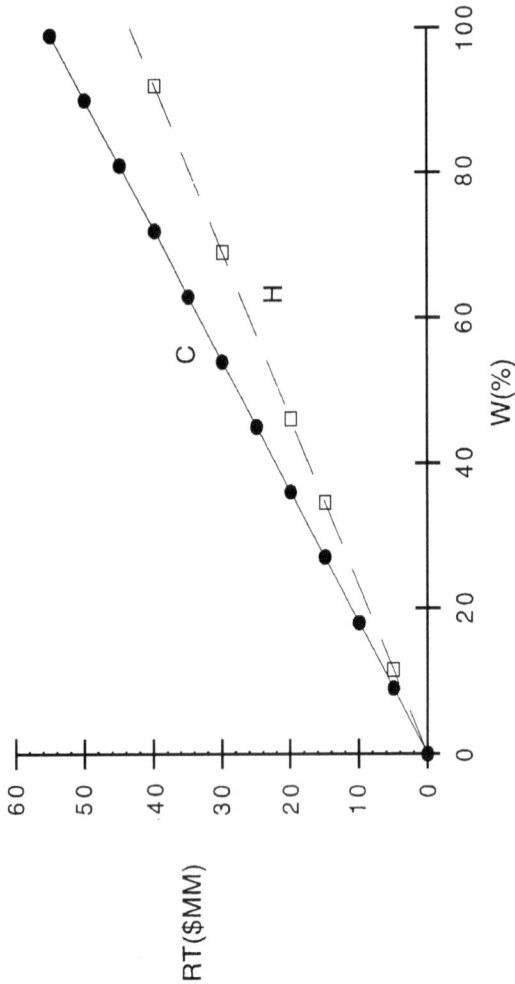

FIGURE 6.3 Plot of the break-even conditions (such that RAV = 0) for risk tolerance RT ($ million) versus working interest W (%) for both the Cozzolino risk-adjustment formula (curve C) and the hyperbolic risk-adjustment formula (curve H) using the data of Fig. 6.1b. RT values higher than the limits at W = 100% imply that 100% working interest should be taken in the project.

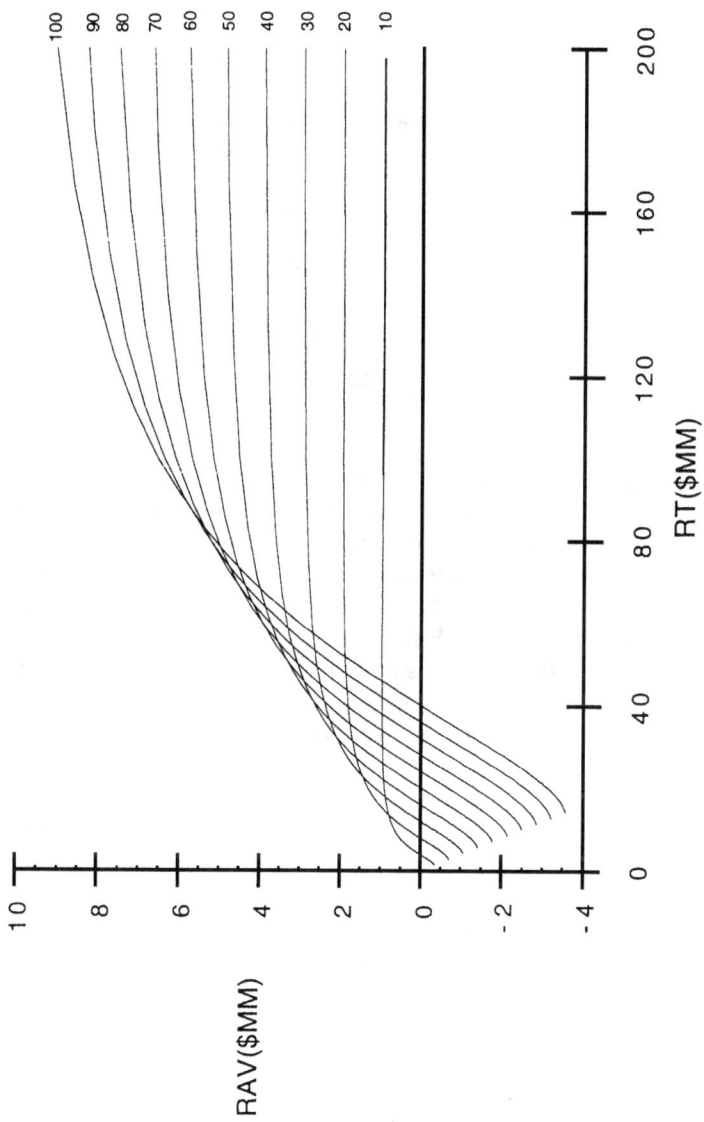

FIGURE 6.4 Risk-adjusted value RAV (in $ million) plotted against risk tolerance RT (in $ million) for various percentage working interests, as labeled on the curves at intervals of 10%. The Cozzolino risk-adjustment formula was used with the data of Fig. 6.1*b*.

increases beyond the maximum estimated value of the project, then the closer is the RAV to $E_1 = p_s V - p_f C = \$10$ million. Thus one can read from Fig. 6.4 what a particular RT corresponds to in terms of a positive or negative RAV. For instance, if the risk tolerance is RT = \$20 million, then it makes sense to accept a working interest of greater than about 25% in order to have a positive RAV.

6.3.2 Results from the Hyperbolic Formula

For the parameters given for the numerical illustration of Fig. 6.1*b*, the hyperbolic RAV formula [Eq. (6.17)] can be written parametrically through

$$\text{RAV} = -10W[x^{-1} \ln(1 + 0.8 \tanh x - 0.2 \tanh 9x)] \qquad (6.22a)$$

with, as before,

$$\text{RT} = \frac{10W}{x} \qquad (6.22b)$$

Fixed Risk Tolerance.   Figure 6.5 gives the estimated RAV values (in \$ million) versus the working interest W (in percent) for different risk tolerances ranging from RT = \$10 million to \$100 million. Again note that a greater positive range of RAV values is obtained as RT increases. Perhaps of interest here is to compare the range of positive RAV values for the risk-aversion model formula of Cozzolino with the results of the hyperbolic model.

Plotted in Fig. 6.3 on the ordinate is the range of RT values against W on the abscissa that produce positive RAV values (for the same W and RT) from the Cozzolino formula (labeled C) and the hyperbolic formula (labeled H), respectively.

Fixed Working Interest.   Figure 6.6 provides the estimated RAV values (in \$ million) versus the risk tolerance RT (in \$ million) as the working interest increases from 10 to 100%. Note that just as for the Cozzolino exponential risk-aversion formula, the hyperbolic formula indicates that the RAV tends to $E_1 W$ as RT increases. The range of W and RT values yielding positive RAV values is plotted in Fig. 6.3 versus the equivalent ranges from the Cozzolino formula.

6.3.3 Comparison of Results

At a fixed working interest, comparison of Fig. 6.4 with Fig. 6.6 shows that the Cozzolino risk-aversion formula is basically more conservative in its estimates of RAV as risk tolerance increases. It takes a higher RT to yield an RAV comparable with that from the hyperbolic formula. The difference for the example given is roughly a factor of 2. The reason for this behavior is that $\tanh(x)$ at large x varies as $1 - 2 \exp(-2x)$, so the x variation is almost twice as rapid as compared with the variation $\exp(-x)$ occurring in the Cozzolino formula.

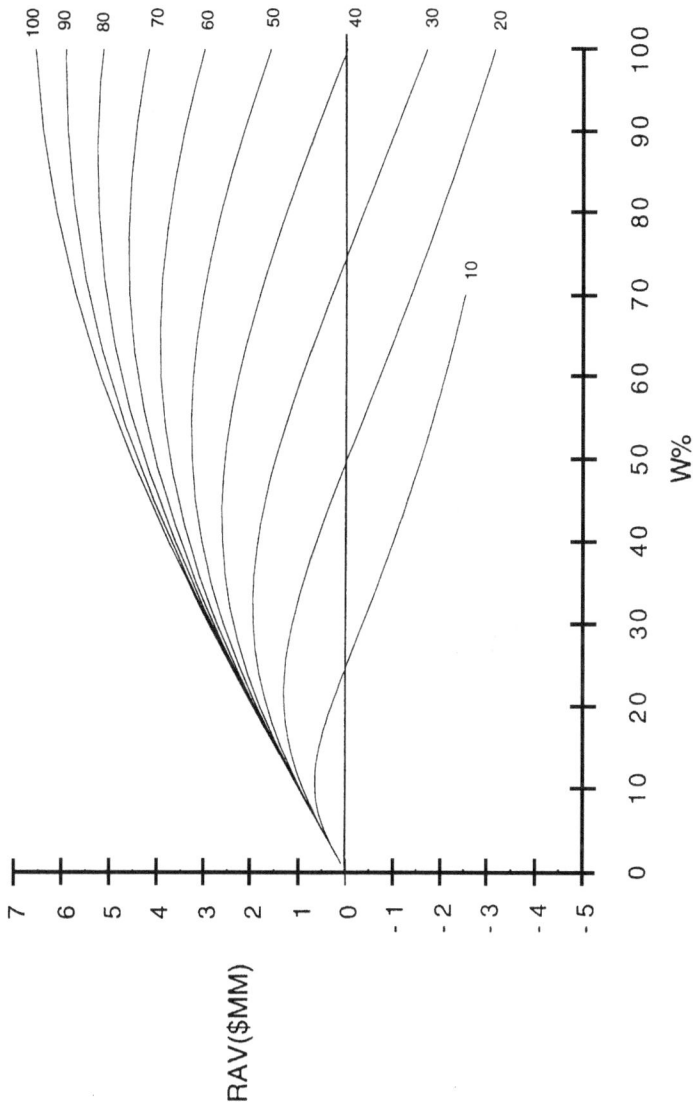

FIGURE 6.5 As for Fig. 6.2 but with the hyperbolic RAV formula. Note that RAV is positive at all $0 \leq W \leq 100\%$ once RT crosses about $42 million, a smaller value than for the Cozzolino formula exhibited in Fig. 6.2.

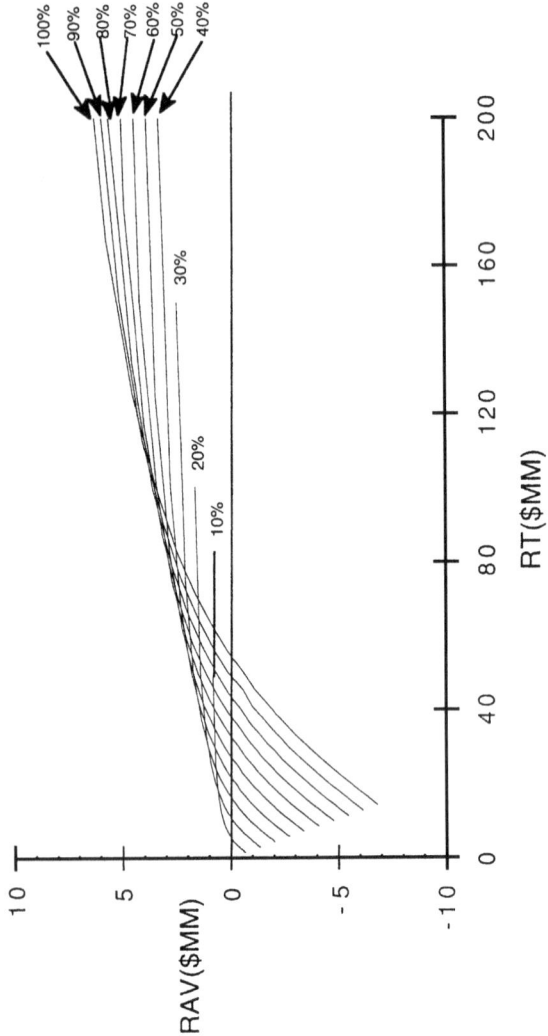

FIGURE 6.6 As for Fig. 6.4 but with the hyperbolic RAV formula.

At a given risk threshold, comparison of Fig. 6.2 with Fig. 6.5 indicates that the Cozzolino formula will require a higher RT to obtain the same range of working interest that yields a positive RAV as compared with the hyperbolic risk-aversion formula. The Cozzolino result accepts only a narrower range of working interest at a fixed risk tolerance, with the corollary that the expected positive RAV values are systematically lower than those obtained using the hyperbolic formula.

The break-even ranges for RT and W that yield RAV ≥ 0 are compared in Fig. 6.3, where, for a given working interest, the Cozzolino formula requires a greater risk tolerance to produce a positive RAV, whereas for a given risk tolerance, the Cozzolino formula would advise a lower working interest be taken to produce a positive RAV than would the hyperbolic risk-aversion formula.

The dominant reason for the systematic offset of results between the Cozzolino exponential-dominated risk-aversion formula and the hyperbolic formula is the functional dependence. Crudely speaking, the hyperbolic tangents vary but slowly in $0 \leq \tanh x \leq 1$ over all x, whereas the exponentials vary without bound. This gentler variation of the hyperbolic contribution implies that the risk-aversion estimates are not as sharply varying as parameter values change—leading to broader ranges of working interest and lower estimates of risk tolerance to produce positive domains of high RAV values.

6.4 NEGATIVE EXPECTED VALUES

The RAV formulas for different modeled behaviors also can be used to assess corporate strategy and philosophy even when there is a negative expected return. For example, consider a project costing $50 million that is *obligatory* so that $W = 1$. There is a 20% chance that the project will make $90 million; otherwise, it will be a $50 million loss. In this case, the expected value at the chance node of Fig. 6.1a is $E_1 = -\$22$ million. It is conventionally taken that the expected value is the risk-neutral cash amount that one should be willing to pay to buy out of the obligation; otherwise, it is more appropriate to undertake the project and take the chance (80%) of a $50 million loss. However, note that the magnitude of the volatility v is 2.5, representing a considerable uncertainty on the estimated mean value, in the sense that within one standard error ($\sigma \equiv vE_1$) one has a range of expected return of $E_1 \pm \sigma$, that is, $E_1(1 \pm v)$, which stretches from $-\$66$ million through to $+\$33$ million. Thus the uncertainty on the mean value should be included in assessing whether the risk-neutral cash amount to buy out of the obligation is really at E_1 or within some tolerance factor around E_1.

The RAV formulas can be used with such problems as follows: Suppose that the corporation is risk averse with a risk tolerance of RT = $50 million; then the RAV using the Cozzolino formula is $-\$39.6$ million, whereas the RAV from the hyperbolic tangent formula is $-\$17.5$ million. Thus the Cozzolino formula tends to emphasize the *negative* fractional uncertainty around the mean value, whereas the hyperbolic tangent procedure tends to emphasize the *positive* fractional uncertainty. In a sense, the Cozzolino formula is pushing more toward the chance of a greater likelihood of encountering the catastrophic loss of $50 million and so suggests that it is better to pay about $39.6 million to buy out of the obligation to be absolutely sure of avoiding the loss.

The hyperbolic tangent RAV recognizes that there is a chance that the uncertainty on the expected value could permit a positive expected value (to within one standard error) and so suggests that one should allow for this chance in assessing RAV and risk-neutral cash settlements of the obligation.

To illustrate the range of behaviors, three cases were run in which the obligation requires $W = 1$, the value $V = \$90$ million, $RT = \$50$ million, and $p_s = 0.2$ but with cost C at the three values $50, $40, and $30 million, respectively, as shown in Table 6.1.

Note that as the negative expected value increases toward zero, the magnitude of the volatility increases to the point where the uncertainty on the expected value is providing a greater chance that there is likely to be a positive *return* (to within one standard error). The hyperbolic tangent formula for RAV uses this chance to suggest that the cost to buy out of the obligation should be tempered by the likelihood of a positive return, whereas the Cozzolino formula for RAV tries to minimize total potential catastrophic damage by suggesting that it is better to buy out of the obligation at a high price, relative to the expected mean value, rather than to gamble on the chance of absolute failure.

Thus the Cozzolino formula tends to be pessimistic, whereas the hyperbolic tangent formula is more optimistic. Essentially, then, the use of individual corporate RAV formulas to assess risk is tied to the corporate philosophy on risk: conservative, neutral, or aggressive. The Cozzolino formula and the hyperbolic tangent formula for RAV reflect this difference in corporate attitude to risk, which was the point of this illustration.

6.5 SUMMARY

It is important to assess quantitatively corporate aversion to risky projects within the framework of a given corporate risk threshold. On the one hand, there is the need to ensure that the corporation does not end up bankrupt if the project fails, but on the other hand, there is the need to figure out what working interest should

TABLE 6.1 Risk-Adjusted Values for Different Cost Estimates

Cost, $ million	E_1, $ million	|Volatility|	Cozzolino RAV, $ million	Hyperbolic tangent RAV, $ million
50	−22	2.5	−39.6	−17.5
40	−14	3.7	−29.8	−14.7
30	−6	8.0	−20.0	−10.8

be taken in the project to maximize potential gains should the project prove profitable.

Clearly, the competing demands of these two corporate positions must be evaluated to assess likely corporate involvement in a project.

Until recently, Cozzolino's (1977, 1978) procedure was the only method readily available for enabling such evaluations to be made. The danger, then, is that the evaluations made of risk-adjusted value and working interest are particularly beholden to the model procedure, without the ability to evaluate the uniqueness, resolution, precision, and accuracy of the results in relation to variations in intrinsic assumptions of the model-dependent RAV.

The advantage to having available alternative functional forms for RAV assessment is that one can then compare and contrast the sensitivity of model results under different assumptions. In this way one can determine those factors which fall within broadly similar ranges for the majority of models versus those factors which are especially sensitive to the intrinsic assumptions of a particular model and so are less likely to reflect accurately corporate risk philosophy.

The other advantage is that different corporations use risk-aversion factors that are not necessarily of the exponential form, as is required by the Cozzolino (1977, 1978) risk-aversion formula. Therefore, the ability to put forward a general formula for risk-aversion values, as done here, means that each corporation can now use its own weighting formula and carry through the corresponding assessments of RAV in relation to risk tolerance and working interest within a corporate required framework.

The particular numerical illustrations presented here address the points of model dependence of results and show the magnitude and distortion of systematic offsets in behaviors for risk and working interest. Any set of model formulas for RAV can now be evaluated for their relative contrasts with the general method given here, and it is this fact that is the main point of this chapter.

CHAPTER 7
LIMITING RISK WITHIN A CONSORTIUM AND IN FOREIGN GOVERNMENT PROJECTS

7.1 INTRODUCTION

At a chance node for an environmental remediation project, such as given in Fig. 7.1, there is a value V, a cost C, and a probability of success of p_s (and a probability of failure $p_f \equiv 1 - p_s$). One of the standard concerns of an environmental corporation is to estimate the working interest W and the risk-adjusted value RAV for the project given a corporate risk tolerance RT. As noted in the preceding chapter, such matters have been researched and developed since the application of utility theory to hydrocarbon exploration opportunities, as illustrated by the Cozzolino (1977, 1978) risk-adjustment formula, which exponentially weights the success and failure branches of the chance-node diagram (see Fig. 7.1) with respect to WV and $-WC$, where W is the fractional working interest. Generalizations to allow for other than exponential weighting and to provide analytical algebraic results for maximum working interest, maximum RAV, risk-neutral RAV, and so on have been developed recently and are presented in Chap. 6.

However, the results are often based on the position that there is no uncertainty in V, C, p_s, or W. In reality, there is uncertainty in all four of these quantities for a variety of reasons: First, the total value V depends on economic models of projected future inflation and escalation costs and allied fiscal factors, whereas at the same time the quantitative environmental scientific estimates, which provide an assessment of transport, storage security, spillage likelihood, leakage chances, and cleanup methods, are themselves uncertain. Thus the conversion of scientific assessment to a cash value V already contains an uncertainty due to the environmental scientific model estimates. Second, the probability of success p_s is related not only to transport and secure storage concerns but also to the probability of uncertainty of spillage and leakage at present and in the future. The estimated project costs C at the time the project is being evaluated depend on assessments of future costs, including uncertainty due to unanticipated handling conditions, site preparation or possible regulations, and cost overruns; the estimated fractional working interest W is uncertain because there is usually flexibility in the values of W that can be optimized to produce a positive RAV for a given risk tolerance, and as the value, cost, and success probability change, the range of working interests yielding a positive RAV also changes. The range also changes as the risk tolerance

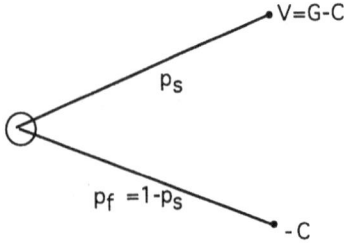

FIGURE 7.1 Chance-node diagram illustrating value V, cost C, probability of success p_s, and probability of failure p_f ($\equiv 1 - p_s$). The expected value at the chance node is $E_1 = p_s V - p_f C$ at a working interest W of 100% and is WE_1 at $W < 100\%$.

varies and also because, even without any uncertainty on the values of p_s, V, and C, there is still an uncertainty, measured by the variance $\sigma^2 \equiv p_s p_f (V + C)^2$, at the chance node of the decision tree of Fig. 7.1 due to the wide difference in value of a success or failure. Thus each of V, C, p_s, and W has some degree of uncertainty. Two questions then become of major concern:

1. How does one quantify the effect of uncertainties on the RAV and on the likelihood of a positive return from a project?

2. Which of the uncertainty factors is most important in influencing RAV so that one has some idea of where to concentrate effort to minimize the uncertainty?

The purpose of this chapter is to address these two concerns.

7.2 RAV ESTIMATES WITH UNCERTAINTIES

7.2.1 General Considerations

The Cozzolino (1977, 1978) formula (exponential weighting) for RAV takes the form

$$\text{RAV} = -\text{RT} \ln \left[p_s \exp\left(-\frac{WV}{RT}\right) + p_f \exp\left(\frac{WC}{RT}\right) \right] \qquad (7.1)$$

whereas the similar hyperbolic tangent weighting formula (MacKay and Lerche, 1996) for *RAV* takes the form

$$\text{RAV} = -\text{RT} \ln \left[1 - p_s \tanh\left(\frac{WV}{RT}\right) + p_f \tanh\left(\frac{WC}{RT}\right) \right] \qquad (7.2)$$

As noted in Chap. 6, a general risk-weighted formula for a function $G(x)$ with a risk-aversion factor proportional to $1 - G(WV/RT)[1 - G(-WC/RT)]$ for the success (failure) branch of the chance-node decision-tree diagram of Fig. 7.1 can be shown (MacKay and Lerche, 1996) to yield

$$\text{RAV} = -\text{RT} \ln\left[1 - p_s G\left(\frac{WV}{\text{RT}}\right) - (1 - p_s) G\left(-\frac{WC}{\text{RT}}\right)\right] \qquad (7.3)$$

7.2.2 Numerical Illustrations

General Observations. Because both the Cozzolino and the hyperbolic tangent RAV formulas are simple algebraic expressions, it is relatively easy to evaluate numerically the expressions for RAV for any given values of V, C, p_s, W, and RT. This simplicity then lends itself to direct Monte Carlo simulations in which individual values of V, C, p_s, W, and RT are selected from predefined distributions (triangular, uniform, normal, log-normal) with ranges specified and the RAV calculated for each set of values. A Monte Carlo random-number selection procedure then can be invoked to choose many input values, and the distribution of RAV values can be computed for the selected input distributions and their ranges. In addition, one obtains the effect of each individual factor contributing to the variance in RAV and so the ability to compute the relative importance of each contribution to the uncertainty in RAV.

The essence of the problem is to have a set of procedures for determining the individual components V, C, p_s, W, and RT and their likely ranges of uncertainty from other considerations. In the case of the value V of a project, the dominant point is to have some scientific method for providing an estimate of transport, storage security, leakage, spillage, and cleanup remediation concerns, as well as a measure of the expected uncertainty in estimates made with scientific models. Once an environmental scientific estimate is made, then economic considerations convert this value to a cash amount V together with increased ranges of uncertainty brought about by the uncertainties in the economic parameters and assumptions.

The basic problem is to assess the environmental scientific aspects and their uncertainty. Fortunately, precisely this problem is the central point of environmental models that estimate the influences of transport, spillage, leakage, storage, and drainage area in determining likely conditions together with uncertainties, as well as the probabilities of being correct in the assessment of one storage site in relation to others. Thus the estimated value V then can be provided, as can the success probability p_s and their likely ranges of uncertainty.

The dominant factors controlling cost estimates and their uncertainties are usually bid prices, seismic survey costs, storage costs, subsurface drilling conditions, and factors having to do with future inflation estimates, taxes, monitoring, etc. While not usually as variable or uncertain as scientific estimates of environmental conditions at present and in the future, nevertheless, project cost estimates are often uncertain by up to 50% historically and occasionally much more.

The variation in risk tolerance that a corporation is prepared to accept is itself a matter quite often of historical precedent within a corporation or often guided by risk thresholds limiting damage to a corporation should a project prove to be unsuccessful. The point here is that cash equivalent values can be provided for V, C, and RT, as well as for their ranges of uncertainty, based on scientific and economic projections of attainable goals.

With respect to working interest, the situation is slightly different. For each set of parameters for V, C, p_s, and RT, it is possible (MacKay and Lerche, 1996) to determine analytically (exactly or approximately) the optimal working interest OWI to maximize RAV, as well as the minimum and maximum working interest values that return a risk-neutral RAV of zero, so that any working interest in the minimum to maximum range will yield a positive RAV (see Chap. 6). However, it also can happen that some range of working interest is a prescribed requirement in order to participate in a project (e.g., an interest greater than 20%). In such cases, there is a prescribed minimum W and obviously also a maximum of $W \leq$ 100%. Within this range, the working interest can vary. Clearly, as the parameters V, C, p_s, and RT vary, the RAV will take on different values for different working-interest prescriptions. Hence there exists the possibility for allowing W to be uncertain within a prescribed range.

In addition, some care has to be exercised in assigning the probability functional behavior for each of the intrinsic variables V, C, p_s, W, and RT. The reasons are that both p_s and W are restricted to the maximum allowable ranges of $0 \leq p_s \leq$ 1 and $0 \leq W \leq 1$, whereas C and RT are each restricted to lie in $\{0, \infty\}$, although value V can range in $\{-C, \infty\}$. Thus a normal distribution for any of the variables cannot be allowed (without truncation) because a normal distribution admits negative values; equally, p_s and W cannot be log-normally distributed (without truncation) because a log-normal distribution covers the range $\{0, \infty\}$, which exceeds the $\{0, 1\}$ range of p_s and W. However, within the framework of these minor restrictions, the variation of V, C, p_s, W, and RT can be as broad as required.

Here the behavior of both the Cozzolino (exponential weighting) and hyperbolic tangent formulas for RAV is investigated under a variety of ranges of parameters to illustrate dominance and relative importance of individual factors contributing to the uncertainty in RAV.

Uniform Probability Distributions.   Consider the situation where environmental waste charge HC estimates to a storage site have been made using scientific techniques, perhaps in the manner of Thomsen and Lerche (1997), yielding ranges of 3 < HC < 9 million container units and with a central estimate of 6 such units. For the purposes of this example, it is given that each container unit can be allocated a profit to the corporation from the governmental contract at a fixed price of $15. At the same time, scientific estimates are also made of the likelihood of the depository being undisturbed after complete fill, suggesting that there is only a combined probability of between 10 and 30% of a seal being in place and of seal integrity being maintained from the present to the end of the lifetime of the toxic waste.

Costs of licenses, monitoring, taxes, and site preparation are set at $5 million if no adverse geologic conditions (such as faults, fractured rocks, etc.) are found but may be as high as $15 million if site conditions (e.g., high fracture density occurring at shallow depths) require special care and the use of advanced studies and technologies. The corporate risk tolerance for similar sites under similar conditions has ranged in previous years from $50 million to $150 million depending on the capital asset and yearly cash-flow picture of the corporation. Thus, on the basis of the preceding information, one would set the ranges

$$\$45 \text{ million} \leq V \leq \$135 \text{ million} \qquad (7.4a)$$

$$\$5 \text{ million} \leq C \leq \$15 \text{ million} \qquad (7.4b)$$

$$0.1 \leq p_s \leq 0.3 \qquad (7.4c)$$

$$\$50 \text{ million} \leq \text{RT} \leq \$150 \text{ million} \qquad (7.4d)$$

Note that the minimum RT ($50 million) is already large compared with the maximum estimated cost ($15 million), so the risk threshold should not be a major factor in influencing probabilistic RAV values. For each of the variables V, C, p_s, and RT, Monte Carlo runs were done to provide probabilistic results, with the random values for each variable being selected from uniform populations, bracketed at their respective minima and maxima by the ranges just given. The working interest was taken to be fixed at 100%.

Figure 7.2a exhibits the relative importance (in percent) of the uncertainties in p_s, C, V, and RT in contributing to the uncertainty for the Cozzolino RAV formula, whereas Fig. 7.2b exhibits the same information for the hyperbolic tangent RAV formula. As anticipated, in both cases RT has the smallest influence on RAV, whereas both p_s and C dominate, contributing a total of about 80% of the uncertainty, with p_s being about 50% of the total uncertainty.

Thus, in terms of relative importance to the uncertainty in the RAV, it is not so much the total charge of container units that dominates in this case but rather the sealing conditions of the storage site after charging that influence the uncertainty on p_s, the probability of success.

In addition, while the range of costs is only a small fraction of the value, the problem here is that the value V has to be multiplied by the success probability ($\approx$ 0.2) so that $p_s V$ is comparable with $(1 - p_s)C$, indicating an expected return (at the center values of each variable) of $p_s V - (1 - p_s)C \approx \$10$ million, which is comparable with the cost values. Hence the relative importance of cost on the RAV in both cases arises because the ratio $p_s V/(1 - p_s)C$ is of order unity.

It is not so much that reducing costs would be the ideal improvement arena but rather that there must be some way to assess whether the range of p_s is really at 0.2 $\pm$ 0.1 or whether p_s is larger or smaller, because the larger the central value of p_s, the smaller is the influence of cost, whereas the smaller the central value of p_s, the less likely is the opportunity to be profitable. If geologic information indicates that there is no further improvement possible in the determination of p_s and its range of uncertainty, then either cost uncertainty should be addressed (possibly through

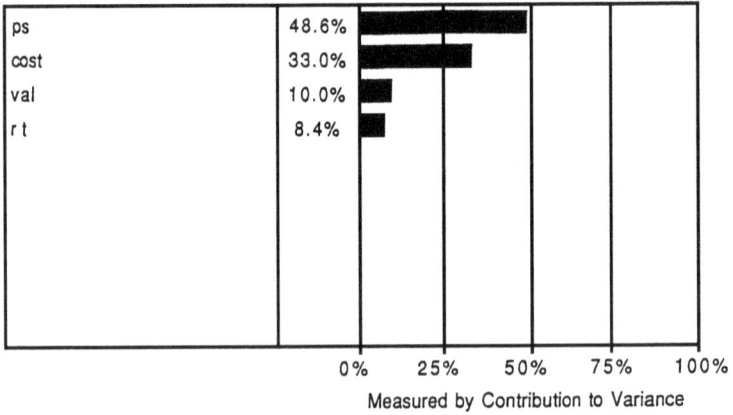

ps	48.6%
cost	33.0%
val	10.0%
r t	8.4%

Measured by Contribution to Variance

(a)

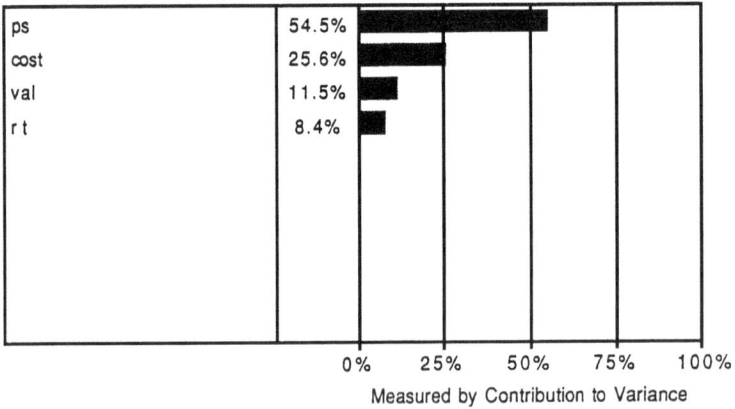

ps	54.5%
cost	25.6%
val	11.5%
r t	8.4%

Measured by Contribution to Variance

(b)

FIGURE 7.2   Relative importance (%) of factors contributing to the uncertainty in RAV:
(a) the Cozzolino formula; (b) the hyperbolic tangent formula for uniform probability distributions of p_s, V, C, and RT with working interest set to $W = 100\%$.

turnkey bids) or value estimates reinvestigated or a hedging position taken to prevent catastrophic loss. In any event, however, the point is that the relative importance of each factor provides an objective way of determining where effort should
be placed to minimize the uncertainty in the RAV.

Apart from the *relative* importance of individual factors contributing to the
uncertainty in the RAV, there is also the *absolute* value of the variance to consider.
The point here is that relative importance informs on *which* factors need to be
addressed for improvement, whereas absolute uncertainty informs on *when* factors
need to be addressed for improvement. If the absolute uncertainty on the RAV is

very much less than the mean value, then there is little uncertainty anyway, even when the relative importance indicates which factors are dominating the variance.

Thus Fig. 7.3a and b, for the Cozzolino formula and the hyperbolic tangent formula for RAV, respectively, shows the cumulative probability of obtaining an RAV that is less than particular amounts. Both curves are approximately log-normally distributed due to the central limit theorem, as can be observed by viewing the figures as histograms.

Two values are of importance: (1) the cumulative probability p_0 of an RAV $\geq$ 0 and (2) the values RAV_{10}, RAV_{50}, and RAV_{90}, which occur at 10%, 50%, and 90% cumulative probability values, respectively, and which measure the range of likely positive returns to be expected given the intrinsic uncertainties.

Reading off from Fig. 7.3a and b, one has

$$p_0 \cong 35\% \qquad \text{(Cozzolino formula)} \qquad\qquad (7.5a)$$

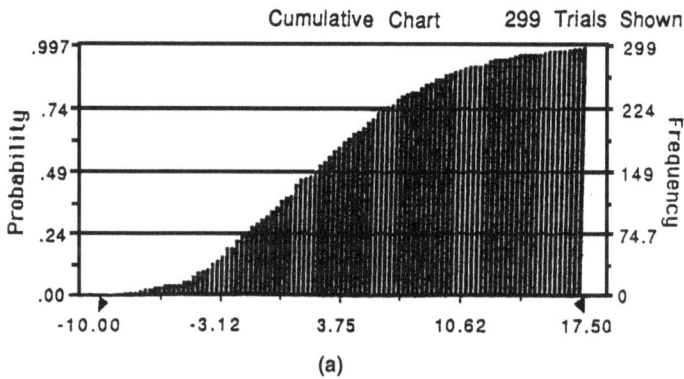

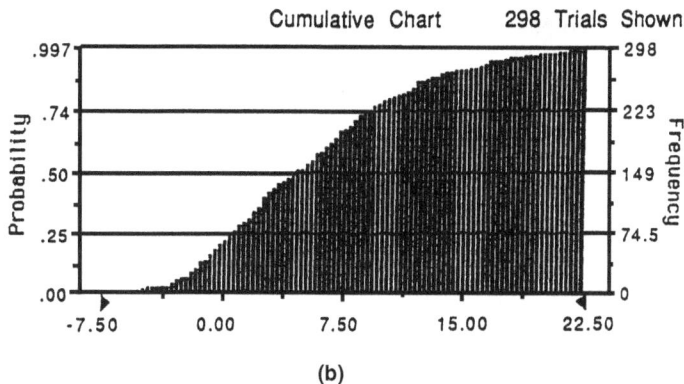

FIGURE 7.3 Cumulative probability plots of RAV for the uniform probability distributions of p_s, V, C, and RT with $W = 100\%$: (a) the Cozzolino formula; (b) the hyperbolic tangent formula.

and

$$p_0 \cong 20\% \qquad \text{(hyperbolic tangent formula)} \qquad (7.5b)$$

so that there is a 65 to 80% chance of a positive RAV no matter which risk-aversion formula is used.

Again reading off from Fig. 7.3a and b, one has

	Cozzolino, $ million	Hyperbolic tangent, $ million
RAV_{10}	−3.9	−1.8
RAV_{50}	2.4	4.6
RAV_{90}	10.3	13.8

The volatility v of the 50% RAV is given by

$$v = \frac{RAV_{90} - RAV_{10}}{RAV_{50}} \qquad (7.6)$$

which takes on the values $v = 5.9$ (Cozzolino) and $v = 3.4$ (hyperbolic tangent), both of which are significantly larger than unity. The volatility provides a quantitative measure of uncertainty on the RAV because if v is small compared with unity, then there is little uncertainty on RAV_{50}, whereas a large value of v ($\gg 1$) provides an indication that there is significant uncertainty on RAV_{50}. Thus, while the estimate RAV_{50} is fairly uncertain, nevertheless, there is between a 2:1 and 4:1 odds-on chance that the project should be undertaken, as measured by the value p_0 where RAV = 0.

In addition to relative importance and probabilities for RAV, two other factors are of relevance: the expected value (not risk-weighted) and the optimal working interest that should be taken—rather than assigning a 100% working interest, as has been done so far.

Referring to Fig. 7.1, the expected value E_1 at the chance node (at $W = 100\%$) is

$$E_1 = p_s V - (1 - p_s) C \qquad (7.7a)$$

and the variance σ^2 is given through

$$\sigma^2 = E_2 - E_1^2 = p_s (1 - p_s) (V + C)^2 \qquad (7.7b)$$

Thus, even for fixed values of V and C, because there is only a probability p_s of success, the expected value has an uncertainty of $\pm\sigma$. Clearly, when V, C, and p_s also have uncertainties, then such add to the total uncertainty on the expected value. Note for reference that if each of p_s, V, and C were to be set at their midpoint values of 0.2, $90 million, and $10 million, respectively, then $E_1 = $10 million and $\sigma = \pm$4 million.

Figure 7.4*a* presents the cumulative probability plot for the Monte Carlo runs indicating a mean value of E_1 = \$9.5 million (within \$0.5 million of the analytic calculation) and a standard error of ±\$8 million, nearly double the value of σ due to the variations in p_s, V, and C around their central values. Note also from Fig. 7.4*a* that the probability of obtaining a positive value for E_1 is about 90% so that, even without risk-weighting the project, the suggestion is that it is likely to be worthwhile financially.

Figure 7.4*b* presents the relative importance (in percent) of the uncertainty in each of p_s, V, and C in contributing to the variance in the expected value. (RT has no importance because E_1 does not depend on RT.) In this case, note that the uncertainty in p_s is contributing almost 50% of the uncertainty, with uncertainty in the value V contributing about 36% and cost uncertainties only about 18%. Again,

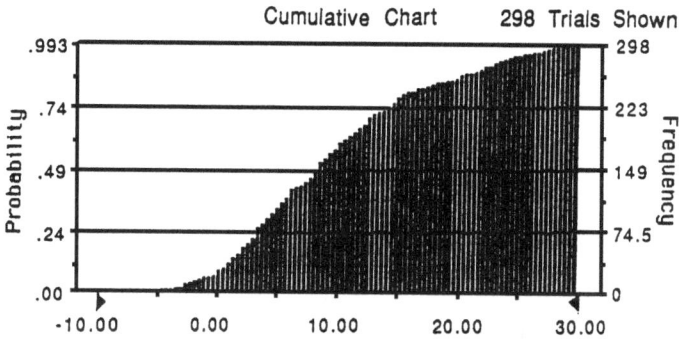

(a)

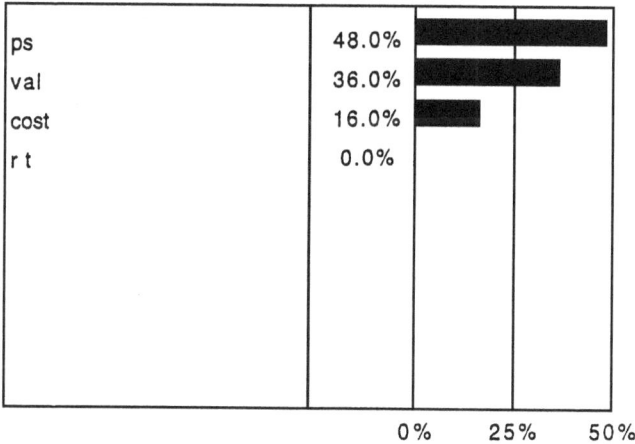

(b)

FIGURE 7.4   (*a*) Cumulative probability plot of expected value E_1 for the uniform probability distributions of p_s, V, C, and RT with $W = 1$. (*b*) Relative importance of factors contributing to the uncertainty in E_1. Note the absence of any effect of RT (because E_1 is independent of RT).

then, the strong indication is that it is the success probability p_s that needs to be addressed in order to improve matters further, but now with a secondary emphasis that suggests that determining better the value V is more important (by a factor 2) than doing a better job on narrowing the cost range of uncertainty.

The optimal working interest OWI that one should take in the project is also of interest. The preceding chapter showed for the Cozzolino RAV formula that OWI was given precisely by

$$OWI = \min\left\{\left(\frac{RT}{C + V}\right) \ln\left[\frac{p_s V}{(1 - p_s) C}\right]; 1\right\} \quad (7.8)$$

in $p_s V > (1 - p_s)C$; otherwise, OWI = 0. At this value of OWI, the maximum RAV is

$$RAV_{max} = V \times OWI - RT \ln\left[p_s\left(1 + \frac{V}{C}\right)\right] \quad (7.9)$$

For the range of values of RT, p_s, V, and C, the Monte Carlo simulations were again used to provide a cumulative probability plot of OWI, as displayed in Fig. 7.5a.

Reading off from Fig. 7.5a, the mean value of OWI is about 62%, whereas the values at 10, 50, and 90% cumulative probability are $OWI_{10} = 5\%$, $OWI_{50} = 69\%$, and $OWI_{90} = 100\%$, respectively, for a volatility of $(OWI_{90} - OWI_{10})/OWI_{50} = 1.4$.

Thus about 60 to 70% optimal working interest is likely the best to take in the project given the uncertainties, but a conservative corporation could take as low as 5%, whereas a more aggressive corporation should aim for 100% interest. Either choice, of course, would change the apparent risk tolerance. Figure 7.5a indicates that there is a 30% probability that an OWI of 100% is appropriate, with an 80% probability that an OWI greater than 25% should be taken.

The relative importance of contributions to the variance in OWI is shown in Fig. 7.5b, indicating that success probability and cost are the dominant contributors, with risk tolerance of lesser importance and with very little dependence on value. The reason for this surprisingly small dependence on value is that the ratio V/C is larger than 3 for the ranges given so that the major sensitivity of Eq. (7.8) is to $\ln[p_s/(1 - p_s)]$, which varies rapidly as p_s varies in $0.1 \leq p_s \leq 0.3$. Thus the dominance of uncertainties in p_s overrides all other contributions to the uncertainty in OWI. In addition, the reason cost C becomes of major importance is the fact that the ratio V/C, while large compared with unity, still varies considerably so that OWI takes on the value unity about 30% of the time. Clearly, the dominant message here is that the costs are sufficiently low, the risk tolerance sufficiently high, and the value sufficiently large that a significant fraction ($\sim$30%) of the Monte Carlo runs opt for 100% OWI, thereby narrowing the probability range to 0 to 70% over which OWI values are sensitive to p_s, C, V, or RT.

In general, the risk-adjusted values place much less emphasis on the value V than does the expected value, whereas the OWI (being roughly linear with respect to cumulative probability) places little emphasis on the value but considerable emphasis on success probability and cost.

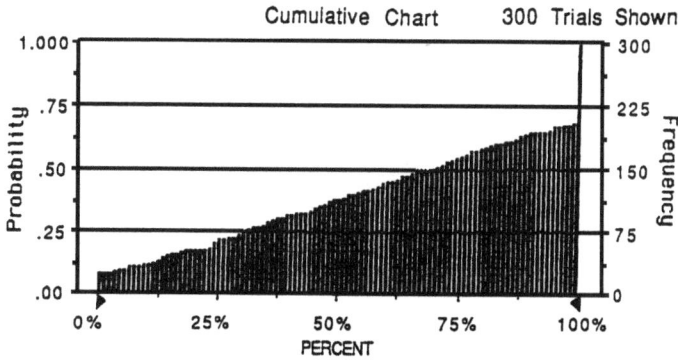

(a)

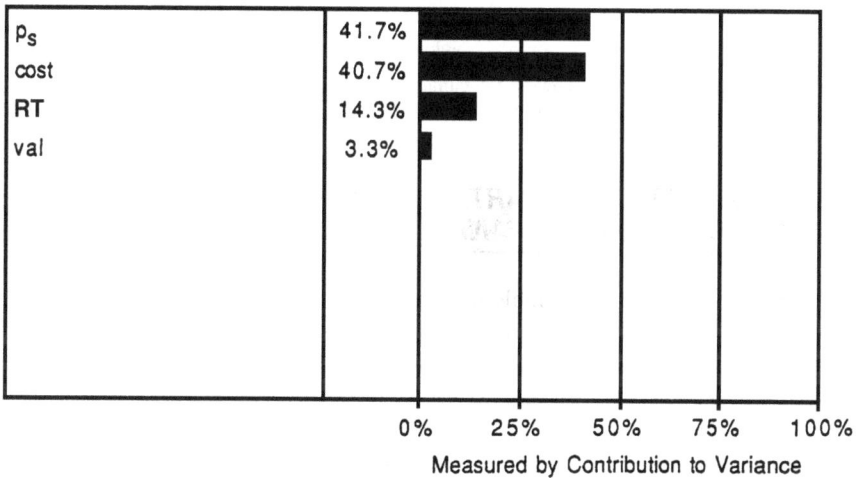

(b)

FIGURE 7.5   Optimal working interest OWI for the Cozzolino RAV formula, which maximizes RAV, for uniform probability distributions of p_s, V, C, and RT with $W = 100\%$. (*a*) Cumulative probability plot of OWI; (*b*) relative importance (%) of factors contributing to the uncertainty in OWI.

Figure 7.6 displays a trend chart illustrating the differences between the estimates for expected value, RAV (Cozzolino), and RAV (hyperbolic tangent) based on cumulative probabilities at different confidence intervals, indicating that the expected value is both more positive and has a wider variance relative to the two RAV plots, with the hyperbolic tangent RAV having a slightly larger variance than the RAV from the Cozzolino method.

While it may be thought that the RAV curves should have a greater variance than the expected value because they involve the extra uncertain variability of RT, the point is that their dependences on RT, as well as on p_s, V, and C, is highly nonlinear, so there is more suppression of the variance than otherwise may have been anticipated.

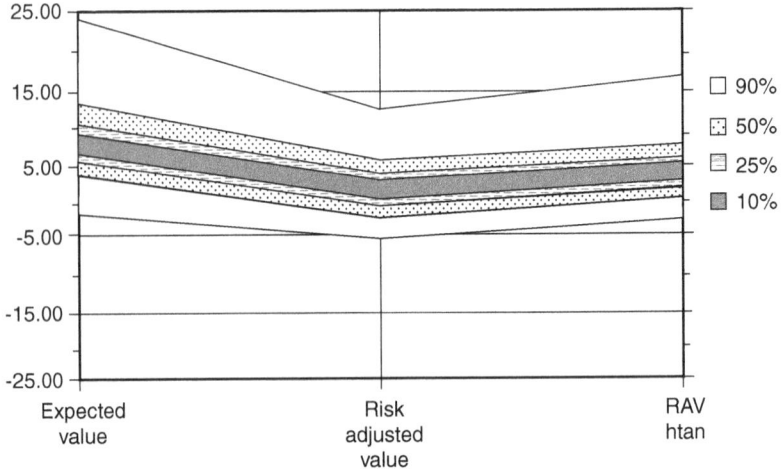

FIGURE 7.6 Trend chart showing the cumulative percentage bands comparison between expected value, RAV (Cozzolino), and RAV (hyperbolic tangent) with centering on the median values for uniform probability distributions of p_s, V, C, and RT with $W = 100\%$.

7.3 EVALUATION OF PARTICIPATION IN A CONSORTIUM OF COMPANIES

The preceding numerical example was predicated on the assumption that a corporation takes 100% working interest in a project. However, in reality, it is often the case that a group of corporations will band together, with each taking less than 100% working interest.

In a general sense, a corporation is offered the choice of some range of fractional participation (greater than zero but less than 100%); the corporation then has to decide what working interest to take.

To illustrate the influence of uncertainty in working interest on RAV, we again consider the situation of uniform probability distributions for V, C, p_s, RT, and W, where the range of each variable is

$$0.1 \leq p_s \leq 0.3 \qquad (7.10a)$$

$$\$5 \text{ million} \leq C \leq \$15 \text{ million} \qquad (7.10b)$$

$$\$45 \text{ million} \leq V \leq \$135 \text{ million} \qquad (7.10c)$$

$$\$50 \text{ million} \leq \text{RT} \leq \$150 \text{ million} \qquad (7.10d)$$

$$0 \leq W \leq 50\% \qquad (7.10e)$$

Again, Monte Carlo runs were done for both the Cozzolino and hyperbolic tangent RAV formulas. Note from Fig. 7.7a and b, which shows the relative importance of individual contributions to the uncertainty in RAV for the two RAV formulas, that the uncertainty in p_s still dominates, but the uncertainties in V and W now jointly contribute about 50 to 60% to the uncertainty in RAV, with risk tolerance RT and cost C providing only small contributions. The Cozzolino RAV formula is more sensitive to uncertainty in value than in working interest (a 1.5:1 ratio), whereas the hyperbolic tangent RAV formula is nearly equally uncertain to both value and working-interest uncertainties. The expected value $E_1 \equiv W[p_sV - (1 - p_s)C]$ is directly proportional to the working interest, and as shown in Fig. 7.7c, this proportionality is reflected in the relative importance contributions, with working interest, success probability, and value all approximately equally important in contributing to the uncertainty in expected value, with contributions from cost uncertainty small and no dependence on RT, of course.

For the cumulative probabilities for RAV from the Cozzolino and hyperbolic tangent formulas, as well as for the expected value, Fig. 7.8 shows the variations; note the influence of working-interest uncertainties on the shapes of the curves. Reading off from Fig. 7.8, the behavior can again be summarized:

	Cozzolino	Hyperbolic tangent	Expected value
RAV_{10} ($ million)	−0.14	0.03	E_{10} ($ million) = 0.04
RAV_{50} ($ million)	1.34	1.64	E_{50} ($ million) = 1.71
RAV_{90} ($ million)	4.67	6.13	E_{90} ($ million) = 6.31
p_0(%) (RAV = 0)	15%	9%	p_0 (E = 0) = 9%
|Volatility|	3.6	3.8	3.7

In this case, all three measures indicate comparable volatilities, but all three also indicate a probability of between 85 and 91% of a positive RAV, whereas the 50% probability RAV value is between $1.3 million and $1.7 million. Thus the addition of a working interest lower than 100% cuts the potential for catastrophic losses while reinforcing the positive aspects of the project.

The question arises as to how the range of working interest allowed ($0 \leq W \leq 50\%$) stands in relation to the optimal working interest OWI formula, which is independent of the input W. Shown in Fig. 7.9a is the cumulative probability of OWI as p_s, V, C, and RT vary, whereas Fig. 7.9b shows that uncertainty in p_s and cost dominate (80%) the uncertainty in OWI, while RT and V uncertainties only jointly contribute about 20% to the uncertainty. Reading off from Fig. 7.9a, there is a 65% chance that a working interest *less than* 100% should be taken in the project but an 80% chance that a working interest *greater than* 30% should be taken.

The upper end of the input range $0 \leq W \leq 50\%$ is met at a cumulative probability of 30%, suggesting that there is a 70% chance that greater than 50% working interest should be taken. Thus, while the opportunity to invest in the project provides a high chance of a positive RAV, the offer of a working-interest fraction less than 50%, while itself a good deal, is not the best deal that could have been accepted if offered; a higher working interest would have been preferred. Note that

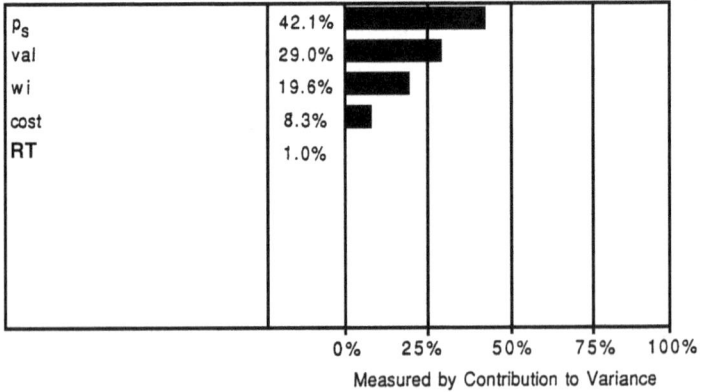

Measured by Contribution to Variance

(a)

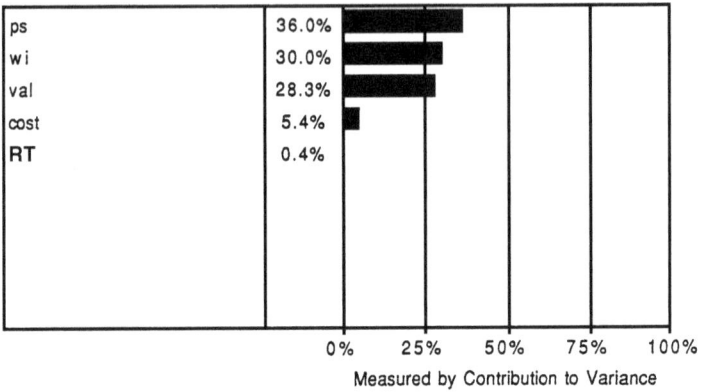

Measured by Contribution to Variance

(b)

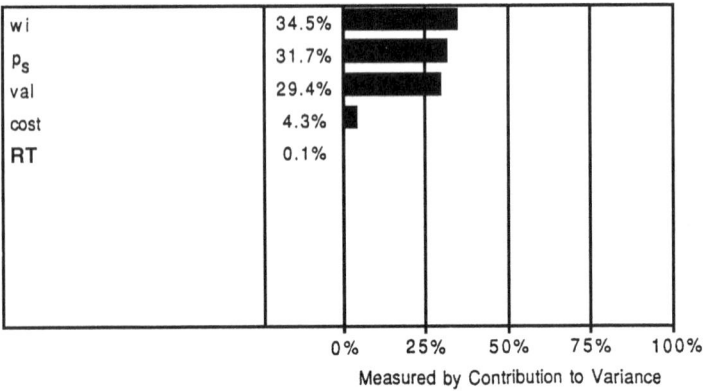

Measured by Contribution to Variance

(c)

FIGURE 7.7   Relative importance (%) of factors contributing to the variance in RAV and expected value when uniform probability distributions are assigned for p_s, V, C, and RT, together with uncertainty in the working interest W, which is also given a uniform probability distribution: (a) RAV (Cozzolino); (b) RAV (hyperbolic tangent); (c) expected value.

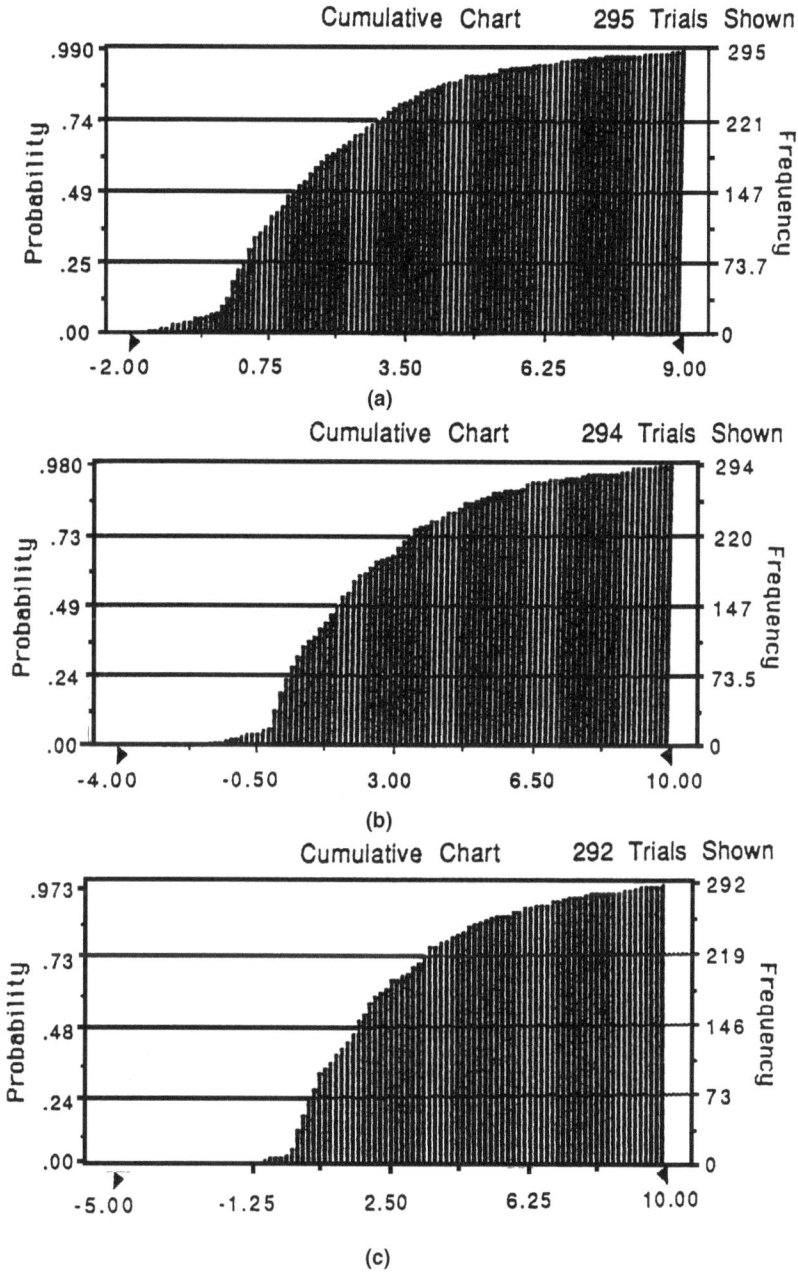

FIGURE 7.8 Cumulative probability plots using the uniform probability distributions for p_s, V, C, RT, and W: (*a*) RAV (Cozzolino); (*b*) RAV (hyperbolic tangent); (*c*) expected value.

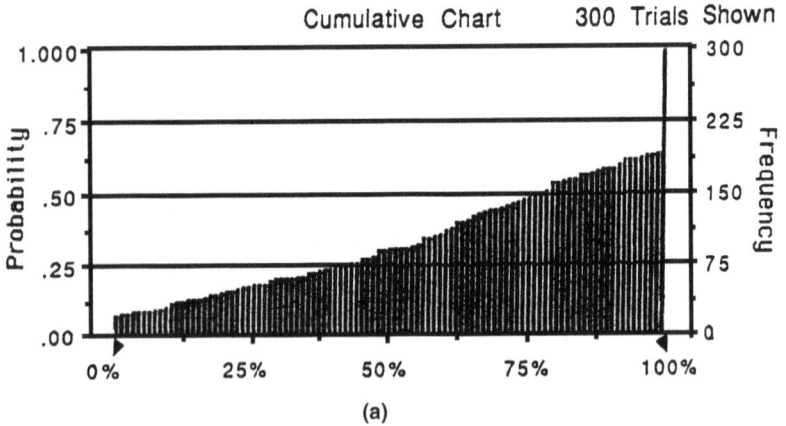

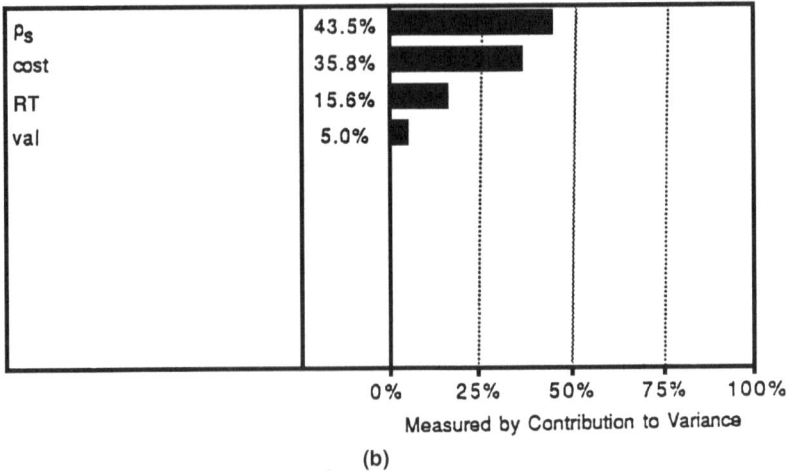

FIGURE 7.9 Optimal working interest OWI for the Cozzolino RAV formula, which maximizes RAV, using the uniform probability distributions for p_s, V, C, RT, and W: (*a*) cumulative probability plot of OWI; (*b*) relative importance (%) of factors contributing to the uncertainty in OWI.

the volatility on the OWI is only 1.2, whereas $\text{OWI}_{50} = 77\%$, so the estimate is ruggedly stable and can be used to bargain for a higher working interest in the project, depending on the conservative, neutral, or aggressive corporate philosophy on risk taking.

7.4 EVALUATION OF PARTICIPATION IN FOREIGN GOVERNMENT PROJECTS

As the costs of a project rise relative to value, both the Cozzolino and the hyperbolic tangent RAV formulas indicate that, if possible, less working interest should be taken. However, often a positive-working-interest commitment in a project is mandated in order to maintain a position in an active area of a country. In such cases, the RAV formulas provide an estimate of the amount that should be spent to buy out of such an obligation rather than face the massive costs that could arise if no success occurs.

For instance, it can (and does) happen that a government agency mandates a company to take a working interest of, say, around 50% as long as the project develops with mounting costs and success uncertain. The fractional working interest allowed to a foreign corporation reduces to 25% if the costs drop, with the project more profitable. The government retains the option to increase the stake of its own national corporation and so decrease a lesser working interest (including zero) that can be retained by the foreign corporation (often predetermined in the contract). Governments recognize that investments with essentially no chance of success are bad for both the foreign company and the national company, so governments are often agreeable to cash payment in lieu of further investment. Thus, with an uncertainty on costs and an uncertainty on mandated working interest, the problem is to figure out the buyout amount that should release the corporation from the obligation.

To illustrate the way in which such problems can be addressed, consider that p_s, V, W, and RT all have the same ranges of uncertainty as in the preceding example, but now increase the costs by a factor of 10, to lie in $50 million $\leq C \leq$ $150 million. All parameters are again taken to be drawn from uniform probability distributions.

A set of Monte Carlo calculations was again made for RAV from the Cozzolino and hyperbolic tangent formulas, as well as for expected value, as shown by the cumulative probability plots of Fig. 7.10.

In this case, the interpretation of Fig. 7.10 is as follows: From Fig. 7.10a, which illustrates the Cozzolino RAV formula results, there is a 90% chance that one should pay at least $3 million to buy out of the obligation, there is a 50% chance that one should pay at least $15 million, but only a 10% chance that one should pay more than $36 million to buy out of the obligation. Likewise, for the hyperbolic tangent RAV (Fig. 7.10b), the 90% chance indicates a buyout price of at least $3 million, a 50% chance one should buy out at more than $12 million, but only a 10% chance that one should buy out at more than $25 million. The expected value probability curve (Fig. 7.10c) indicates a 90% chance that one should buy out at more than $3 million, a 50% probability of a buyout at more than $14 million, and only a 10% probability that one should buy out at more than $31 million. If the government buyout amount is set higher than these ranges, then it is preferable to complete the project. Depending on corporate strategy, some value in the range between $3 million and $30 million will satisfy the corporate decision makers.

The question of which components of uncertainty are relatively important in contributing to the buyout ranges of RAV is illustrated in Fig. 7.11 for the Cozzolino RAV, the hyperbolic tangent RAV, and expected value. As anticipated, the working-interest mandate dominates at over 70% of the uncertainty in RAV,

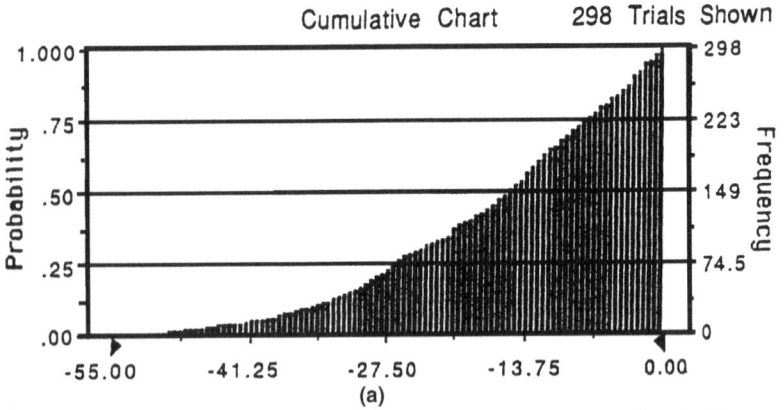

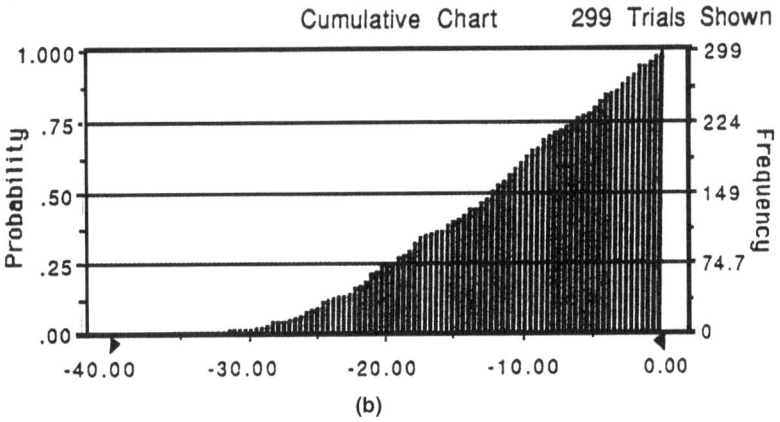

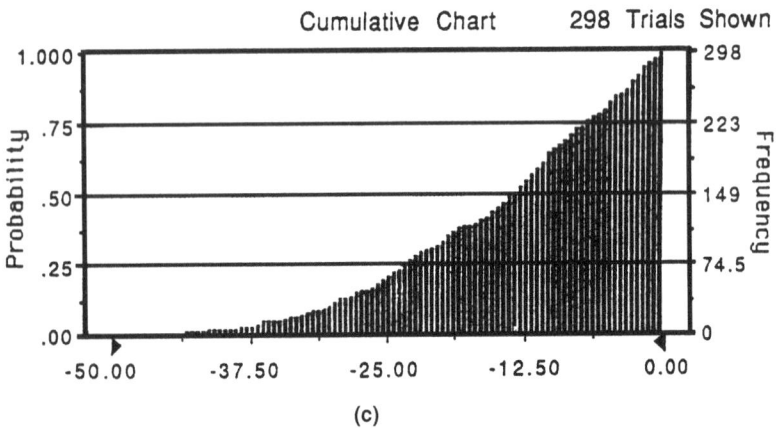

FIGURE 7.10 Cumulative probability plots of RAV and expected value for the buyout situation of high cost when uniform probability distributions are assigned for p_s, V, C, RT, and W: (a) RAV (Cozzolino); (b) RAV (hyperbolic tangent); (c) expected value.

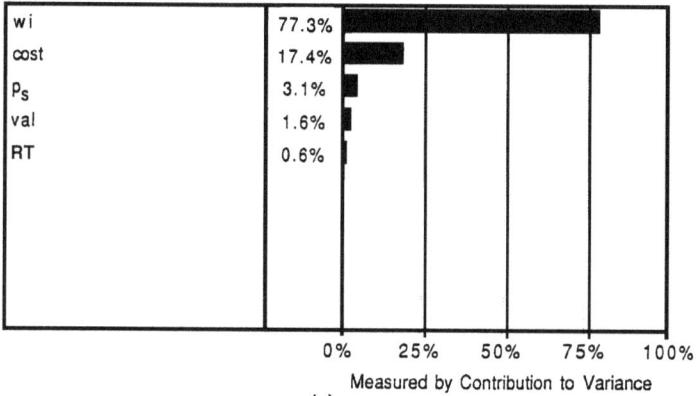

wi	77.3%
cost	17.4%
p_s	3.1%
val	1.6%
RT	0.6%

0%   25%   50%   75%   100%

Measured by Contribution to Variance

(a)

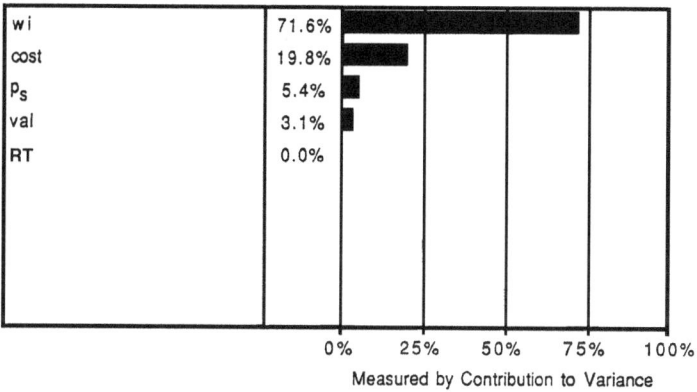

wi	71.6%
cost	19.8%
p_s	5.4%
val	3.1%
RT	0.0%

0%   25%   50%   75%   100%

Measured by Contribution to Variance

(b)

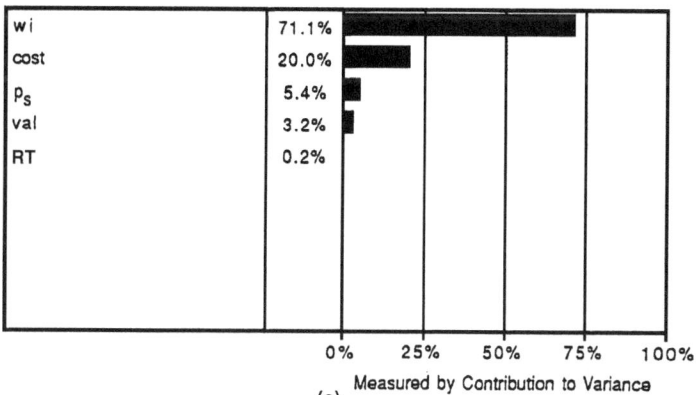

wi	71.1%
cost	20.0%
p_s	5.4%
val	3.2%
RT	0.2%

0%   25%   50%   75%   100%

Measured by Contribution to Variance

(c)

FIGURE 7.11   Relative importance (%) of factors contributing to the uncertainty in buyout amount using the uniform probability distributions for p_s, V, C, RT, and W: (a) RAV (Cozzolino); (b) RAV (hyperbolic tangent); (c) expected value.

with the uncertainty in cost picking up about 20%. The uncertainties in value, success probability, and risk tolerance have but small roles to play. Clearly, the aim here is to lower the mandated working interest by negotiation and to narrow the range of its uncertainty; in addition, costs need to be kept to the bare bones. However, considerably more effort (3.5 times as much) should be spent on negotiation with the government to arrange a different working-interest-mandated contribution in order to lower the anticipated buyout value to somewhere less than the $3 million to $30 million range (90 to 10% probability) indicated by the mandates in place.

7.5 SUMMARY

The importance has long been recognized of providing quantitative measures of risk aversion in relation to value, cost, success probability, corporate risk tolerance, and estimating working-interest fractions for environmental remediation projects. The Cozzolino (1977, 1978) risk-adjusted value formula (which uses exponential weighting) was the predominant tool used, particularly in the oil exploration industry, to assess risk until recently, when generalizations to allow for any functional form of weighting have been developed (MacKay and Lerche, 1996). Thus the intercomparison of different corporate model results for RAV can now be made so that one is no longer so beholden to the specifics of a single available formula.

The use of RAV formulas is often predicated on precise values being available for the input parameters p_s, V, C, W, and RT. Different values, of course, can be inserted into the RAV formulas so that some idea of stability, sensitivity, and accuracy of results to changes in the input parameters is available.

However, the fact that the estimated values of p_s, V, C, W, and RT all carry some degrees of uncertainty, from both geologic, environmental, and economic conditions, and the fact that the probability distributions for the uncertainties are not too well known mean that the RAV assessments from any and all models also should incorporate these uncertainties. In this way, probabilistic ranges of uncertainty in RAV estimates, and the relative importance of factors contributing to the uncertainty, can be investigated to determine where and when to focus effort to improve matters.

What has been done here is to emphasize how one can bring to bear statistical considerations in assessing uncertainty in RAV estimates and the importance of determining probabilities for positive RAV and of probable ranges of buyout values for a high-cost project.

The numerical illustrations indicate how the RAV probabilities depend on the model functions (Cozzolino, hyperbolic tangent) used to provide RAV estimates. In addition, a mandated range of working interest can be addressed as an extra variable contributing to the probabilistic range of RAV, whereas negative RAV values for a high-cost project can be used to assess the probable buyout amount one should be prepared to pay depending on corporate risk philosophy.

The integrated use of environmental modeling methods—geochemistry, geology, seismic processing, economic costs, and fiscal risk—can have an impact on the determination of economic worth and so of risk-adjusted value. The methods described here show how and when the individual subdisciplines have to be addressed to improve the situation. Perhaps one of the more interesting facets of such assessments is to have objective, reproducible methods expressed in simple form to demonstrate *why* the improvements are needed and the level of increased resolution that needs to be obtained. At the same time, the exigencies of reality must be kept firmly in mind in terms of work cost, personnel, and reporting deadlines. In addition, in order for a composite group of personnel to interact together positively, it is almost mandatory to have some objective measures of need for improvement of individual factors influencing profit estimates; the alternative is subjective measures, with the attendant inferences for selective bias by individuals in what they believe needs improving rather than what can be shown to need improving.

Effectively, what is being risked here is the ability to protect the corporation from massive financial ruin. If the expected value is not too negative, then there is always a chance of not losing money and so one recommends that the environmental remediation project be carried out, but as the expected value becomes even more negative, the prospect of making any money becomes so remote that it is better to pay the buyout amount—unless it, too, is an extremely large amount and it is a less expensive loss to carry on with the project. However, if there is any chance of bankruptcy occurring, then it is always the case that the buyout will be paid or the working interest adjusted to a smaller value to minimize the bankruptcy chance. And this is, indeed, what risk tolerance is all about in the negative-expected-value situation.

In summary, this chapter may be used by an environmental decision-making group to analyze the risk-adjusted value of a project under uncertain conditions in the economic and scientific estimates of a project and under different models (and hence assumptions) describing the variation of these estimates. The material of this chapter also can assist in the determination of a corporation's involvement in a project during negotiations for the establishment of a consortium or with a foreign government that reserves the option to alter its financial interest in a project.

CHAPTER 8
CORPORATE INVOLVEMENT IN MULTIPLE ENVIRONMENTAL PROBLEMS

8.1 INTRODUCTION

Because environmental remediation projects can be involved in high liability claims, they present a high risk in a corporation's fiscal success. Thus procedures for limiting corporate economic exposure are always involved in the decision to pursue certain available opportunities and in decisions concerning the working-interest fraction W that a particular corporation would prefer to take in a given opportunity.

However, while each opportunity presented to or available to a corporation can be evaluated in isolation (see Chaps. 4, 6, and 7), the difficulty is that most corporations do not have an unlimited budget, so they cannot take the optimal working interest OWI in each opportunity; the reason is corporate liability, in the sense that if each project were to fail with probability p_{fi} for the ith project, with a total cost of C_i, and if OWI_i were to be taken in each opportunity, then for N such projects, the corporate liability on cost *exposure* would be $\sum_{i=1}^{N} C_i OWI_i$. In the unlikely event that a corporation is operating on a continuing steady cash-flow basis, it might choose to limit corporate liability to cost *expenditure*, given as $\sum_{i=1}^{N} C_i p_{fi} OWI_i$. This situation is rare because of the long delays (3 or more years usually) before a project can yield a positive cash flow. Most corporations insist that working interests W_i be taken such that $C_0(W) \equiv \sum_{i=1}^{N} C_i W_i$ is less than or equal to the available budget B. In such cases, one cannot take the optimal working interest in each opportunity. The problem, then, is to find some procedure to optimize the total risk-adjusted value RAV from the N opportunities while staying within the mandated corporate budget B; i.e., the portfolio of opportunities is balanced. The classic method of resolving the various optimal working interests is with a linear optimization scheme.

The primary purpose of this chapter is to provide an analytic technique that can be used without extensive computational requirements for determining the portfolio balancing with a fixed total cost exposure (or cost expenditure) limit for determined input parameter values for each opportunity. In addition, if the individual input variables for each opportunity are themselves uncertain then, as will be shown, it is possible to generalize the procedure to yield a probabilistic interpretation that illustrates how to determine the probability for optimizing the total RAV and the probable range of working interest for each opportunity.

8.2 DETERMINISTIC PORTFOLIO BALANCING

Consider a sequence of opportunities $i = 1, 2, \ldots, N$ in which risk-aversion values RAV(i; W_i) have been computed for each opportunity i as a function of working interest W_i for a given corporate risk tolerance RT. The RAV values can be computed from any of the risk-aversion formulas given in previous chapters. The optimal working interest OWI(i) for each opportunity is also calculated so that the maximum RAV, $\text{RAV}_{\text{max}}(i, \text{OWI})$, is also known. In making these estimates, it is assumed that value V_i, costs C_i, success probability p_{si}, and risk tolerance RT are all *precisely* known. The expected worth of an opportunity described by Fig. 6.1a is then $E_i = p_{si}V_i - p_{fi}C_i$ (with $p_{fi} = 1 - p_{si}$), while the standard error σ_i of the expected worth is $\sigma_i = |(V_i + C_i)|(p_{si}\, p_{fi})^{1/2}$ [see Eqs. (6.1) and (6.2)] so that the volatility v_i of the expected worth is $v_i = \sigma_i/|E_i|$.

8.2.1 Relative Importance

The relative importance RI_i of the ith opportunity in relation to all others is usually defined either as

$$\text{RI}_i\,(W_i) = \frac{\text{RAV}(i;\, W_i)}{\sum\limits_{j=1}^{N} \text{RAV}(j;\, W_j)} \qquad \text{(unweighted)} \qquad (8.1a)$$

or, if weighted inversely with respect to volatility, as

$$\text{RI}_i\,(W_i) = \frac{\text{RAV}(i;\, W_i)\, /v_i}{\sum\limits_{j=1}^{N} [\text{RAV}(j;\, W_j)\, /v_j]} \qquad (8.1b)$$

Here, RAV(i; W) is positive or zero only. In either event, the relative importance of each opportunity is available.

8.2.2 Profitability

The contribution of each opportunity to the total profitability P is then usually defined through

$$P_i = \text{RAV}(i;\, W_i) \qquad \text{(unweighted)} \qquad (8.2a)$$

or

$$P_i = \frac{\mathrm{RAV}(i; W_i) / v_i}{\displaystyle\sum_{j=1}^{N} 1/v_j} \qquad \text{(weighted)} \qquad (8.2b)$$

with

$$P = \sum_{j=1}^{N} P_j \qquad (8.2c)$$

8.2.3 Costs

Maximum costs of each opportunity, for a fractional working interest W_i, are $C_i W_i$ and have to be borne by the corporation. If the opportunity does not succeed, then no future revenues are ever available against which to replenish the corporate reserves, so a net cost of $C_i W_i$ is the fiscal drain against the corporation. If the opportunity succeeds, then future revenues allow later replenishment of the corporate outlay. However, at the time of committing to the project, the corporation only has a budget B to use with the estimated costs that it knows it must bear. Two limiting cases are available:

1. *Cost exposure,* defined as total costs

$$C_0 = \sum_{i=1}^{N} C_i W_i \qquad (8.3a)$$

which measures the authorization for expenditure and participation in each and every opportunity.

2. *Cost expenditure,* defined as total probable costs

$$C_E = \sum_{i=1}^{N} C_i W_i p_{fi} \qquad (8.3b)$$

which measures the likely amount of cost that one would have to bear on a long-term *continuing cash-flow* basis once the failure probability of each opportunity is allowed for. In most situations, it is the cost-exposure limit with which a corporation has to deal against budget, and portfolio balancing will be carried through here on a cost-exposure basis. Cost expenditure is often used for strategic planning over the 5- to 10-year scale, whereas cost exposure is more usually appropriate for tactical planning on the 1- to 3-year scale.

8.3 BUDGET CONSTRAINTS

A total budget B is available. The problem is to figure out what fraction of the budget should go to each opportunity in order to maximize the total portfolio of profitability from each. Two cases are available.

8.3.1 High Budget

Suppose first that the budget is sufficiently high that participation in each opportunity could be at the OWI_i for each opportunity. Each project is then maximized in terms of its RAV. The total cost exposure is

$$C_0(\text{max}) = \sum_{i=1}^{N} C_i OWI_i \qquad (8.4a)$$

so the budget should be at

$$B \geq C_0(\text{max}) \qquad (8.4b)$$

which, for parabolic risk aversion [see Eq. (8.7)], is

$$C_0(\text{max}) = RT \sum_{i=1}^{N} \frac{C_i}{v_i^2 E_i} \qquad (8.4c)$$

If the budget funds remain $[B - C_0(\text{max})]$, the funds are returned to the corporation for other projects.

8.3.2 Low Budget

If the total budget B is less than total cost exposure $C_0(\text{max})$, then optimal working interest cannot be taken in each opportunity. One has to settle for less than optimal profitability. The question is to figure out a procedure for balancing the portfolio of opportunities so as to maximize profitability, recognizing that the profitability will be less than optimal. It is this question that is addressed now.

8.4 FINDING THE BEST WORKING INTERESTS

To show how the procedure works at maximizing portfolio balancing, in the next section of this chapter several numerical illustrations of increasing complexity are considered. The Cozzolino (1977, 1978) exponential risk-adjusted value formula for the problem of Fig. 6.1 can be used, with

$$RAV = -RT \ln\left(p_s \exp\left(-\frac{WV}{RT}\right) + p_f \exp\left(\frac{WC}{RT}\right)\right) \quad (8.5)$$

which has a maximum at the optimal working interest

$$OWI = RT(V + C)^{-1} \ln\left(\frac{p_s V}{p_f C}\right) \quad (8.6a)$$

and at $W = OWI$,

$$RAV_{max} = V\, OWI - RT \ln\left[p_s\left(1 + \frac{V}{C}\right)\right] \quad (8.6b)$$

A useful approximation to the exponential RAV formula is the parabolic RAV formula

$$RAV \approx WE(1 - \frac{1}{2} Wv^2 \frac{E}{RT}) \quad (8.7)$$

where the mean value is $E = p_s V - p_f C$ (taken to be positive), and the volatility v is given through

$$v^2 = \frac{p_s V^2 + p_f C^2}{E^2} - 1 \quad (8.8)$$

The parabolic RAV formula has a maximum of $RAV_{max} = \frac{1}{2} RT/v^2$ occurring on $W_{max} = v^{-2} RT/E$, provided $W_{max} < 1$ and $RAV = 0$ on $W = 0$ and on $W = 2W_{max}$ (by definition of a parabola). As can be seen from the preceding formulas, RAV_{max} and W_{max} are simpler to compute with the parabolic formula, yielding essentially the same results as with the more complex exponential formula.

Here, the unweighted procedure will be used together with the parabolic RAV formula for illustrative purposes only, because analytically exact expressions can be developed for the working interests and the total RAV. It will then be shown how to generalize the procedure for any functional form of RAV.

The parabolic RAV formula is more risk averse at high working interest values than is the exponential RAV formula but is almost identical at low working interests. This property of the parabolic RAV formula parallels real decisions where managers are often quick to farm out an opportunity to around 50% involvement, even when the risk tolerance is high, but are much less willing to take small working interests in an opportunity. If attempts to model this diversification preference are attempted by lowering the risk tolerance in the exponential formula, then the entire range of working-interest involvement is affected. However, if one uses the parabolic RAV formula at the same risk tolerance, only the high end of the range of working interests is influenced (by reduction)—precisely what is required to correct for this managerial preference.

8.4.1 The Parabolic RAV Formula

For each opportunity $i = 1,\dots, N$, there is some best W_i. The task is to obtain an explicit formula describing W_i. The procedure for so doing is as follows: The total RAV of the N projects is

$$\text{RAV} = \sum_{i=1}^{N} \text{RAV}_i(W_i) \tag{8.9}$$

where it is taken that $W_i \geq 0$ and $W_i \leq 2W_{\text{max},i}$. It is also taken that the total budget is constrained, that is, $B < \text{RT} \sum_{i=1}^{N} [C_i/(E_i \, v_i^2)]$, so that optimal working interest in each and every opportunity *cannot* be taken.

There are N values, $W_1,\dots, W_N$, to obtain from Eq. (8.9). If no constraints were emplaced, then each W_i would be at its optimum of OWI_i. However, there is the single constraint of

$$B = \sum_{k=1}^{N} C_k W_k \tag{8.10}$$

which can be used to write the jth opportunity working interest as

$$W_j = C_j^{-1}\left(B - \sum_{\substack{k=1 \\ k \neq j}}^{N} C_k W_k\right) \tag{8.11}$$

where $k \neq j$ means that the term $k = j$ is to be omitted in the summation. Then rewrite Eq. (8.9) as

$$\text{RAV} = \sum_{\substack{L=1 \\ L \neq i \\ L \neq j}}^{N} \text{RAV}_L(W_L) + \text{RAV}_i(W_i) + \text{RAV}_j\left[C_j^{-1}\left(B - \sum_{\substack{k=1 \\ k \neq j}}^{N} C_k W_k\right)\right] \tag{8.12}$$

where in the first sum the terms $L = i$ and $L = j$ are omitted and in the second one the term $k = j$ is omitted.

Inspection of Eq. (8.12) shows that only the two end terms involve W_i. Then the maximum of RAV with respect to W_i occurs when

$$C_i^{-1}\frac{\partial \text{RAV}_i(W_i)}{\partial W_i} - C_j^{-1}\frac{\partial \text{RAV}_j(x)}{\partial x} = 0 \tag{8.13}$$

where $x = C_j^{-1}(B - \sum_{\substack{k=1 \\ k \neq j}}^{N} = C_k W_k)$. For the parabolic RAV formula one has

$$\frac{\partial \text{RAV}_i(W_i)}{\partial W_i} = E_i \left(1 - \frac{W_i v_i^2 E_i}{RT}\right) \tag{8.14}$$

Hence Eq. (8.13) becomes

$$E_i C_i^{-1}\left(1 - \frac{W_i v_i^2 E_i}{RT}\right) - E_j C_j^{-1}\left[1 - \frac{v_j^2 E_j}{RT} C_j^{-1}\left(B - \sum_{\substack{k=1 \\ k \neq j}}^{N} C_k W_k\right)\right] = 0 \tag{8.15a}$$

that is,

$$E_i C_i^{-1}\left(1 - \frac{W_i v_i^2 E_i}{RT}\right) = E_j C_j^{-1}\left(1 - \frac{W_j v_j^2 E_j}{RT}\right) \tag{8.15b}$$

But i and j are arbitrary choices, so each side of Eq. (8.15b) must be a constant, independent of i or j. Let this constant be H so that

$$W_i = (1 - C_i H E_i^{-1}) \frac{RT}{v_i^2 E_i} \qquad i = 1,\dots, N \tag{8.16}$$

Then use the fact that

$$\sum_{i=1}^{N} C_i W_i = B \tag{8.17}$$

to determine H, leading to

$$W_i = RT (E_i v_i^2)^{-1} \left\{ 1 - \frac{C_i}{E_i} \frac{\displaystyle\sum_{j=1}^{N} (C_j/v_j^2 E_j) - B/RT}{\displaystyle\sum_{j=1}^{N} (C_j^2/v_j^2 E_j^2)} \right\} \tag{8.18}$$

provided that $0 \leq W_i \leq \min\{2\text{OWI}_i, 1\}$ and under the low budget constraint. Inspection of Eq. (8.18) shows that the budget constraint is just the requirement that

$$\sum_{j=1}^{N} \frac{C_j}{v_j^2 E_j} > \frac{B}{RT} \tag{8.19}$$

and when this inequality is in force, then each $W_i \leq \text{OWI}_i$.
    The requirement that $W_i \geq 0$ is that

$$\alpha_i \equiv \sum_{j=1}^{N} \frac{C_j}{v_j^2 E_j} - \left(\sum_{j=1}^{N} \frac{C_j^2}{v_j^2 E_j^2}\right) \frac{E_i}{C_i} \leq \frac{B}{RT} \tag{8.20}$$

Hence order the summations in Eqs. (8.18), (8.19), and (8.20) with respect to α_i, with $\alpha_1 < \alpha_2 < \cdots < \alpha_N$.

Then, as the budget B is systematically decreased, $W_N = 0$ when $B = \text{RT}\alpha_N$, and the Nth opportunity is removed from consideration. As B systematically decreases, in turn, W_{N-1}, W_{N-2}, etc. reach zero, and those opportunities are discarded. Thus at any given budget it is relatively simple to determine which opportunities should be invested in, as well as the working interest that should be taken.

The analytic exact formula for determining W_i, as given through Eq. (8.18), then maximizes the total RAV of all the N opportunities under the fixed budget constraint so that optimal working interest cannot be taken in each and every opportunity.

To draw a close parallel with the single parametric representation occurring with the exponential RAV formula for obtaining best working interests to balance the portfolio of opportunities, one also can write Eq. (8.16) in the form

$$W_i = \text{OWI}_i \left(1 - \frac{C_i H}{E_i} \right)$$

with

$$H = \frac{\sum\limits_{i=1}^{N} C_i \text{OWI}_i - B}{\sum\limits_{i=1}^{N} C_i^2 \text{OWI}_i/E_i} \geq 0$$

Then note that $W_i = 0$ on $H = E_i/C_i$. Thus, if the summations are organized in order $E_1/C_1 > E_2/C_2 > \cdots > E_N/C_N$, then as B decreases, so H increases. Then, as H crosses E_N/C_N, the Nth opportunity has $W_N = 0$ and so is discarded. The remaining $N - 1$ opportunities are then used to recalculate the summations from $i = 1$ to $N - 1$, and H is increased (B is decreased) until $W_{N-1} = 0$; the process is then repeated. The last opportunity to be discarded is when $H = E_1/C_1$, when $B = 0$. Thus, as the budget decreases from the maximum of $B_{\max} = \sum\limits_{i=1}^{N} C_i \text{OWI}_i$ (at which optimal working interest can be taken in each and every opportunity) to $B = 0$, the various opportunities are steadily discarded in order of their values of E/C; this parameter is similar to the economic evaluation tool of risked present worth investment (PWI).

8.4.2 The Exponential RAV Formula

The procedure used with the parabolic RAV formula to obtain an analytic expression for each W_i in the portfolio also can be used with the exponential RAV formula, although in this case a closed-form analytic solution for W_i is not possible, but a single parametric representation does obtain. Proceeding through the differentiations of the preceding section, the difference arises at the point where the explicit functional form of the RAV is used to compute $\partial RAV(W)/\partial W$. Instead of the parabolic RAV(W) formula of Eq. (8.7), if one uses the exponential RAV formula of Eq. (8.5), one obtains

$$W_j = RT(V_j + C_j)^{-1} \ln \frac{p_{sj} p_{fj}^{-1} (V_j/C_j - H)}{1 + H} \qquad (8.21)$$

where the single parameter H is to be determined from

$$B = RT \sum_{j=1}^{N} C_j(V_j + C_j)^{-1} \ln \frac{p_{sj} p_{fj}^{-1} (V_j/C_j - H)}{1 + H} \qquad (8.22)$$

Note that the exponential RAV formula yields

$$OWI_i = RT(V_i + C_i)^{-1} \ln\left(p_{si} p_{fi}^{-1} \frac{V_i}{C_i} \right) \qquad (8.23)$$

which occurs on $H = 0$.

Also note from Eq. (8.21) that

$$W_j = 0 \qquad on \qquad H = \frac{E_j}{C_j} = \frac{p_{sj} V_j - 2 p_{fj} C_j}{C_j} \qquad (8.24a)$$

Thus, if the terms in the summation in Eq. (8.22) are ordered with respect to decreasing values of E_j/C_j, with $E_1/C_1 > E_2/C_2 > \cdots > E_N/C_N$, then as H increases from zero, W_N will first go to zero, and thereafter, at higher H, the Nth opportunity is not considered. As H increases further, W_{N-1} will next go to zero, and so on. If one proceeds in this manner, the final term to drop is W_1, at which point $B = 0$. Thus, increasing $H \geq 0$ corresponds to a steadily decreasing budget, with $H = 0$ corresponding to a budget so that OWI can be taken in each and every opportunity. A practical strategy for determining H given the budget B is as follows: Rewrite Eq. (8.22) in the form

$$\ln(1 + H) = \left(a - \frac{B}{RT} \right) b^{-1} + b^{-1} \sum_{j=1}^{N} \frac{C_j}{C_j + V_j} \ln \left(\frac{V_j}{C_j} - H \right) \qquad (8.24b)$$

where

$$a = \sum_{j=1}^{N} (C_j + V_j)^{-1} C_j \ln \frac{p_{sj}}{p_{fj}} \qquad (8.25a)$$

and

$$b = \sum_{j=1}^{N} \frac{C_j}{(C_j + V_j)} \tag{8.25b}$$

and under the conditions that only those terms in the respective sums are to be included for which $H < E_j/C_j$.

Then, for a given budget B, Eq. (8.24) can be iterated to provide H, with the (n + 1)th iteration given through

$$\ln(1 + H_{n+1}) = \left(a - \frac{B}{RT}\right) b^{-1} + b^{-1} \sum_{j=1}^{N} C_j (C_j + V_j)^{-1} \ln\left(\frac{V_j}{C_j} - H_n\right) \tag{8.26}$$

starting from $H_0 = 0$ on the zeroth iteration, $n = 0$.

Thus, at each iteration one checks to see if $W_j \geq 0$ [from Eq. (8.21)] with the respective H_n. If $W_j < 0$, then that term in the summations of Eqs. (8.25a), (8.25b), and (8.26) is ignored, and the iteration is repeated. Pragmatically, convergence is extremely rapid, with no more than 5 to 10 iterations ever needed. In this way H is determined for a fixed budget B.

The procedure is clear: For a given budget, steadily increase H until the appropriate solution to Eq. (8.22) is obtained, *including* dropping terms as H crosses each E_j/C_j. Then, with the H so obtained, merely insert H into Eq. (8.21) to obtain the working interests that optimize the portfolio of opportunities.

Thus portfolio optimization for the exponential formula is reduced to a single parametric determination for a given budget. Clearly, for *any* functional form chosen for RAV(W), the procedure is applied as illustrated here.

This general argument, although explicitly developed here for unweighted RAV formulas, is also appropriate if any weighting is done. This procedure is developed in Appendix 8A.

8.5 DETERMINISTIC NUMERICAL ILLUSTRATIONS

Consider three opportunities A, B, and C with the following characteristic parameters:

Opportunity A: $V = \$110$ million; $p_s = 0.5 = p_f$; $C = \$10$ million; $RT = \$30$ million

Opportunity B: $V = \$200$ million; $p_s = 0.5$; $p_f = 0.5$; $C = \$100$ million; $RT = \$30$ million

Opportunity C: $V = \$300$ million; $p_s = 0.4$; $p_f = 0.6$; $C = \$120$ million; $RT = \$30$ million

Opportunities A and B each have a mean value of $E = \$50$ million, whereas opportunity C has $E = \$48$ million. Using the parabolic formula, opportunity A has $\text{RAV}(A)_{\text{max}} = \$10.417$ million at an $\text{OWI}(A) = 0.417$, whereas opportunity B has $\text{RAV}(B)_{\text{max}} = \$1.667$ million at $\text{OWI}(B) = 0.067$, and opportunity C has $\text{RAV}(C)_{\text{max}} = \$0.816$ at $\text{OWI}(C) = 0.034$, for a total possible maximum RAV of $\$12.9$ million. Note that $E(A)/C(A) = 5$, $E(B)/C(B) = 0.5$, and $E(C)/C(C) = 0.3$ so that, as the budget decreases, eventually opportunity C is expected to be least worthwhile taking any fraction of, followed by opportunity B.

Optimization of the total RAV at different budgets has been carried out under two conditions: (1) using the analytic formula for parabolic RAV given in the preceding section and (2) comparing the results obtained from the analytic formula with those arising from a Simplex solution solver, which has been the standard numerical method to analyze this type of problems in the oil industry. Both procedures yielded identical results under all conditions addressed, but the analytic method presented here is much faster numerically than the Simplex search method.

8.5.1 A Budget of $20 million

The value of $C \times \text{OWI}$ is $\$4.16$ million for opportunity A, $\$6.667$ million for opportunity B, and $\$4.082$ million for opportunity C, for a total cost of $C_i\text{OWI}_i = \$14.915$ million. Thus the budget of $\$20$ million exceeds the total costs at the optimal working interest for each opportunity, so OWI is taken in each at a cost of $\$14.915$ million; a budget return of $\$5.085$ million to the corporate coffers is then made.

8.5.2 A Budget of $11 million

In this case the budget is less than the total needed to invest in each opportunity at its OWI but is large enough so that no single opportunity should be discounted. Both the parabolic analytic formula and the Simplex method return the optimal values of

$W(A) = 0.4033$	$\text{RAV}(A) = \$10.406$ million	$C(A)W(A) = \$4.0328$ million
$W(B) = 0.0452$	$\text{RAV}(B) = \$1.4946$ million	$C(B)W(B) = \$4.5248$ million
$W(C) = 0.0204$	$\text{RAV}(C) = \$0.6847$ million	$C(C)W(C) = \$2.4424$ million

for a total RAV of $\$12.585$ million. In this case, note that the optimization of total RAV has kept the working interest in opportunity A very close to its OWI and has kept the involvement in opportunity B at about three-quarters of the OWI but has dropped the involvement in opportunity C to about two-thirds of its OWI—in line with E/C being smallest for opportunity C so that the lower budget is forcing the working interest in opportunity C closer to zero first.

8.5.3   A Budget of $4 million

In this case the threshold value of $E/C = 0.3$ has been crossed for opportunity C, which is therefore discounted completely. The budget is then split between opportunities A and B in the proportions

$W(A) = 0.3765$ $RAV(A) = \$10.3$ million $C(A)W(A) = \$3.7647$ million

$W(B) = 0.0024$ $RAV(B) = \$0.115$ million $C(B)W(B) = \$0.2353$ million

for a total RAV of $10.415 million. In this case, because opportunity B has a lower E/C ($= 0.5$) than opportunity A ($= 5$), the lower budget forces less involvement in opportunity B. The total maximum RAV for all three opportunities is $12.9 million, so portfolio balancing yields 97.6% (at $B = \$11$ million) and 80.9% (at $B = \$4$ million) of the maximum by optimizing the budget fractions allocated to each opportunity.

8.6   PROBABILISTIC PORTFOLIO BALANCING

For ranges of uncertainty on value V, cost C, success probability p_s, and risk tolerance RT, for each opportunity, the cumulative probability distribution for RAV at different values of working interest W can be constructed. A measure of the worth of the ith opportunity is taken to be the average value for RAV, written $<RAV(i, W)>$, where the angular brackets denote average values; a measure of uncertainty on this value is taken to be the volatility with

$$v_i(W) = \frac{RAV_{90}(i, W) - RAV_{10}(i, W)}{<RAV(i, W)>} \qquad (8.27)$$

where RAV_{90} and RAV_{10} are the values of RAV at 90% and 10% cumulative probability, respectively. It is also taken that $<RAV(i, W)> > 0$.

In this case, volatility-weighted relative importance $RI(i, W_i)$ is given through

$$RI(i, W_i) = \frac{<RAV(i, W_i)> / v_i(W_i)\,]}{\sum_{j=1}^{N} [<RAV(j, W_j)> / v_j(W_j)]} \qquad (8.28)$$

whereas an unweighted relative importance can be given through

$$RI(i, W_i) = \frac{<RAV(i, W_i)>}{\sum_{j=1}^{N} <RAV(j, W_j)>} \qquad (8.29)$$

The contribution P_i to the total probable profit P at the average cumulative value is

$$P_i = <RAV(i, W_i)> \quad \text{(unweighted)} \quad (8.30a)$$

or

$$P_i = \frac{<RAV(i, W_i)> /v_i(W_i)}{\sum_{j=1}^{N} 1/v_j(W_j)} \quad \text{(weighted)} \quad (8.30b)$$

with

$$P = \sum_{i=1}^{N} P_i \quad (8.30c)$$

The problem, then, is to determine the fractional working interests appropriate to maximize likely probability of profit given the uncertainties and given a fixed budget B.

Consider the deterministic expressions given in Sec. 8.2 for fixed values of p_{si}, V_i, and C_i for each opportunity ($i = 1, N$) together with the fixed values of RT and budget B. Corporate budget and corporate risk tolerance are better established than the variations known to occur in estimates of p_{si}, V_i, and C_i for each opportunity. Thus, if p_{si}, V_i, and C_i are allowed to vary within preset ranges for each opportunity and with known distributions, then Monte Carlo simulations can be run using the deterministic formulas of Sec. 8.2 and constructions made of the probable ranges of each W_i, together with the probable RAV distribution for each opportunity and for the total of all opportunities. One merely has to record frequency of occurrence of events and also the individual variations of the intrinsic parameters that contribute the largest degrees of uncertainty to the probabilistic outcomes, in a manner similar to that done elsewhere (MacKay and Lerche, 1996) for a different problem.

In this way, not only can portfolio balancing be done in a deterministic manner, but also the balancing can be done in a probabilistic framework. In addition, cost exposure is used to balance a fixed budget.

8.7 PROBABILISTIC NUMERICAL ILLUSTRATIONS

8.7.1 Variable Value

The value V ascribed to an opportunity is usually somewhat uncertain due to the two dominant factors of environmental scientific assessments and potential future liabilities. As a consequence, there is some fluctuation to be expected in the optimal working interest that should be taken in each opportunity and in the total RAV.

To illustrate the effect of uncertainties in value V on the optimization, the values for the three opportunities presented in the preceding section were used and allowed to vary uniformly by $\pm 10\%$ around central values. Thus, for opportunity A, \$100 million $\leq V \leq$ \$120 million, with a central value of \$110 million; for opportunity B, \$180 million $\leq V \leq$ \$220 million, with a central value of \$200 million; for opportunity C, \$270 million $\leq V \leq$ \$330 million, with a central value of \$300 million. The Monte Carlo procedure then selected values at random from each of the three distributions in constructing cumulative probability distributions for total RAV and working interest for each opportunity using the analytic formulas for parabolic working interest given in Sec. 8.2. Approximately 1500 trials were run on a small PC in approximately 6 minutes (the Simplex solution solver was not used in this exercise because it was far too slow except to spotcheck individual runs). The computations were done for a budget of \$11 million, a case for which all three opportunities have some working interest at the center values, as exhibited in Sec. 8.3.

Shown on Fig. 8.1 is the cumulative probability plot for total RAV with the abscissa in units of \$ million. In this case P_{68} (an approximate estimate of the mean) occurs at $RAV_{68} = \$12.78$ million, whereas P_{90} is at $RAV_{10} = \$13.06$ million and P_{10} is at $RAV_{10} = \$12.13$ million, so the volatility of the 68% chance is only $\nu = 7.2\%$, indicating a high degree of confidence in an overall RAV of around \$12.78 million. There is a two-thirds chance that the total RAV will be less than \$12.78 million but only a 10% chance of *less* than \$12.13 million and only a 10% chance of *greater* than \$13.06 million. Hence one can write RAV = $\$12.78^{+0.28}_{-0.65}$ million to display directly the uncertainty at the 10 and 90% levels. Of interest also is the range of uncertainty on working interest for each opportunity, exhibited in Fig. 8.2 for opportunities A, B, and C. Note from Fig. 8.2a for opportunity A that P_{68} occurs at $W_{68}(A) = 0.41$, whereas P_{90} and P_{10} occur, respectively, at $W_{90}(A) = 0.43$ and $W_{10}(A) = 0.38$, so the volatility on working interest is $\nu(A) = 12.2\%$; equally, from Fig. 8.2b, one has $W_{68}(B) = 0.045$, $W_{90}(B) = 0.0474$, and $W_{10}(B) = 0.0425$, for a volatility of $\nu(B) = 13.1\%$, whereas from Fig. 8.2c, one has $W_{68}(C) = 0.022$, $W_{90}(C) = 0.025$, and $W_{10}(C) =$

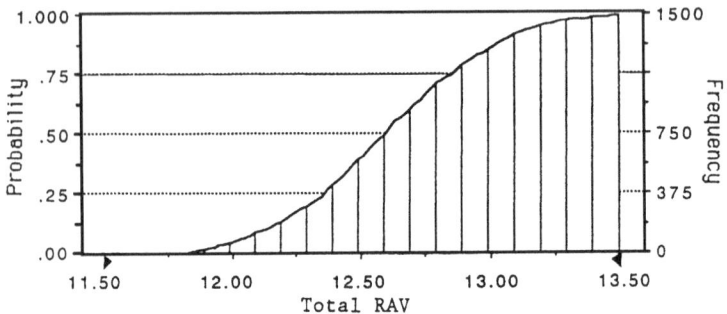

FIGURE 8.1 Cumulative probability distribution plot for total RAV when values for each of the three opportunities vary by $\pm 10\%$ around their central values, for a budget of \$11 million. The abscissa unit is \$ million.

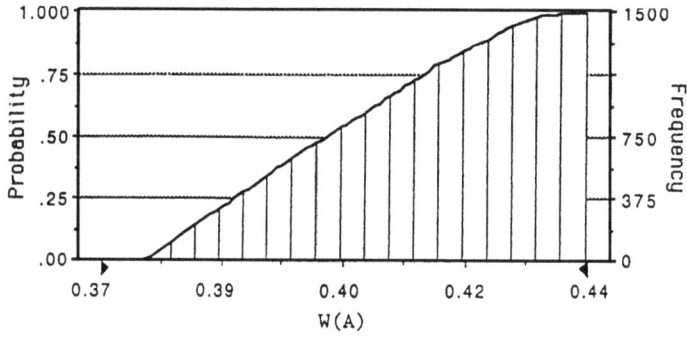

(a)

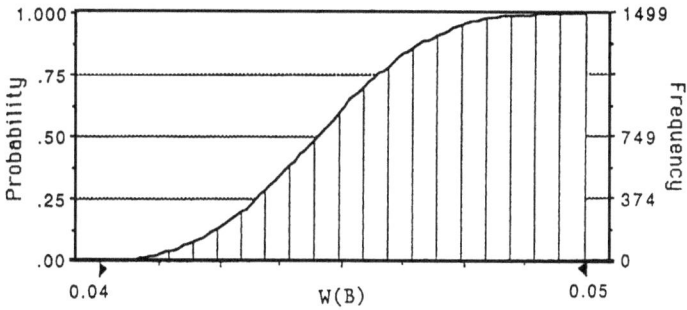

(b)

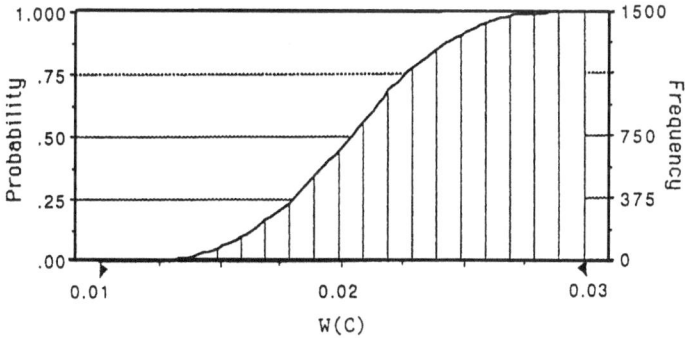

(c)

FIGURE 8.2 Cumulative probability distribution plot for working interest optimizing the total RAV when values for each of the three opportunities vary by ±10% around their central values, for a budget of $11 million. The abscissa unit is fractional working interest (*a*) for opportunity *A*; (*b*) for opportunity *B*; (*c*) for opportunity *C*.

0.015, for a volatility of $v(C) = 45\%$. The increasing volatilities of the different opportunities reflect directly the fact that the closer the budget is to crossing over the threshold at which $H = E/C$ for each opportunity, the lower is the working interest to be taken in that opportunity. Because opportunity C is the closest to this threshold, this fact is picked up in the greater degree of volatility on the likely working interest to be taken.

8.7.2   Variable Value and Cost

In addition to the value of an opportunity being uncertain, one may encounter the situation where cost estimates are also uncertain. To illustrate how variations in cost influence total RAV and working interest in the portfolio, the costs of each opportunity also were allowed to vary randomly with a uniform distribution centered on the deterministic values used in Sec. 8.3 and with 10% variation around the central values. Thus, for opportunity A, $9 million $\leq C \leq$ $11 million; for opportunity B, $90 million $\leq C \leq$ $110 million; and for opportunity C, $108 million $\leq C \leq$ $132 million. The values for each opportunity also were taken to vary uniformly within 10% of their central values as for the preceding example. Again, a Monte Carlo set of 1500 random runs was made using a small PC with the following results: As shown in Fig. 8.3, the total RAV is now changed due to the extra "spread" arising from the cost uncertainty. Thus RAV_{68} = $12.8 million, RAV_{90} = $13.25 million, and RAV_{10} = $12.0 million, so the volatility is now $v = 9.8\%$, an increase of 2.6% relative to the preceding case of variable values only.

Equally, the probable working interests for each opportunity also reflect the extra uncertainty caused by cost fluctuations. Shown in Fig. 8.4 for opportunities A, B, and C are the cumulative probability distributions of working interest. Reading off from Fig. 8.4, one has

$W_{68}(A) = 0.41$ $W_{90}(A) = 0.43$ $W_{10}(A) = 0.38$ $v(A) = 12.2\%$

$W_{68}(B) = 0.05$ $W_{90}(B) = 0.06$ $W_{10}(B) = 0.04$ $v(B) = 40\%$

$W_{68}(C) = 0.022$ $W_{90}(C) = 0.029$ $W_{10}(C) = 0.015$ $v(C) = 64\%$

Thus there is a significant increase produced in the range of uncertainty of the probable working interests for opportunities B and C by the 10% fluctuations in costs of each opportunity, representing the fact that low value and high cost, when taken together, both drive E/C to a smaller value and so influence the closeness of the working interest to the point where $H = E/C$ demands no involvement in the opportunity. Put another way, near the peaks of the individual RAV curves there is less influence of variations in input uncertainties because there is no slope at the peak of RAV with W. However, the steepness of the slopes away from the peaks, particularly for opportunity C, cause a rapid slide away from significant investment in that opportunity relative to less steeply sloping opportunities.

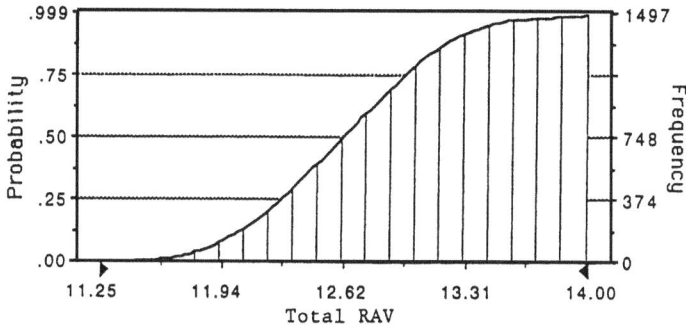

FIGURE 8.3   As for Fig. 8.1 but now including the cost of each of the opportunities being uncertain by ±10% around their central values. The budget is at $11 million.

8.7.3   Variable Value, Cost, and Success Probability

Apart from uncertainties in values and costs, it is also the case that the chance of success p_s is uncertain. To illustrate the additional effects on RAV and working-interest probabilities caused by uncertainty in success probability, and to keep a balance with the prior two examples, a 10% uncertainty on p_s for each opportunity, drawn from a uniform population, is now included with the 10% uncertainties on V and C. Thus, for opportunities A and B, $0.45 \leq p_s \leq 0.55$, and for opportunity C, $0.36 \leq p_s \leq 0.44$, with each centered at the deterministic values of Sec. 8.3. A suite of 1500 Monte Carlo runs was again done, with results for cumulative RAV probability shown in Fig. 8.5 and for working-interest probability in Fig. 8.6 for opportunities A, B, and C. The budget was kept at $11 million.

To be noted from Fig. 8.5 is that $RAV_{68} = \$13.63$ million, $RAV_{90} = \$14.87$ million, and $RAV_{10} = \$10.74$ million, for a volatility of $v = 30\%$—a considerable increase compared with the prior two examples and reflecting the relative importance of uncertainty in success probability in determining RAV.

Reading off from Fig. 8.6, the working interests for each opportunity are

$$W_{68}(A) = 0.42 \qquad W_{90}(A) = 0.45 \qquad W_{10}(A) = 0.36 \qquad v(A) = 9.4\%$$

$$W_{68}(B) = 0.05 \qquad W_{90}(B) = 0.06 \qquad W_{10}(B) = 0.03 \qquad v(B) = 60\%$$

$$W_{68}(C) = 0.023 \qquad W_{90}(C) = 0.032 \qquad W_{10}(C) = 0.01 \qquad v(C) = 96\%$$

Thus the 10% fluctuations in success probability have a considerable effect on the uncertainties in working interest for the more risky opportunities B and C, with opportunity C clearly dominated by being closer to the critical threshold of no involvement.

8.7.4   Very Low Budget

The prior three examples were all carried out for a budget of $11 million, at which partial involvement could be taken in all three opportunities. Consider, now, the

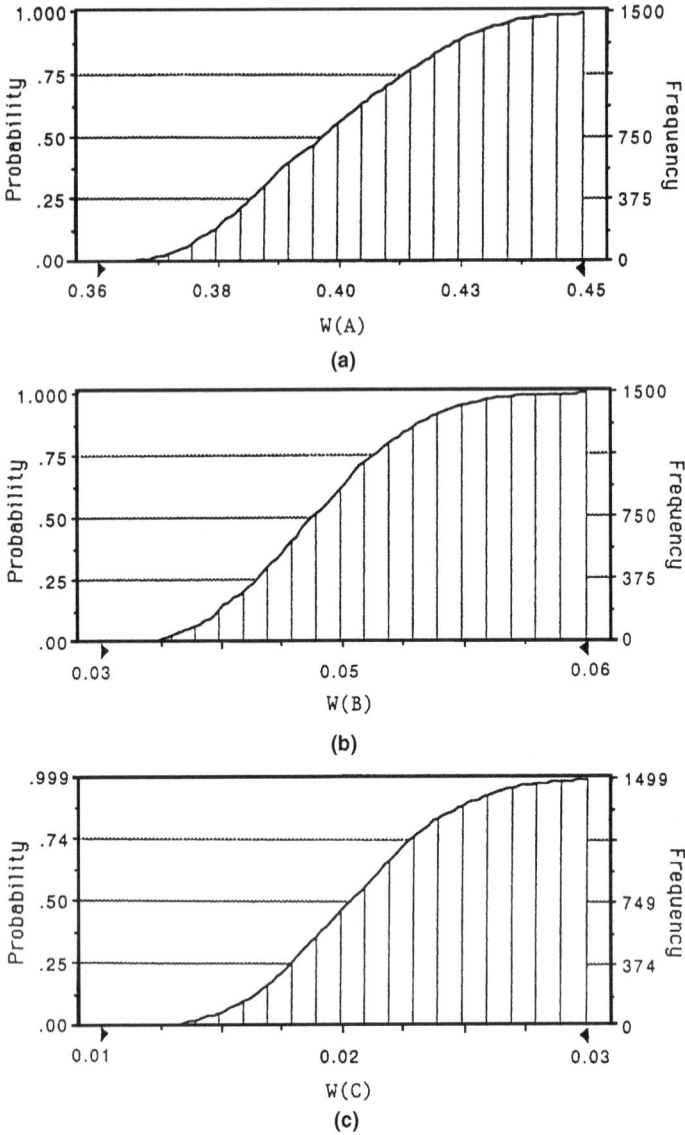

FIGURE 8.4 As for Fig. 8.2, but now including the cost of each of the opportunities being uncertain by ±10% around their central values; the budget is at $11 million. (*a*) For opportunity *A*; (*b*) for opportunity *B*; (*c*) for opportunity *C*.

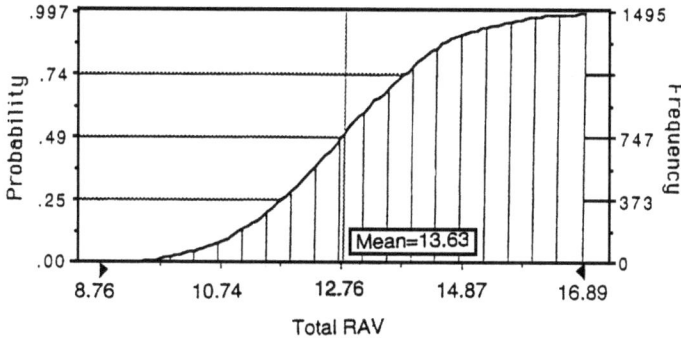

FIGURE 8.5   As for Fig. 8.3, but now also including the probability of success for each of the opportunities being uncertain by ±10% around their central values. The budget is at $11 million.

same uncertainty ranges on values, costs, and success probabilities for each oppor-tunity but when the budget is lowered to $4 million. This situation was shown in Sec. 8.3 (the deterministic case) to lead to no involvement in opportunity C, and the budget was partitioned only between opportunities A and B. Again, a suite of 1500 Monte Carlo runs was done, with the results exhibited in Fig. 8.7 for total RAV probability and in Fig. 8.8 for cumulative working-interest probability for opportunities A, B, and C.

Reading from Fig. 8.7 indicates that $RAV_{68} = \$11.3$ million, $RAV_{90} = \$12.45$ million, and $RAV_{10} = \$8.74$ million, with a volatility $v = 33.5\%$, whereas Fig. 8.8 indicates that it is still the case that no involvement should be taken in opportunity C. Figure 8.8a and b indicates that

$$W_{68}(A) = 0.39 \qquad W_{90}(A) = 0.42 \qquad W_{10}(A) = 0.33 \qquad v(A) = 23\%$$

$$W_{68}(B) = 0.048 \qquad W_{90}(B) = 0.055 \qquad W_{10}(B) = 0 \qquad v(B) = 115\%$$

Thus most of the budget is now committed to opportunity A, and the volatility of opportunity B makes that opportunity one of higher risk.

8.7.5   Low to High Budget Comparison

As the budget varies for given ranges of uncertainty on values, costs, and success probabilities, the relative importance is needed of each uncertainty in influencing the range of variation of the total RAV. The importance of this information lies in the fact that if one wishes to determine better the investment in each opportunity, then it is crucial to be aware of which parameter values and their ranges of uncer-tainty are causing the largest degree of variability in RAV. In this way the decision can be made as to which parameters to focus on in order to narrow their ranges of uncertainty. The relative importance RI of the contribution of each parameter to

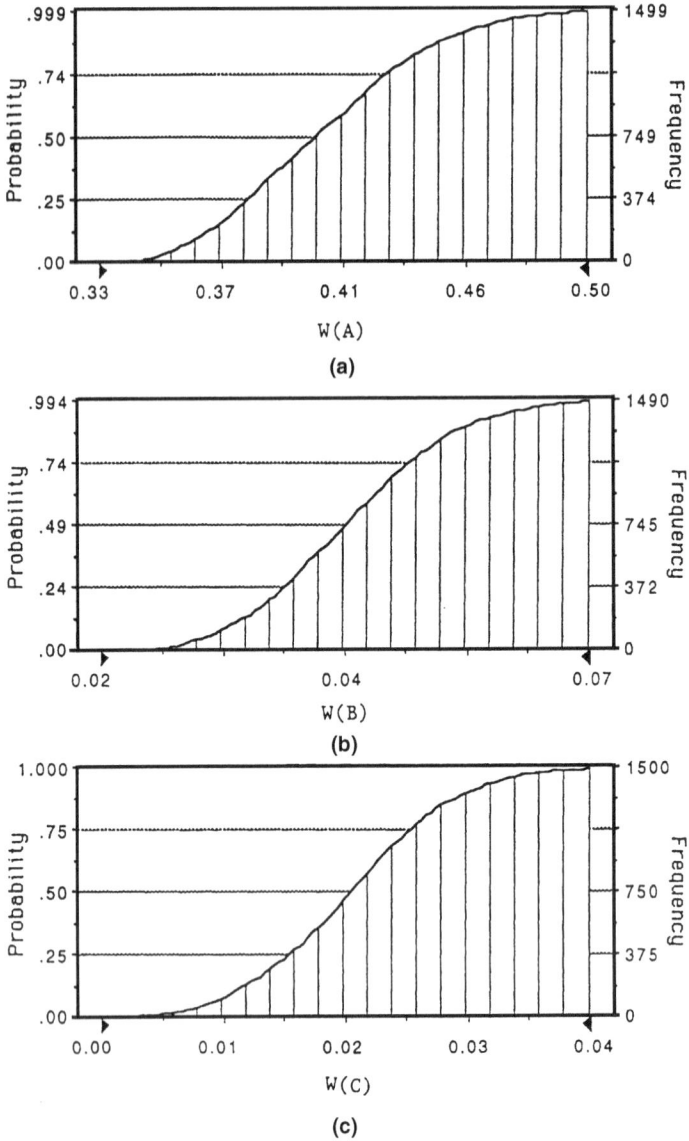

FIGURE 8.6 As for Fig. 8.4, but now also including the probability of success for each of the opportunities being uncertain by ±10% around their central values. The budget is at $11 million. (a) For opportunity A; (b) for opportunity B; (c) for opportunity C.

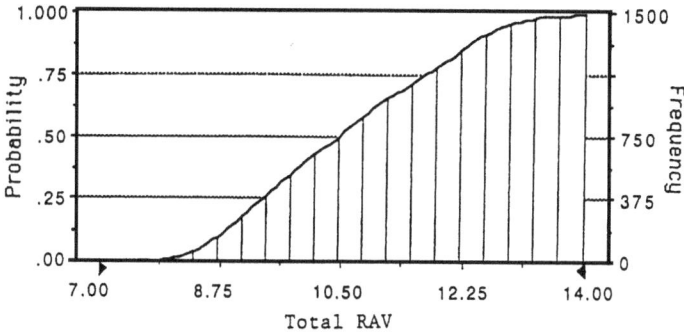

FIGURE 8.7   As for Fig. 8.5, but with the budget reduced to $4 million.

uncertainty in the total RAV cumulative probability distribution provides the requisite measure. Note that this procedure only prioritizes the relative dominance of individual parameters. Thus one has a measure that *if* one wishes to reduce uncertainty ranges, then the RI provides the information on where to expend effort to narrow uncertainty; the volatility provides a measure of *when* one needs to improve uncertainty ranges.

To illustrate the effect of each of the ranges of variability of values, costs, and success probabilities, Fig. 8.9*a* and *b* plots the relative importance (in percent) of each uncertainty factor used in the preceding numerical examples for the cases of a budget of $11 million (Fig. 8.9*a*) and $4 million (Fig. 8.9*b*), respectively. Because investment can be taken in all three opportunities in the case of a budget of $11 million (albeit with only small RAV values for opportunities *B* and *C* compared with opportunity *A*), Fig. 8.9*a* shows that there is some relative contribution to the total RAV variance of each uncertainty. However, what is clear is that the uncertainty in success probability for opportunity *A* is absolutely dominant in causing the majority (85%) of the variance in RAV, with about 7% caused by uncertainty in success probability for opportunity *B*. The remaining factors add only 8% cumulatively. Clearly, if the uncertainty on total RAV is to be more limited in the $11 million budget case, then most effort (85%) has to be put on narrowing the range of uncertainty of success probability for opportunity *A*; all other factors are secondary to this goal.

Equally, lowering the budget to $4 million removes opportunity *C* as a contributor to the total RAV, as reflected in Fig. 8.9*b*, where there is no contribution to the uncertainty in total RAV from parameter variations of opportunity *C*. Because a greater fraction of the budget now goes into opportunity *A*, the dominance of the uncertainty in success probability for opportunity *A* is increased in relative importance to 94%, with the uncertainties in the value of *A* and the cost of *A* adding another 5.7%. The uncertainties in the success probability and cost for opportunity *B* are now of extremely small relative importance. Thus, in this low-budget situation, even more effort should be put into attempting to narrow down the range of uncertainty of the chance of success for opportunity *A*.

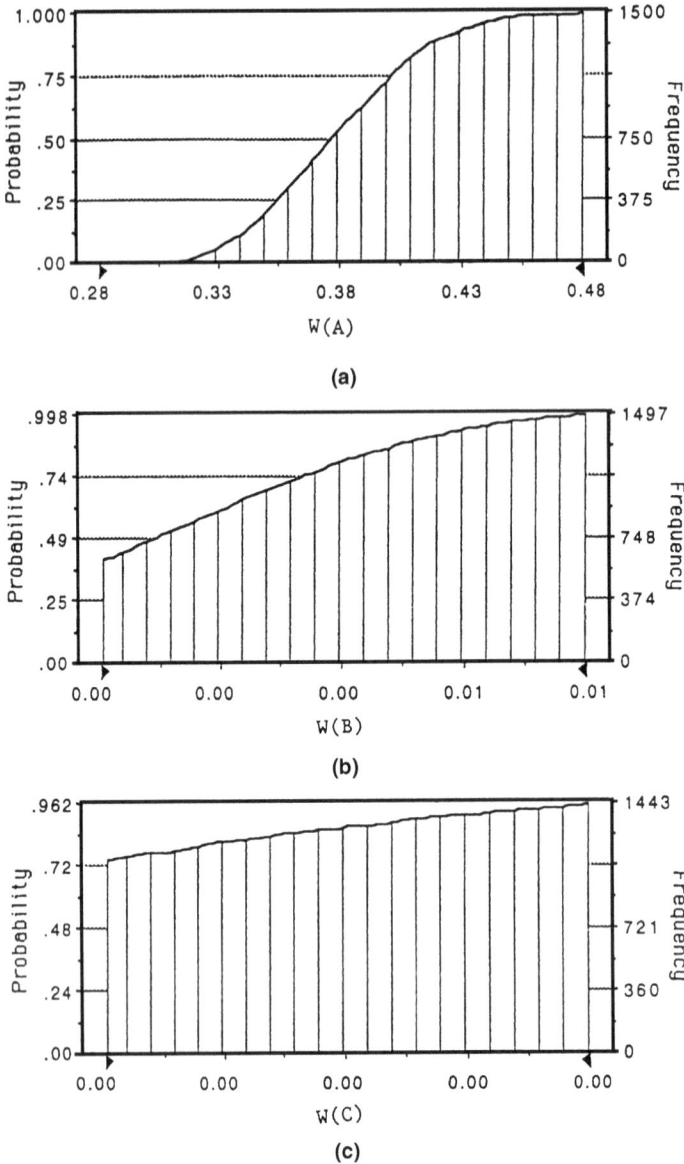

FIGURE 8.8 As for Fig. 8.6, but with the budget reduced to $4 million: (*a*) opportunity *A*; (*b*) opportunity *B*; (*c*) opportunity *C*.

The advantage to having available the ability to determine which factors are causing the greatest uncertainty in estimates of total RAV in a portfolio is that one can quickly determine where to place effort to improve matters without spending effort on factors that do not play a significant role. It is this fact that the relative importance figures provide in simple graphic form.

$P_s(A)$	85.2%		
$P_s(B)$	6.4%		
$V(A)$	1.8%		
$C(A)$	1.5%		
$P_s(C)$	1.5%		
$V(B)$	1.2%		
$C(C)$	0.8%		
$V(C)$	0.8%		
$C(B)$	0.6%		

0%   25%   50%   75%   100%

Measured by Contribution to Variance

(a)

$P_s(A)$	94.0%		
$V(A)$	3.5%		
$C(A)$	2.2%		
$C(B)$	0.2%		
$P_s(B)$	0.1%		
$V(C)$	0.0%		
$P_s(C)$	0.0%		
$V(B)$	0.0%		
$C(C)$	0.0%		

0%   25%   50%   75%   100%

Measured by Contribution to Variance

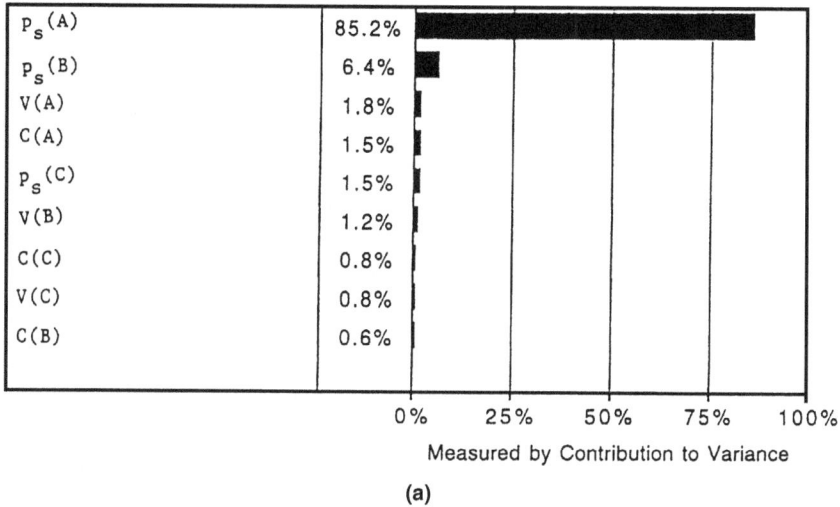

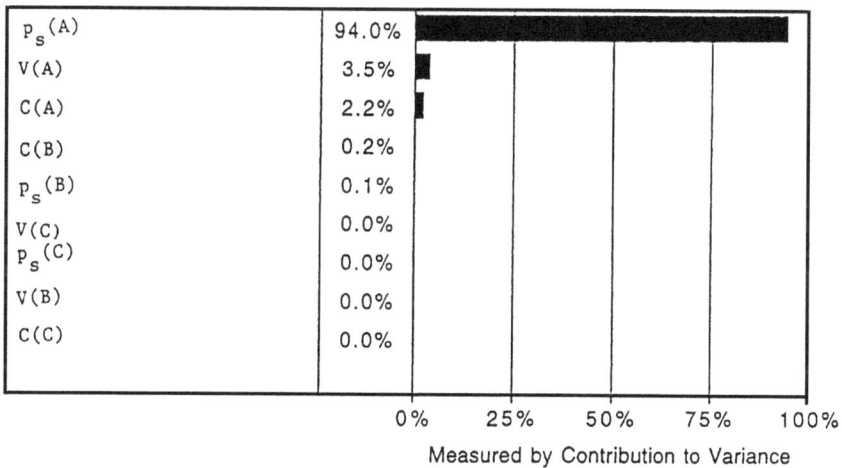

FIGURE 8.9   Relative importance diagram for individual parameters influencing the uncertainty on the total RAV. Note the dominance of the uncertainty in probability of success for opportunity A. (*a*) For a budget of $11 million; (*b*) for a budget of $4 million.

8.8   COMPARISON OF PARABOLIC AND EXPONENTIAL RAV

8.8.1   Deterministic Results

The three opportunities in Secs. 8.3 and 8.5 were arranged with opportunity A always dominant compared with opportunities B and C, so it was easy to visualize the widely separated parameter ranges that showed the relative impact of changes influencing RAV results. However, it is most often the case that the parameters describing different opportunities are much closer together than the simple examples used in the preceding sections for illustrative purposes. Accordingly, in this section of the chapter a discussion is given of three opportunities with parameters as shown in Table 8.1, together with a fixed risk tolerance of RT = $300 million and a budget B = $100 million, so that OWI cannot be taken in each and every opportunity. Because the functional forms are different for the exponential and parabolic RAV formulas, different OWI values ensue, as shown in Table 8.2.

At a budget of $100 million and RT = $300 million, the portfolio is optimized as shown in Table 8.3. There is, then, little to choose between the values of RAV or working interest from either the parabolic or exponential formulas. Perhaps the point to be made is that the two formulas for RAV reflect an insensitivity of the optimal RAV for the total portfolio to particular choices of functional RAV forms for each opportunity. Such stability of results is of great use when planning optimization strategies because the results are not volatile, implying a statistical sharpness to financial decisions for involvement in the different opportunities.

The major part of the reason for this close parallel between the optimizations of the parabolic and exponential risk-adjusted formulas is that the parabolic RAV formula is a Taylor series expansion (to quadratic order) in working interest of the Cozzolino exponential formula. Because all optimization values have positive slopes for $\partial RAV(W)/\partial W$, each RAV formula yields a working interest less than OWI, precisely the domain where the parabolic and exponential RAV formulas are very close to each other.

8.8.2 Statistical Results

Because of the close parallel between the parabolic and exponential RAV formulas in the region where $W <$ OWI, it follows that one also anticipates a close parallel in probabilistic ranges of output when values, costs, and success probabilities

TABLE 8.1 Parameters for Opportunities A, B, and C

Opportunity	V, $ million	C, $ million	p_s	p_f	E, $ million	E/C
A	100	80	0.6	0.4	28	0.35
B	200	100	0.5	0.5	50	0.5
C	300	120	0.4	0.6	48	0.4

TABLE 8.2 Optimal Working Interests for Opportunities A, B, and C

Opportunity	Parabolic		Exponential	
	OWI	C × OWI, $ million	OWI	C × OWI, $ million
A	1.0	80.000	1.00	80.000
B	0.67	66.667	0.693	69.315
C	0.34	40.816	0.365	43.785
TOTAL		187.483		193.090

TABLE 8.3 Portfolio Optimization for Opportunities *A*, *B*, and *C*

Opportunity	Parabolic			Cozzolino		
	W	*C* × *W*, $ million	RAV	*W*	*C* × *W*, $ million	RAV
A	0.4793	38.345	10.443	0.4495	35.963	10.422
B	0.4071	40.706	14.139	0.419	41.937	14.470
C	0.1746	20.949	6.229	0.184	22.100	6.473
TOTAL			30.811			31.365

are allowed to vary for each opportunity. To demonstrate that such a close parallel does in fact occur, 1500 Monte Carlo simulations were run for both the parabolic and exponential RAV formulas, respectively, when ±10% uncertainty is permitted on each of V, p_s, and C for each of the three opportunities, with uniform probability distributions for each parameter and with centered values as for the deterministic case given earlier. The risk tolerance RT was held fixed at $300 million, and the budget B was held constant at $100 million for all runs.

Figure 8.10 shows the cumulative probability distributions for RAV from the parabolic (Fig. 8.10*a*) and exponential (Fig. 8.10*b*) formulas. There is only a minor difference of less than 2% between the two plots, reinforcing the anticipation of only minor shifts.

Figure 8.11 shows the cumulative probability distributions for working interest in the three opportunities using both the parabolic (Fig. 8.11*a–c*) and exponential (Fig. 8.11*d–f*) expressions. Again, there is but little difference in the working-interest probability distributions (less than about 2%), reflecting the close parallel of the two formulas for RAV in the domain where $W <$ OWI.

The relative importance of the variations in the input parameters V, C, and p_s for each opportunity in contributing to uncertainty on the optimal RAV of the portfolio also reflects the close parallel of the parabolic (Fig. 8.12*a*) and exponential (Fig. 8.12*b*) formulas. Both Fig. 8.12*a* and *b* indicates that it is the uncertainty in success probabilities of opportunities B and A that dominate the relative importance, with about 34% from $p_s(B)$ and 25% from $p_s(A)$, for a total of around 59% of the total variance in RAV for the portfolio. The uncertainty in working interest for any opportunity is also closely parallel for both formulas, as exhibited in Fig. 8.13 for opportunity A, with $p_s(A)$, $C(A)$, and $p_s(B)$ contributing about 75% of the variance in both Fig. 8.13*a* (parabolic RAV) and Fig. 8.13*b* (exponential RAV). Similar parallelism also occurs for opportunities B and C (not shown here).

The conclusion is clear: The expectation that the parabolic and exponential RAV formulas provide closely parallel results for portfolio optimization is substantiated in both the deterministic and statistical treatments. Either formula will lead to a portfolio optimization that is ruggedly stable and insensitive to the precise details of the functional form, making for trustworthy participation decisions in multiple opportunities and allowing identification of the sensitivity and relative importance of the various input variables in influencing both the uncertainty of the total RAV and the uncertainty of the individual working interests for each oppor-

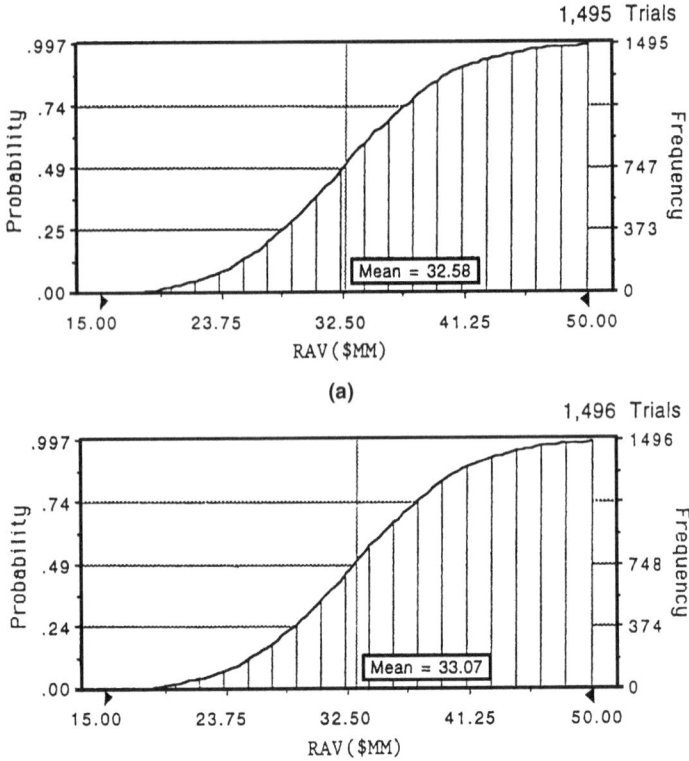

FIGURE 8.10 Cumulative probability plot of total RAV for the three opportunities of Sec. 8.6: (*a*) for the parabolic RAV formula; (*b*) for the Cozzolino formula. The abscissa is RAV in $ million.

tunity. In this way, priority can be assigned to improving individual factors influencing investment decisions without the concern that such priorities are RAV model-dependent to a high degree—such is not the case.

8.9 TRANSPORT AND BURIAL OF WASTES

Of concern in this section is evaluation of the complex situation where a corporation is not assessing its potential involvement for a single environmental cleanup or disposal project but has the opportunity to participate in many projects at the same time. To set the development within a familiar framework (see Chap. 4 and Fig. 4.1), we focus attention here on the particular problem of transport and long-term storage and monitoring of chemical, industrial, toxic, or radioactive wastes, although the general sense of the argument applies equally well to any environmental project or, indeed, to a large class of engineering problems (such as bridge

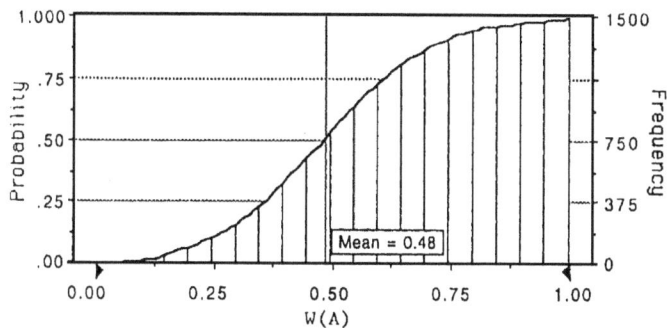

(a)

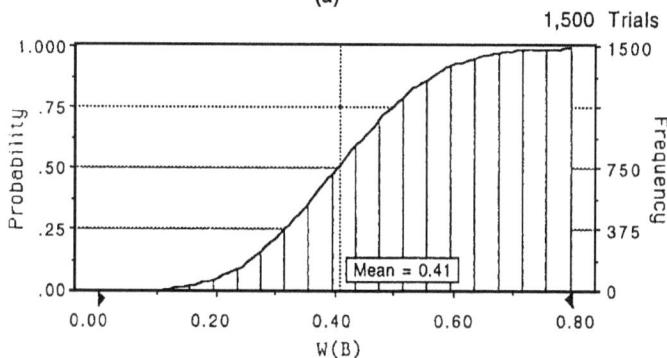

(b)

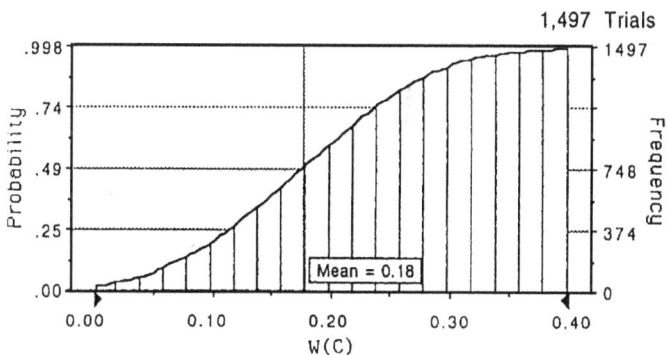

(c)

FIGURE 8.11 Cumulative probability plot of fractional working interest for the three opportunities of Sec. 8.5. The parabolic RAV formula was used to obtain the plots for (*a*) opportunity *A*, (*b*) opportunity *B*, and (*c*) opportunity *C*, whereas the Cozzolino RAV formula was used to obtain the plots for (*d*) opportunity *A*, (*e*) opportunity *B*, and (*f*) opportunity *C*.

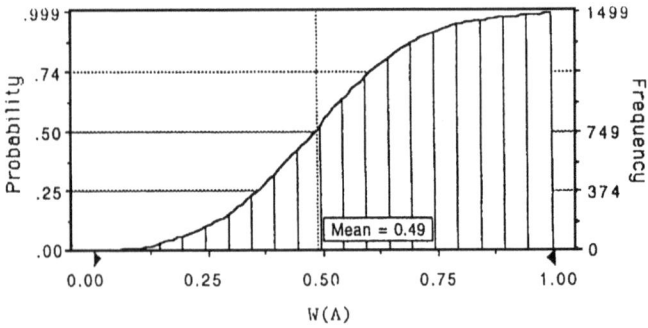

(d)

1,499 Trials

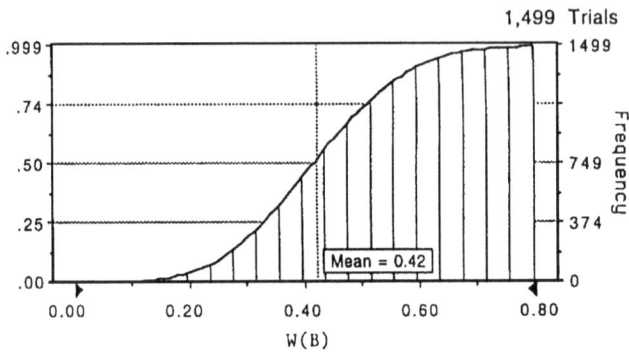

(e)

1,500 Trials

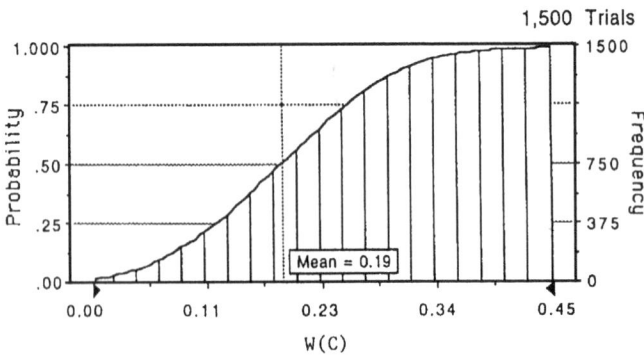

(f)

FIGURE 8.11   (*Continued*)

building, dam impoundment, etc.). For any project, we distinguish four dominant scenarios: (1) transport without spillage from the location of origin and storage without leakage, (2) transport without spillage but leakage at the burial site, (3) spillage during transport of the waste but no further contamination problems at the

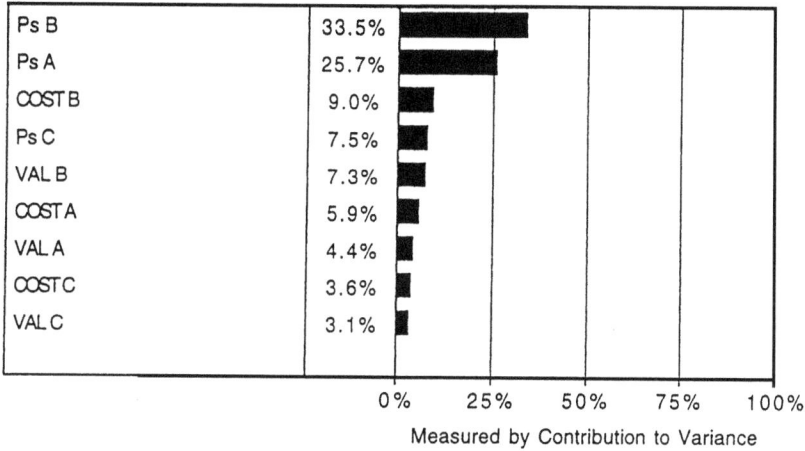

(a)

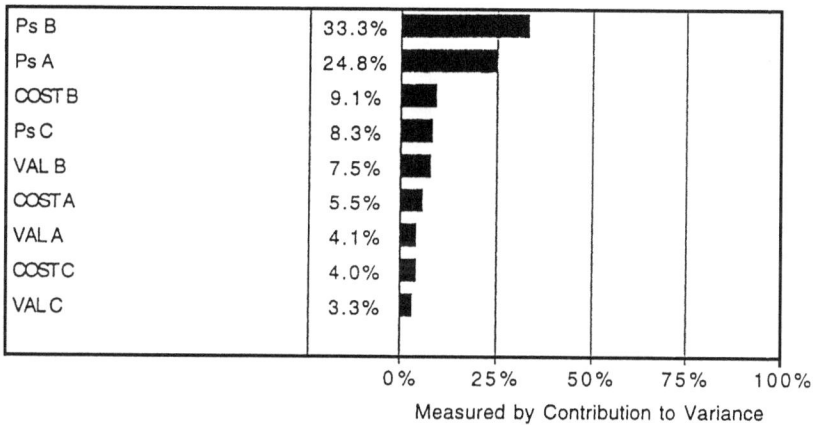

(b)

FIGURE 8.12   Relative importance diagram of ±10% variations in V, C, and p_s for each opportunity in contributing to the uncertainty on the total RAV of the portfolio: (*a*) for the parabolic RAV formula; (*b*) for the Cozzolino RAV formula.

storage location, and (4) transport with spillage and leakage at the repository site. Each project has its own characteristic parameters for costs and probabilities of success and failure, along the different possible pathways, and each has its own particular monitoring and regulatory requirements. Spillage during transport may occur because of human error or accident associated with the particular mode of transportation and/or failure of some of the containers used for transportation of the waste. Once the waste material reaches the repository site, it is assumed that a separate procedure for isolation of the waste is implemented. Such a procedure

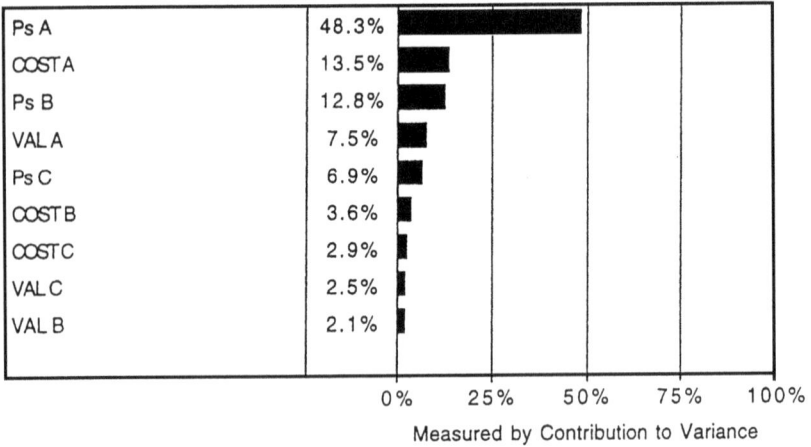

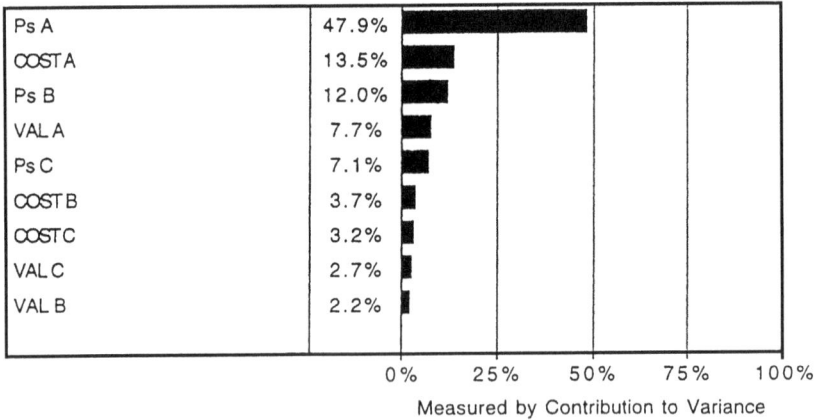

FIGURE 8.13   Relative importance diagram of ±10% variations in V, C, and p_s for each opportunity in contributing to the uncertainty on working interest for opportunity A: *(a)* for the parabolic RAV formula; *(b)* for the Cozzolino RAV formula.

relies on engineering measures (and physical barriers) that include, but are not limited to, the insulating properties of the waste containers (Sandia Report, 1994; Westinghouse Savannah River Company Report, 1996). Hence transport and burial of the wastes are treated as independent (disjoint) events, with the probability of success (or failure) of the transport component not affecting the probability of success (or failure) of burial.

The decision-tree diagram of Fig. 8.14 depicts these four events for the ith such project, together with the associated probabilities and total anticipated costs $C_{1,i}$, $C_{2,i}$, $C_{3,i}$, and $C_{4,i}$, respectively. Thus $C_{1,i}$ is the estimated expenditure under optimal conditions; $C_{2,i}$ includes, additionally, the costs of excavation of leaking containers, repackaging, remediation, and increased monitoring at the operating facility; $C_{3,i}$ consists of (in addition to $C_{1,i}$) the costs of collecting and transporting spilled waste as well as any other contaminated material from a spillage location; and $C_{4,i}$ encompasses all previous financial obligations (Paleologos and Lerche, 1999). On the other side of an environmental opportunity is the funding agency for waste control that offers a contract at a fixed price G_i for transport, burial, and monitoring for a given time. Usually the contract price G_i will be less than the costs for the worst-case scenario $C_{4,i}$, and also may be less than $C_{3,i}$ and $C_{2,i}$. The values of each scenario (contract price minus costs) are shown in Fig. 8.14.

A corporation would like to maximize its involvement in all such projects in order that they provide the largest return to the company. On the other hand, the more projects a company is involved in, the greater is the chance that at least one will undergo a catastrophic event, and the greater also is the likelihood of future liability claims. Both cases cause the corporation major financial loss. What a corporate decision maker would like to do is to estimate the best working interest for each project so that the total profit returned by all projects is maximized. Additionally, a corporation would like to limit liability for each project so that, again, the total liability for all projects is no more than some fixed amount that a corporation believes it can afford should an extreme, high-cost catastrophic event occur in one or more of the projects. The objective of this section is to address these issues within the context of decision analysis.

8.10 MULTIPLE PROJECTS

Consider a set ($i = 1, 2,..., N$) of N projects, each of the forms described in the preceding section and each with its own contract price G_i. An amount CE (or risk-adjusted value RAV in oil industry terminology) is termed the *certainty equivalent*

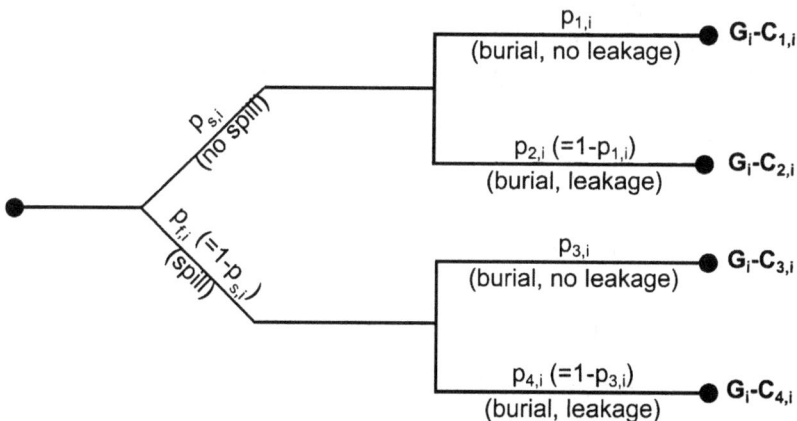

FIGURE 8.14 Four-path decision tree for a transport and storage environmental problem.

of an opportunity involving two equiprobable random outcomes of value A and B, respectively, if one is indifferent between CE and the rights to this opportunity. In business situations, most individuals exhibit risk-averse behavior, that is, CE < (A + B)/2; otherwise, when CE > (A + B)/2 or CE = (A + B)/2, the behavior is described as risk-seeking or risk-neutral, respectively (Keeney and Raiffa, 1993; Raiffa, 1997). A utility function U encodes the strength of subjective preferences and risk attitudes toward specific uncertain outcomes of an opportunity, with the shape of the utility function approximating a decision maker's attitude toward risk. Thus U is concave when CE < (A + B)/2, convex when CE > (A + B)/2, and linear when CE = (A + B)/2, respectively (Krzysztofowicz, 1986; Keeney and Raiffa, 1993; Clemen, 1996).

Based on an exponential utility model,

$$U(x) = 1 - e^{-x/\mathrm{RT}} \tag{8.31}$$

where x stands for the dollar value of an opportunity, and RT stands for the risk tolerance (the maximum amount a corporation is prepared to lose without endangering its cash-flow situation or total assets). Chapter 4 presented the certainty-equivalent amount CE for a single project of the form presented in Fig. 8.14. For the ith project from a total number N of projects, when a working fractional interest W_i ($0 \leq W_i \leq 1$) is taken at a corporate risk tolerance RT_i, the certainty amount was given (see Chap. 4) by

$$\mathrm{CE}_i(W_i) = -\mathrm{RT}_i \ln \left\{ p_{s,i} \left[p_{1,i} \exp \left(\frac{-W_i(G_i - C_{1,i})}{\mathrm{RT}_i} \right) + p_{2,i} \exp \left(\frac{-W_i(G_i - C_{2,i})}{\mathrm{RT}_i} \right) \right] \right.$$

$$\left. + p_{f,i} \left[p_{3,i} \exp \left(\frac{-W_i(G_i - C_{3,i})}{\mathrm{RT}_i} \right) + p_{4,i} \exp \left(\frac{-W_i(G_i - C_{4,i})}{\mathrm{RT}_i} \right) \right] \right\} \tag{8.32}$$

Equation (8.32) was obtained using the exponential model [Eq. (8.31)] to derive the expected utility for a working fraction of the project in Fig. 8.14 and the definitions of the CE and the utility function, from which it follows that the expected utility of a project equals the utility value of the certainty equivalent. Also derived in this chapter was a second-order (quadratic) approximation to Eq. (8.32) that is easier to analyze to changes in the values of parameters and which, when applied to project i, yields

$$\mathrm{CE}_i(W_i) = a_i W_i^2 + b_i W_i \qquad a_i = -\frac{v_i^2 E_i^2}{2\mathrm{RT}_i}, \qquad b_i = E_i \tag{8.33}$$

with the expected value E_i of any four-branch system (see Fig. 8.1) given by

$$E_i = G_i - p_{s,i}(p_{1,i}C_{1,i} + p_{2,i}C_{2,i}) - p_{f,i}(p_{3,i}C_{3,i} + p_{4,i}C_{4,i}) \tag{8.34}$$

the variance σ_i^2 calculated by

$$\sigma_i^2 = p_{s,i}p_{1,i}(1 - p_{s,i}p_{1,i})\, C_{1,i}^2 + p_{s,i}p_{2,i}\,(1 - p_{s,i}p_{2,i})\, C_{2,i}^2$$

$$+ p_{f,i}p_{3,i}(1 - p_{f,i}p_{3,i})\, C_{3,1}^2 + p_{f,i}p_{4,i}(1 - p_{f,i}p_{4,i})\, C_{4,i}^2$$

$$- 2\,[\,p_{s,i}^2 p_{1,i}p_{2,i}C_{1,i}C_{2,i} + p_{f,i}^2 p_{3,i}p_{4,i}C_{3,i}C_{4,i}$$

$$+ p_{s,i}p_{f,i}(p_{1,i}C_{1,i} + p_{2,i}C_{2,i})\,(p_{3,i}C_{3,i} + p_{4,i}C_{4,i})\,]\qquad(8.35)$$

and the volatility v_i defined by

$$v_i = \frac{\sigma_i}{E_i}\qquad(8.36)$$

Thus, for a set $(i = 1,\dots, N)$ of such projects the total CE is given by

$$\mathrm{CE} = \sum_{i=1}^{N} \mathrm{CE}_i(W_i)\qquad(8.37)$$

and it is Eq. (8.37) that one wishes to maximize. If each project is independently assessed, and if no other corporate constraints are in force, then $\partial \mathrm{CE}/\partial W_i = \partial \mathrm{CE}_i/\partial W_i$ for all $i = 1, 2,\dots, N$, and the maximum total CE is obtained when the working interest in each project is at $W_i = W_{\mathrm{max},i}$, where $W_{\mathrm{max},i}$ is the fractional working interest that maximizes the certainty-equivalent amount of each project i. However, it is an extremely rare circumstance when the preceding situation holds. There is always one or more additional corporate criteria enforced, and these are discussed next.

8.11 EXPOSURE AND EXPENDITURE CONSTRAINTS

Application of different RT_i for each project i is done here to allow for the situation where projects represent different sectors of the environmental market (with different risk characteristics, potential future liabilities, etc.) and/or considerations involving corporate strategy. Distinct risk tolerances also may be used because of interest to increase market share in a particular environmental sector or corporate presence in a particular national or international market. As well as setting a risk-tolerance limit RT_i for each project, a corporation also usually takes a global perspective on the total amount B it could afford to lose without compromising its operational viability should one or more catastrophic situations occur. There are two common ways of setting a limit: corporate exposure and corporate expenditure. Each is considered in turn.

8.11.1 Corporate Exposure

In this situation, a corporation calculates that if the worst-case scenario occurs for the ith project, where working interest W_i has been undertaken and with catastrophic costs for the task equaling $C_{4,i}$, then the corporation would be liable for the amount $W_i C_{4,i}$. The cost-exposure limit B is set so that the working interests in all projects are adjusted to maintain the relation

$$\sum_{i=1}^{N} W_i C_{4,i} \leq B \qquad (8.38)$$

In this way, a corporation limits total exposure to catastrophic costs to no more than B. The problem to address in this cost-exposure situation is to maximize the total risk-adjusted value CE of Eq. (8.37) subject to the constraint of Eq. (8.38).

8.11.2   Cost Expenditure

In this case, a corporation takes the viewpoint that it is unlikely that the worst-case scenario will occur in all projects. Accordingly, the corporation often argues for an expected-value loss, which is obtained for each project by calculating the expected value E_i as though there were no worth to the contract, that is, $G_i = 0$. Thus from Fig. 8.14 one has E_i given by Eq. (8.34), and the expected-value loss L_i can be calculated by

$$L_i = p_{s,i} \, (p_{1,i} C_{1,i} + p_{2,i} C_{2,i}) + p_{f,i} \, (p_{3,i} C_{3,i} + p_{4,i} C_{4,i}) \qquad (8.39)$$

Thus the corporation sets a cost-expenditure limit B such that the working interests in all projects are subject to the constraint

$$\sum_{i=1}^{N} W_i L_i \leq B \qquad (8.40)$$

Because $L_i \leq C_{4,i}$ for each opportunity, it follows that the working interest one could take under a cost-expenditure corporate philosophy is larger than under cost-exposure considerations.

8.12   OPTIMAL WORKING INTEREST

In either a cost-exposure or a cost-expenditure approach, the general constraint can be written as

$$\sum_{i=1}^{N} W_i X_i = fB \qquad (8.41)$$

where X_i stands for either L_i or $C_{4,i}$. Here, f is the fraction of the maximum limit a corporation will allow, enabling one to use the equality sign in Eq. (8.38) or Eq. (8.40). For each opportunity one then computes $CE_i \, (W_i)$ from Eq. (8.32) or (8.33) and retains only those projects for which there is at least some range of

W_i in $0 \le W_i \le 1$ for which the certainty equivalent amount is positive. The residual opportunities are eliminated because they have no capability of generating profit.

In classical optimization, there are two well-known approaches for solving this type of problem: (1) the substitution method and (2) lagrangian analysis. Under the substitution method, constraint (8.41) can be solved for one of the decision variables W_k in terms of fB, X_k, and the other $N - 1$ variables such that $W_k = (fB - \Sigma W_j X_j)/X_k$, where the summation is over $j = 1, 2,..., N$ and $j \ne k$. Substituting the expression for W_k in Eq. (8.37) allows the total CE to be written in terms of $N - 1$ decision variables. The maximum total CE is obtained by setting the partial derivatives of CE with respect to W_i (where $i = 1, 2,..., N$ and $i \ne k$) equal to zero. The resulting $N - 1$ nonlinear equations can be solved by numerical solvers that are based on a variation of Newton's method, with the concave exponential or quadratic utility functions ensuring globally optimal solutions (Press et al., 1987). Under the lagrangian approach, expression $L = \Sigma CE_i(W_i) - \lambda(\Sigma W_i X_i - fB)$ is maximized. Here, λ is an undetermined lagrangian multiplier, and summation is over $i = 1, 2,..., N$. This leads to the solution of $N + 1$ partial derivatives $\partial L/\partial W_i = 0$, $i = 1, 2,..., N$, and $\partial L/\partial \lambda = 0$, which again can be accomplished with the use of numerical solvers.

The two methods can be seen easily to result to the same conditions. Regard the constraint (8.41) as yielding a working interest W_k for the kth project, which according to the substitution method can be expressed as

$$ W_k X_k = fB - \sum_{\substack{j=1 \\ j \ne k}}^{N} W_j X_j \tag{8.42} $$

Substituting W_k from Eq. (8.42) into Eq. (8.37) yields

$$ \mathrm{CE} = \mathrm{CF}_k \left(\frac{fB - \Sigma W_j X_j}{X_k} \right) + \Sigma \mathrm{CE}_j(W_j) \tag{8.43} $$

where the summation is over $j = 1, 2,..., N$ but $j \ne k$. Now the certainty-equivalent amount in Eq. (8.43) has a maximum with respect to variations in $W_i (i = 1, 2,..., N$ but $i \ne k)$ when

$$ \frac{\partial \mathrm{CE}}{\partial W_i} = \frac{\partial \mathrm{CE}_i}{\partial W_i} + \frac{\partial \mathrm{CE}_k}{\partial W_k} \frac{\partial W_k}{\partial W_i} = \frac{\partial \mathrm{CE}_i}{\partial W_i} + \frac{\partial \mathrm{CE}_k}{\partial W_k} \left(-\frac{X_i}{X_k} \right) = 0 \tag{8.44} $$

which can be rearranged to yield

$$ X_i^{-1} \frac{\partial \mathrm{CE}_i}{\partial W_i} = X_k^{-1} \frac{\partial \mathrm{CE}_k}{\partial W_k} \tag{8.45} $$

But i and k are arbitrary choices. The only way Eq. (8.45) can be satisfied for any and all choices of i and k is if each side equals a constant H that is independent of the parameters of projects i and k.

Thus the requirement for a global maximum with respect to W_i [Eq. (8.44)] leads to the condition

$$\frac{\partial CE_i/\partial W_i}{X_i} = H \qquad \text{for } i = 1, 2, ..., N \tag{8.46}$$

which relates the optimal working interest of any project to H. The global optimality condition of Eq. (8.46) is known in economics as the *equimarginal principle* and underlies the optimal allocation of a scarce resource among competing demands. Taking the derivative of Eq. (8.32) for CE_i with respect to W_i and combining with Eq. (8.46) leads to

$$(G_i - C_{1,i} - HX_i)A_1 + (G_i - C_{2,i} - HX_i)A_2 + (G_i - C_{3,i} - HX_i)A_3$$
$$+ (G_i - C_{4,i} - HX_i)A_4 = 0 \tag{8.47a}$$

where

$$A_1 = p_{s,i}P_{1,i}\exp(W_iC_{1,i}/RT_i) \qquad A_2 = p_{s,i}P_{2,i}\exp(W_iC_{2,i}/RT_i)$$

$$A_3 = p_{f,i}P_{3,i}\exp(W_iC_{3,i}/RT_i) \qquad A_4 = p_{f,i}P_{4,i}\exp(W_iC_{4,i}/RT_i) \tag{8.47b}$$

Equation (8.47a) relates the working interests W_i ($i = 1, 2, ..., N$) that maximize the total certainty-equivalent amount to the constant H, which has yet to be determined. In Eqs. (8.47a) and (8.47b) the parameters of each project (probabilities, costs, risk tolerances) are known, the only unknowns being W_i and H. Thus the working interests that provide the maximum total return are each a function of the parameters of each project and the unknown constant H, that is, $W_i = W_i(H, q_i)$, where q_i is the vector of all parameters for the ith opportunity. Equations (8.47a) and (8.47b) are then combined with Eq. (8.41) in order to constrain the working interests of the projects to the limiting cost fB a corporation is prepared to expend. Thus one obtains a relation between the constant H and the limiting cost fB in the form

$$\sum_{i=1}^{N} W_i(H, q_i) X_i = fB \tag{8.48}$$

Note that using the substitution method, one can develop $N + 1$ equations [N equations from Eqs. (8.46) and (8.48)]. Furthermore, these equations are exactly what one would obtain through the lagrangian method. Indeed, Eq. (8.46) results directly from $\partial L/\partial W_i = 0$, $i = 1, 2, ..., N$, and Eq. (8.48) from the relation $\partial L/\partial \lambda = 0$. In this respect, H coincides with the lagrangian multiplier λ and represents the marginal change in the objective function (8.37) per unit change in the level of the constraint.

As an alternative to nonlinear programming techniques (Hillier and Lieberman, 1980), the preceding optimization problem [Eqs. (8.47a) and (8.47b)], subject to constraint (8.48), can be solved simply as follows: (1) Select an arbitrary value of H. Equations (8.47a) and (8.47b) can now be used, given the parameters of each project, to calculate the unknowns W_i for all projects $i = 1, 2,..., N$. (2) Insert these working interests into Eq. (8.48), thereby providing the value of the limiting cost fB that corresponds to the chosen value of H. (3) Vary H, and repeat this procedure. A single curve of fB versus H is then obtained from which one can read off the value of the constant H corresponding to the fB value mandated by the corporation. (4) Having established the value of H, one can evaluate the optimal involvements W_i from Eqs. (8.47a) and (8.47b), the corresponding optimal CE_i from Eq. (8.32), and the maximum total certainty-equivalent amount from Eq. (8.37) for all projects $i = 1, 2,..., N$. In this way the relevant working interest of each opportunity is determined that satisfies both the corporate risk tolerance RT_i and the corporate limit to total risk fB and which also maximizes the total return from all projects. This procedure can be easily implemented numerically and is usually much faster than the numerical searches for global maxima (Press et al., 1987).

8.13   PARABOLIC UTILITY FUNCTION

For the exponential utility function, it is not possible to obtain an analytic expression describing the behavior in multiple opportunities because the combination of exponential variations in Eq. (8.32) does not lend itself to analytic evaluation. However, for the quadratic (in W) parabolic formula of Eq. (8.33), one can provide analytic representations. In addition, a Taylor series expansion of the exponential expression [Eq. (8.32)] around $W = 0$ has the parabolic formula as its first two terms, so both certainty-equivalent formulas are very similar before their respective peaks. Thus, not only is it simpler to work with the parabolic expression, but it also provides results that are almost identical to those arising from the exponential function. This section develops the procedure for optimal involvement in multiple projects based on a parabolic utility function.

Differentiating Eq. (8.33) with respect to W_i and inserting into Eq. (8.46) yield

$$\frac{\partial CE_i}{\partial W_i} = 2a_iW_i + b_i = HX_i \qquad \text{for } i = 1, 2,..., N \qquad (8.49)$$

which on substitution of the expressions for a_i and b_i and rearrangement of terms give

$$W_i = \frac{(E_i - HX_i)RT_i}{(v_iE_i)^2} \qquad \text{for } i = 1, 2,..., N \qquad (8.50)$$

Equation (8.50) relates W_i and the constant H explicitly in a linear form where the expected value E_i and the volatility v_i are determined from Eqs. (8.34) and (8.36), respectively, and X_i stands for either L_i or $C_{4,i}$. Substituting Eq. (8.50) into Eq. (8.33) yields the optimal certainty-equivalent amount of each project

$$
\text{CE}_i = \frac{\text{RT}_i}{2v_i^2}\left[1 - \left(\frac{HX_i}{E_i}\right)^2\right] \qquad \text{for } i = 1, 2, \dots, N \tag{8.51}
$$

Equation (8.51) can be used provided that the optimal working interest W_i evaluated from Eq. (8.50) lies in the range $0 \le W_i \le 1$, which translates into the requirements

$$
\frac{E_i}{X_i} > H \qquad \text{corresponding to } W_i \ge 0 \tag{8.52a}
$$

and

$$
\frac{E_i}{X_i} - \frac{(v_i E_i)^2}{X_i \text{RT}_i} \le H \qquad \text{corresponding to } W_i \le 1 \tag{8.52b}
$$

Combining Eqs. (8.52a) and (8.52b) gives the allowable range for the constant H that corresponds to working interests in the range $0 \le W_i \le 1$ as

$$
\text{lb}_i = \frac{E_i}{X_i} - \frac{(v_i E_i)^2}{X_i \text{RT}_i} \le H \le \frac{E_i}{X_i} = \text{ub}_i \tag{8.52c}
$$

If the value of H exceeds the upper bound of Eq. (8.52c), E_i/X_i, then this corresponds to a negative working interest, and hence no working interest should be taken in the ith project ($W_i = 0$). If the value of H is smaller than the lower bound of Eq. (8.52c), then this corresponds to $W_i > 1$, and hence the working interest in the ith project should be required to equal 1 ($W_i = 1$), with the corresponding optimal certainty-equivalent amount given by

$$
\text{CE}_i = E_i\left(1 - \frac{v_i^2 E_i}{2\text{RT}_i}\right) \tag{8.52d}
$$

Suppose now that a particular value of H is chosen initially. Then use of Eqs. (8.52a) and (8.52b) enables one to identify the opportunities for which $0 \le W_i \le 1$, as well as those for which $W_i = 1$. Let the opportunities be ordered according to their increasing values of $\text{ub}_i = E_i/X_i$ with $\text{ub}_1 > \text{ub}_2 > \text{ub}_3 \cdots$. For a particular H value, let there be K projects with $\text{ub}_i \ge H$ and M projects for which $W_i = 1$. Then the relation between H and the corporate liability fB is given by

$$
\sum_{i=1}^{K} W_i X_i + \sum_{j=1}^{M} X_j = fB \tag{8.53}
$$

which can be written, using Eq. (8.50), in the form

$$fB = \left(\sum_{j=1}^{M} X_j + \sum_{i=1}^{K} \frac{X_i RT_i}{E_i v_i^2} \right) - H \sum_{i=1}^{K} \frac{X_i^2 RT_i}{E_i^2 v_i^2} \tag{8.54}$$

Note that as the value of H increases, the value of fB decreases because as H crosses to a higher value than $ub_j = E_j/X_j$, the jth opportunity is discarded because it would have $W_j < 0$. Further, as H is increased, it becomes easier to satisfy Eq. (8.52b), that is, $lb_i \leq H$, so the number of opportunities in which 100% working interest could be taken decreases. Hence the corporate exposure to liability fB decreases. In other terms, as the corporation lowers its acceptable risk level fB to involvement in potential catastrophic payouts, the parameter H rises until, in the end, H will exceed ub_1 when no involvement in any projects would be recommended. The determination of H in terms of fB (or of fB in terms of H), then, allows one to write the total certainty-equivalent amount for all projects as [Eqs. (8.33) and (8.52d)]:

$$CE = \sum_{i=1}^{K} E_i W_i \left(1 - \frac{E_i W_i v_i^2}{2RT_i} \right) + \sum_{j=1}^{M} E_j \left(1 - \frac{E_j v_j^2}{2RT_j} \right) \tag{8.55}$$

Equations (8.54) and (8.55) indicate that as H increases, fB decreases, and the total CE also decreases.

8.14 APPLICATION

We provide now a numerical illustration of the previous procedure using the parabolic utility function for ease of presentation. Consider three potential projects with the risk-tolerance values $RT_3 = 2RT_2 = 4RT_1$ assigned to each project. Let the projects be arranged so that $E_3 = 2E_2 = 4E_1$ with volatilities $v_3^2 = 2v_2^2 = 4v_1^2$. Evaluate the optimal working interest for each project assuming a cost-expenditure basis with $L_3 = 2L_2 = 4L_1$. Note that the projects have been arranged in an increasing order of RT, E, v, and L. Thus project 1 has the lowest risk tolerance, expected return, volatility, and potential average catastrophic loss of all projects, and project 3 exhibits the highest values in all four attributes. The total corporate budgetary constraint against loss [Eq. (8.40) or (8.41)] is given by

$$W_1 + 2W_2 + 4W_3 = \frac{fB}{L_1} \tag{8.56}$$

Then Eq. (8.50) yields (for $0 \leq W_i \leq 1$)

$$W_1 = \frac{(E_1 - HX_1)RT_1}{(v_1 E_1)^2} \tag{8.57a}$$

$$W_2 = \tfrac{1}{2}\,W_1 \qquad \text{and} \qquad W_3 = \tfrac{1}{4}\,W_1 \qquad\qquad (8.57b)$$

Substituting Eqs. (8.57a) and (8.57b) into Eq. (8.56) provides the value of the constant H:

$$H = \frac{E_1}{L_1}\left(1 - \frac{1}{3}\,\frac{v_1^2\,E_1}{RT_1 L_1}\,fB\right) \qquad\qquad (8.58)$$

Let us analyze now the working interest for each project according to different regions of H values:

1. $H < E_1/L_1$ because in Eq. (8.58) the second term within the parentheses is positive; the reverse ($H > E_1/L_1$) would have corresponded to a negative working interest, leading to the decision of not participating in any project.

2. When $E_1/L_1 > H > (E_1/L_1)\,(1 - v_1^2\,E_1/RT_1) > 0$, corresponding to $fB < 3L_1$, the working interest in each project is given by Eqs. (8.57a) and (8.57b).

3. When $(E_1/L_1)(1 - v_1^2 E_1/RT_1) > H > (E_1/L_1)\,(1 - 2v_1^2\,E_1/RT_1)$, the range of fB values is given by $5L_1 > fB > 3L_1$. Substitution of Eq. (8.57b) into Eq. (8.56) yields $W_1 = fB/3L_1$. For the range of fB values considered, this makes W_1 greater than one; hence $W_1 = 1$. Substitution of Eq. (8.57b), with respect to W_2, into Eq. (8.56) yields $W_2 = (fB/L_1 - 1)/4$ and, from Eq. (8.57b), $W_3 = W_2/2$.

4. When $(E_1/L_1)(1 - 2v_1^2\,E_1/RT_1) > H > (E_1/L_1)\,(1 - 4v_1^2\,E_1/RT_1)$, the range of fB values is $7L_1 > fB > 5L_1$. With $W_1 = fB/3L_1$ and $W_2 = (fB/L_1 - 1)/4$, one must take $W_1 = W_2 = 1$. From Eqs. (8.57b) and (8.56), one has $W_3 = (fB/L_1 - 3)/4$.

5. When $(E_1/L_1)(1 - 4v_1^2\,E_1/RT_1) > H$, which corresponds to $fB > 7L_1$, then $W_3 > 1$ and hence one takes $W_1 = W_2 = W_3 = 1$.

Thus in this simple illustration one takes a progressively greater working interest first in project 1, then in project 2, and finally in project 3 as the ability of the corporation (measured through fB) to afford catastrophic losses increases.

8.15 SUMMARY

Most environmental companies are involved in many projects at the same time. Such involvement presents the opportunity for diversification to a company in several areas of the environmental market but also increases the likelihood of exposure to uncertain events that can result in major financial loss immediately and/or make the company liable to future claims. Accordingly, each corporation has some predetermined global risk tolerance, expressed by a maximum total amount the company is prepared to lose without compromising its cash-flow viability and total assets. This total maximum liability represents the primary financial constraint for a company in participating in multiple environmental projects. A secondary financial constraint arises from the risk tolerance a corporation attaches to each particular project, according to each project's characteristics to profit bearing and range of uncertainty. The overarching objective for a corporation is to optimize its involvement in all such projects in order that the total return is maximized. Thus each company may limit potential exposure to the negative aspects of a project by taking less than 100% working interest in a contract. In terms of the global objective, this translates into determining the working interests for each and all projects, the combination of which will satisfy both the preceding financial constraints to risk and the requirement to maximize the total profit.

This chapter has served two purposes: First was the need to obtain analytic expressions for optimizing total risk-adjusted value for a portfolio of opportunities in the face of a constrained budget, and second was the need to determine the effect of ranges of uncertainties of values, costs, and success chances for each opportunity on the total RAV and to figure out a procedure to identify which parameters were causing the greatest uncertainty.

With respect to the first purpose, it was shown for a parabolic profile of RAV versus working interest for each opportunity that a closed-form expression could be written down exactly for the working interest that should be taken in each opportunity in order to maximize the total RAV under a constrained budget condition. It also was shown for the exponential RAV formula that the same procedure allows one to write down an analytic but parametric relation between working interest and total budget in terms of a single parameter. The procedure developed operates with any functional form chosen for RAV versus working interest. Numerical illustrations of the procedure indicated the pragmatic operation of the method (which is extremely fast numerically compared with the more conventional Simplex solution solver methods that are used traditionally).

With respect to the second purpose, the ability to provide a rapid investigation of ranges of uncertainty of values, costs, and chances of success for each opportunity on the total RAV probability and probable working interest ranges for each opportunity means that one can quickly focus on which parameters need to be addressed if one is to narrow the uncertainty on total RAV for a given budget or, indeed, for a range of possible budgets.

The methods presented here can be used with any functional form of RAV dependence on working interest to provide simple expressions, either in closed form or involving only a single parametric representation, for the working interest to be taken to optimize a portfolio of opportunities. It is this fact that both the deterministic and probabilistic numerical illustrations have been designed to exhibit.

This methodology was applied to assess the involvement in three projects arranged at increasing order of expected return, tolerance to risk, uncertainty, and potential to financial loss. Depending on the ability of a corporation to afford catastrophic losses, involvement in the projects increased in a conservative manner; i.e., a greater working interest always was taken in the least uncertain (but least profitable) project than in the riskier projects. Riskier opportunities were considered only after full participation had been achieved in safer projects.

APPENDIX 8A: WEIGHTED RAV OPTIMIZATION

In the body of this chapter the total RAV was maximized, with each opportunity having its RAV added arithmetically to the total. However, corporations often weight the relative RAV contributions. The purpose of this appendix is to show that optimization of the portfolio can be carried through equally with weighting factors.

Thus, if the weights assigned are σ_i (>0) ($i = 1,\ldots, N$) for $RAV_i(W_i)$, then the total weighted RAV is

$$RAV = \sum_{i=1}^{N} \frac{\sigma_i RAV_i(W_i)}{\sum_{j=1}^{N} \sigma_j} \tag{8.59}$$

The budget constraint also could be weighted if done more on a cash-flow basis than a fixed-budget basis. Let that weighting be $\rho_i > 0$ ($i = 1,\ldots, N$) in the sense that the constraint to be applied is

$$\sum_{i=1}^{N} C_i \rho_i W_i = B \tag{8.60}$$

Then, following the procedure of Sec. 8.2, the optimal RAV occurs when

$$\frac{\sigma_i}{\rho_i C_i} \frac{\partial RAV_i(W_i)}{\partial W_i} = H = \text{constant, all } i \tag{8.61}$$

For the parabolic RAV formula,

$$\frac{\partial RAV_i(W_i)}{\partial W_i} = E_i\left(1 - \frac{W_i}{OWI_i}\right) \tag{8.62}$$

so Eqs. (8.61) and (8.62) yield

$$W_i = OWI_i\left(1 - \frac{H\rho_i C_i}{\sigma_i E_i}\right) \tag{8.63}$$

with

$$\sum_{i=1}^{N} C_i \text{OWI}_i \rho_i - H \sum_{i=1}^{N} \frac{\text{OWI}_i \rho_i^2 C_i}{\sigma_i E_i} = B \qquad (8.64)$$

which determines H in terms of the budget B. A similar analysis can be carried through for any weighting applied to any functional form chosen for the dependence of RAV on working interest.

CHAPTER 9
APPORTIONMENT OF COST OVERRUNS TO HAZARDOUS WASTE PROJECTS

9.1 INTRODUCTION

One of the more interesting challenges facing a corporation in environmental projects arises when the estimated costs of each project end up being less than the actual costs incurred. In 1997, the average cost for private-sector environmental remediation projects in the United States was 25 to 50% over the initial budget (Al-Bahar and Crandell, 1990; Diekmann and Featherman, 1998). Cost overruns related to major military, energy, and information-technology projects in the United States, the United Kingdom, and a number of developing countries were between 40 and 500% (Morris and Hough, 1987). Cost overruns arise primarily from (1) external risks due to modifications in the scope of a project and changes in the legal, economic, and technologic environments, (2) technical complexity of a project due to size, duration, or technical difficulty, (3) inadequate project management manifested in the control of internal resources, poor labor relations, and low productivity, and (4) unrealistic cost estimates because of improper quantification of the uncertainties involved (Yeo, 1990; Minato and Ashley, 1998).

Two possibilities can then occur depending on the way the original contract was written: Either the corporation absorbs the extra costs of the project because the contract was for a fixed price, or it renegotiates the additional cost overruns as part of an addendum to the original contract. It is also possible within a single project that a total cost overrun occurs and one renegotiates the extra costs. In either event, the corporation then has the job of apportioning the costs to the projects. Clearly, from a corporate perspective there is the ongoing need to maximize the profit the corporation can make. However, there is also the need to ensure that extra costs are not assigned to a project or part of a project that is only marginally profitable; otherwise, one can easily arrange for a particular project to lose money.

For instance, suppose a new computer system is part of an unanticipated expense. The computer system will be used for two projects. If one of the projects is highly successful at generating revenue for the corporation, while the other project is only marginally profit-making, then assigning *all* the computer system costs to the highly profitable venture will lower that project's profitability. Equally, assigning *all* the computer costs to the marginal project may drive its worth negative. There must, then, be a way of assigning extra costs to a set of projects in such a way as to maximize the total profit available to the corporation.

The same sense of argument is operative for a single project in which various parts of the project provide profit and each of which have costs; one again wishes to apportion the extra costs so as to maximize the total overall profit to the corporation. The purpose of this chapter is to illustrate how one goes about such an apportionment.

9.2 PROJECT COSTS AND COST OVERRUNS

Consider, for ease of illustration only, two projects A and B (although it is extremely easy to generalize the analysis to N projects). Let project A (B) have an expected value E_A (E_B) and estimated direct costs C_A (C_B). And take it that the corporation has a total working interest of 100% in each project. If the corporation has unlimited risk tolerance, then a measure of profitability is

$$M_1 = \frac{E_{A0}}{C_A} + \frac{E_{B0}}{C_B} \tag{9.1}$$

where $E_{A0}(E_{B0})$ is the expected value in the absence of any costs at all and so is independent of costs. If the corporation has a finite risk tolerance, then the risk-adjusted value RAV is used for each project rather than the expected value. In this case a measure of profit is

$$M_2 = \frac{\text{RAV}_{A0}}{C_A} + \frac{\text{RAV}_{B0}}{C_B} \tag{9.2}$$

where RAV_{A0} (RAV_{B0}) is the risk-adjusted value in the absence of any costs at all; it too is then independent of cost. In either case, what is being used as a profitability index is the return to the corporation for each dollar spent. Clearly, a value of each component on the right-hand side of Eq. (9.1) or Eq. (9.2) greater than unity implies that net gains exceed costs for each project so that each is then profitable. Then $M_1 > 2$ and $M_2 > 2$ follow automatically. However, one should be careful *not* to use the requirements $M_1 > 2$ and $M_2 > 2$ on their own as measures of the same profitability. The reason, of course, is that it is possible to have E_{A0}/C_A, say, much larger than unity, whereas E_{B0}/C_B could be much smaller than unity, but the sum could still yield $M_1 > 2$.

Suppose, now, that an extra cost δC is incurred either because of new equipment (e.g., computer system) being purchased or because of a cost overrun on unanticipated parts of each project (e.g., the cost of drums to bury the waste rises beyond the maximum price anticipated at the start of the project). Then one wishes to apportion this extra cost δC in such a way as to keep the profit measure M_1 (or M_2) at its maximum possible value.

The way to do this is as follows: Let a fraction $\sin^2 \theta$ of the extra cost δC be assigned to project A and, correspondingly, a fraction $\cos^2 \theta$ be assigned to project B. The corresponding measure of worth is now reduced to

$$M(\theta) = \frac{E_{A0}}{(C_A + \delta C \sin^2 \theta)} + \frac{E_{B0}}{(C_B + \delta C \cos^2 \theta)} \qquad (9.3)$$

If $M(\theta)$ is to be maximal, then differentiating Eq. (9.3) with respect to θ and setting the derivative equal to zero yield

$$\frac{\partial M}{\partial \theta} = 0 = \sin \theta \cos \theta \left[\frac{E_{A0}}{(C_A + \delta C \sin^2 \theta)^2} - \frac{E_{B0}}{(C_B + \delta C \cos^2 \theta)^2} \right] \qquad (9.4)$$

Hence θ takes on one of three values, θ_1, θ_2, or θ_3, with either

$$\theta_1 = 0 \qquad \text{(all costs are placed on project } B\text{)} \qquad (9.5a)$$

$$\theta_2 = \frac{\pi}{2} \qquad \text{(all costs are placed on project } A\text{)} \qquad (9.5b)$$

or

$$\delta C \sin^2 \theta_3 = \frac{(C_B + \delta C)(E_{A0}/E_{B0})^{1/2} - C_A}{1 + (E_{A0}/E_{B0})^{1/2}} \qquad (9.5c)$$

Now if $\theta_1 = 0$ is used, then the value of $M(\theta)$ is

$$M(\theta = 0) = \frac{E_A}{C_A} + \frac{E_B}{C_B + \delta C} \qquad (9.6)$$

whereas if $\theta_2 = \pi/2$ is used, then

$$M\left(\theta = \frac{\pi}{2}\right) = \frac{E_A}{C_A + \delta C} + \frac{E_B}{C_B} \qquad (9.7)$$

whereas if θ_3 is used, then

$$M(\theta = \theta_3) = (C_B + \delta C \cos^2 \theta_3)^{-1} (E_A E_B)^{1/2} \left[1 + \left(\frac{E_A}{E_B}\right)^{1/2} \right] \qquad (9.8)$$

However, $\sin^2 \theta_3$ from Eq. (9.5c) can be in the range $0 \le \sin^2 \theta_3 \le 1$ only when the following two conditions are *both* satisfied:

$$\text{(i)} \quad (C_B + \delta C) E_A^{1/2} > C_A E_B^{1/2} \qquad (\sin^2 \theta_3 > 0) \qquad (9.9a)$$

and

$$\text{(ii)} \quad \delta C > C_B \left(\frac{E_A}{E_B} \right)^{1/2} - C_A \qquad (\sin^2 \theta_3 < 1) \qquad (9.9b)$$

Suppose first that at least either Eq. (9.9a) or Eq. (9.9b) is *not* satisfied. Then either $\theta_1 = 0$ or $\theta_2 = \pi/2$ are the only choices. In this situation one has $M(\theta = 0) > M(\theta = \pi/2)$ whenever $E_A/C_A > E_B/C_B$, and conversely, $M(\theta = 0) < M(\theta = \pi/2)$ whenever $E_B/C_B > E_A/C_A$. In either of these cases, one assigns all the extra costs to project A (whenever $E_A/C_A > E_B/C_B$) or project B (whenever $E_B/C_B > E_A/C_A$).

Suppose, however, that both of Eqs. (9.9a) and (9.9b) *are* satisfied. Then one also could have Eq. (9.8) as an optimal value. In this case, suppose that $M(\theta = 0) > M(\theta = \pi/2)$; then one only has to check whether $M(\theta = \theta_3)$ exceeds $M(\theta = 0)$ to determine which apportionment of costs is the best.

There is also a second sort of optimization possible for cost apportionment. Often taxes and royalty charges are assessed to a corporation based on the amount earned per dollar spent. Clearly, in such a case, the aim is to apportion extra costs to *minimize* $M(\theta)$. Fortunately, all one has to do in this case is insert θ_1, θ_2, or θ_3 into Eq. (9.3) to see which provides the *smallest* value of $M(\theta)$.

9.3   NUMERICAL ILLUSTRATIONS

9.3.1   Identical Projects

As a simple first illustration, consider the case of two identical environmental projects with $E_{A0} = E_{B0}$ and $C_A = C_B$. Write the cost overrun δC as fC_B, where the dimensionless number f is in the range $0 < f < \infty$. In this case, Eq. (9.5c) yields $\sin^2 \theta_3 = 1/2$ (i.e., $\theta_3 = \pi/4$), indicating that cost overruns should be equally split between both projects. Evaluation of $M(\theta)$ at $\theta = 0$, $\pi/2$, and $\theta_3 = \pi/4$ yields the result

$$M(\theta = 0) = M\left(\theta = \frac{\pi}{2} \right) = \frac{E_{B0}}{C_B} \frac{2 + f}{1 + f} \qquad (9.10a)$$

while

$$M\left(\theta_3 = \frac{\pi}{4} \right) = \frac{E_{B0}}{C_B} \frac{4}{2 + f} \qquad (9.10b)$$

It is simple to show from Eq. (9.10) that $M(\theta_3 = \pi/4)$ is always less than $M(\theta = 0)$. Thus, in this situation, the result is that one should *only* apportion the cost overruns equally to the identical projects if one is interested in minimizing $M(\theta)$; if one wishes to *maximize* $M(\theta)$, then cost overruns should be assigned 100% to one of the two projects. Depending on the corporate aim (highest profit measure or lowest profit measure to avoid taxation), cost overruns will be assigned very differ-ently to the two projects.

9.3.2 Unequal Projects

As a second simple illustration, consider the situation where $E_{A0} = 2E_{B0}$ and $C_A = 2C_B$ so that the ratios E_{A0}/C_A and E_{B0}/C_B are identical, just as they were in the preceding example. But the introduction of the cost overrun $\delta C (\equiv fC_B)$ breaks this symmetry because δC cannot, simultaneously, be the *same* fixed fraction of C_B and C_A because $C_A = 2C_B$.

In this example, direct evaluation of $M(\theta)$ at $\theta = 0$, $\pi/2$, and θ_3 yields

$$M(\theta = 0) = \frac{E_{B0}}{C_B} \frac{2+f}{1+f} \qquad (9.11a)$$

$$M\left(\theta = \frac{\pi}{2}\right) = \frac{E_{B0}}{C_B} \frac{4+f}{2+f} \qquad (9.11b)$$

and

$$M(\theta = \theta_3) = \frac{E_{B0}}{C_B} \frac{(3+2^{3/2})\,2^{-1/2}}{3+f} \qquad (9.11c)$$

Here, θ_3 exists provided only that

$$f > 2^{1/2} - 1 \qquad (9.11d)$$

and is given through

$$\sin^2 \theta_3 = 2^{1/2}(2^{1/2} - 1)\,[1 - f^{-1}\,(2^{1/2} - 1)] \qquad (9.11e)$$

For low cost overruns, so that $0 < f < 2^{1/2} - 1$, only Eqs. (9.11a) and (9.11b) provide accessible measures of worth. In such a situation, $M(\theta = \pi/2)$ always exceeds $M(\theta = 0)$ for all f. For $f > 2^{1/2} - 1 \cong 0.41$, it can be shown that $M(\theta = \theta_3)$ never exceeds $M(\theta = \pi/2)$. In fact, $M(\theta = \theta_3)$ is always less than $M(\theta = 0)$ for $f > 2^{1/2} - 1$. Thus, in this case, if the corporate need is to maximize the measure of worth, then $M(\theta = \pi/2)$ should be chosen so that all cost overruns are assigned to project A. However, if the corporate aim is to minimize apparent profit, then $M(\theta = 0)$ should be chosen for low cost overruns ($f < 2^{1/2} - 1$) so that all cost overruns are assigned to project B, whereas for high cost overruns ($f > 2^{1/2} - 1$), one should choose $M(\theta = \theta_3)$ in order to minimize apparent measures of successful operation.

In this case one sees that the cost overrun apportionment between the two projects is highly dependent on how much the cost overrun is relative to the initial estimated costs of project B. For less than about 41% [$(2^{1/2} - 1) \times 100\%$] overrun, the costs are attributed 100% either to project A or to project B depending on corporate need to minimize or maximize apparent profitability, respectively. But for higher ($\geq 41\%$) cost overruns, it is more advantageous to partition the extra costs between projects A and B whenever the corporation is interested in minimizing the apparent success ratios of the two projects.

9.4 ARBITRARY NUMBERS OF PROJECTS

If N projects are undertaken, with the jth such project having total costs C_j and an expected value in the absence of all costs of E_j, then cost overruns of δC are apportioned fractionally, with the jth project picking up the fraction $f_j \delta C$ of the total. In this case, the corresponding measure of worth M_N is

$$M_N = \sum_{j=1}^{N} \frac{E_j}{C_j + f_j \delta C} \qquad (9.12)$$

subject to the constraints $f_j \geq 0$ for all j and that the fractional apportionments sum to unity, that is,

$$\sum_{j=1}^{N} f_j = 1 \qquad (9.13)$$

One can use the constraint (9.13) as follows: Rewrite Eq. (9.13) as

$$f_k = -f_j + \left(1 - \sum_{\substack{m=1 \\ m \neq k,j}}^{N} f_m \right) \qquad (9.14)$$

where in the summation in Eq. (9.14) the terms $m = j$ and $m = k$ are omitted.

To find the extrema of Eq. (9.12), one differentiates with respect to f_j and sets $\partial M_N / \partial f_j$ equal to zero. Substituting the constraint (9.14) in Eq. (9.12), only the terms involving f_j and f_k are present in $\partial M_N / \partial f_j$. Thus at the extremal points of M_N one obtains

$$\frac{\partial M_N}{\partial f_j} = \frac{-E_j \delta C}{(C_j + f_j \delta C)^2} + \frac{E_k \delta C}{(C_k + f_k \delta C)^2} \equiv 0 \qquad (9.15)$$

One can rewrite Eq. (9.15) in the form

$$\frac{E_j}{(C_j + \delta C f_j)^2} = \frac{E_k}{(C_k + \delta C f_k)^2} = H^2 \qquad (9.16)$$

But j and k were arbitrary choices. The only way Eq. (9.16) can be satisfied for any and all choices of j and k is if both sides are a positive constant, written as H^2 in Eq. (9.16). It then follows that

$$f_k = \frac{(E_k^{1/2}/H) - C_k}{\delta C} \tag{9.17}$$

If f_k is to be positive, then

$$H \leq \frac{E_k^{1/2}}{C_k} \equiv \alpha_k$$

If f_k is to be less than unity, then

$$H \geq \frac{E_k^{1/2}}{C_k + \delta C} \equiv \beta_k$$

Suppose that H satisfies these requirements for the N projects, that is, $a_k \geq H \geq \beta_k$, $k = 1,\ldots, N$. Then insert Eq. (9.17) into the constraint Eq. (9.13) and so determine H directly from

$$H = \frac{\displaystyle\sum_{j=1}^{N} E_j^{1/2}}{\displaystyle\sum_{j=1}^{N} C_j + \delta C} \tag{9.18}$$

Equation (9.18) specifies H in terms of the parameters of each project and of the cost overrun δC. Alternatively, one can regard H as given and then write the cost overrun as

$$\delta C = H^{-1}\left(\sum_{i=1}^{N} E_i^{1/2} - H \sum_{i=1}^{N} C_i \right) \tag{9.19}$$

Note that either of Eqs. (9.18) and (9.19) imply that the absolute upper limit H_{max} of H is given through

$$H_{\text{max}} = \frac{\displaystyle\sum_{j=1}^{N} E_j^{1/2}}{\displaystyle\sum_{j=1}^{N} C_j} \tag{9.20}$$

whereas the lower limit is $H \to 0$, corresponding to $\delta C \to \infty$. Thus $0 \leq H \leq H_{\text{max}}$. The one assumption in this derivation is that *all* the N projects satisfy $a_k \geq H \geq$

β_k, $k = 1,\ldots, N$. We investigate this point next. Organize the N projects in order of increasing α_k ($\equiv E_k^{1/2}/C_k$) values, with $\alpha_1 \le \alpha_2 \le \cdots \le \alpha_N$.

Then suppose that a value of H is chosen with $\alpha_p \ge H \ge \alpha_{p-1}$. Then none of the projects from $j = 1$ to $j = p - 1$ can be made to bear any fraction of the cost overrun δC, only those from $j = p$ to $j = N$. Hence the general procedure is clear. Calculate a sequence of values

$$H_p = \frac{\displaystyle\sum_{j=p}^{N} E_j^{1/2}}{\displaystyle\sum_{j=p}^{N} C_j + \delta C} \qquad p = 1,\ldots, N \qquad (9.21)$$

for specific project parameters and a given cost overrun δC. For each H_p, check where it lies within the sequence $\alpha_p \ge H_p \ge \alpha_{p-1}$. Then check that $H_p \ge \beta_p p$. Hence one determines that value of p that allows partitioning of the total cost overrun. Such linear sorting problems are most easily accomplished with a spreadsheet program on a personal computer. The principles, however, remain the same as for the simple example of the two-project situation.

The development in this section followed the solution technique of two well-known approaches in classic optimization: the substitution method and lagrangian analysis (for an introduction to optimization, see any text in operations research, e.g., Hillier and Lieberman, 1980). Furthermore, the global optimality condition of Eq. (9.16) is known in economics as the *equimarginal principle* and underlies the optimal allocation of a scarce resource among competing demands.

9.5 SUMMARY

The problem addressed in this chapter is one often faced by corporations involved in environmental projects. Often one does not know the exact conditions of, say, an environmental storage system prior to undertaking construction of the system and also later storage of the waste material. However, one provides a bid for the project that is based on the best estimates that one could make. As site construction proceeds, it becomes clear that costs will rise significantly relative to the original estimates made, perhaps because of unanticipated formation permeability in the case of underground storage. If the contract is a fixed-price offer with no chance of renegotiation, then the company must decide whether to default on the project (and so be required, often, to declare bankruptcy) or to absorb the extra overrun costs within its ongoing projects. Clearly, by apportioning costs, the net gain-to-cost ratio for each project is affected. The total balance of effective gains to losses can be either increased or decreased by judicious choice of the fraction of the cost overruns one assigns to each project. A corporate position is to lower the effective gain-to-cost ratio for projects such that the sum for *all* projects is minimized; in this way, tax liability can be reduced. An alternate corporate position is to apportion cost overruns in order to maximize the gain-to-cost ratio for all projects when summed over each project. The two strategies for apportioning cost overruns lead to very different apportionment fractions, as shown by the numerical illustrations with two projects.

Precisely how a particular corporation acts and reacts to the event of cost overruns depends on its long-term aims of, say, tax avoidance versus profit enhancement. There is considerable merit to either strategy, perhaps accounting for the different ways individual corporations have operated.

This chapter has attempted to provide a procedure for evaluating rationally objective measures of gains to costs including apportionment of cost overruns. The simple numerical examples were designed to illustrate, as sharply as possible, these points.

CHAPTER 10
BAYESIAN UPDATING OF TOXIC LEAKAGE SCENARIOS

10.1 INTRODUCTION

One of the major unknowns in storage of toxic waste is the leakage potential after burial, presumably a fundamental reason that ongoing monitoring of stored canisters is carried out. Yet, even when a monitor indicates leakage occurrence, there arises the problem of what remedial action to take. Perhaps a simple illustration is of value. Prior to burial, take it that many different scenarios of possible causes of leakage have been investigated. The aim of providing the leakage probabilities for such potential causes is not only to determine the best way to package toxic waste against probable leakage conditions but also to minimize further detrimental conditions after burial that would contribute to the leakage chance.

Suppose, however, that a leak is found. The question arises as to whether the unequivocal cause of the leakage can be ascertained so that the appropriate remedial action can be taken. Clearly, any ambiguity in the determination of the cause of the leak can lead to the wrong remedial action being taken, which, in turn, could lead to the hazardous situation not only not being fixed but even being exacerbated. Thus leakage scenarios and their probabilities of being correct need to be updated constantly as leaks are discovered. This chapter shows, initially, with a simple example, how such updating can be done effectively. Later sections develop these concepts in a more advanced fashion.

10.2 THEORETICAL CONSIDERATIONS

The difficulty one faces stems from the actual burial undertaken: Either the burial is successful in having no leakage or it is not. In the case of a successful situation, a postburial review is undertaken to compare the actual parameters with those predicted; in the case of an unsuccessful situation, a postburial review is undertaken to determine the causes of the leakage. In either event, new information is available that is used to update the risk assessment of the situation in relation to decisions to (1) continue with the prior strategy of evaluation, (2) modify the strategy

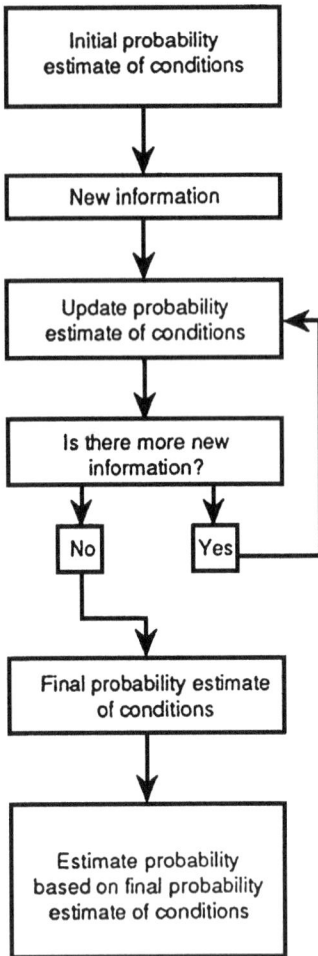

FIGURE 10.1 Flow diagram for estimating probability based on bayesian updating of a priori estimates using additional information.

based on the information uncovered, or (3) perform remedial action on the burial site. Thus the whole process of risk analysis for the burial site is reevaluated based on the monitor information. Based on the later information acquired, the updating changes prior assessments of probabilities and so has an impact on all the geologic, toxic production, and economic assessments. This sort of updating is an ongoing process as a site is prepared, developed, and used for burial (Fig. 10.1). At all stages of investigation of an opportunity, the updating continues. This use of a posteriori information to update prior assessments is usually referred to as *bayesian updating,* after Bayes, who formulated the basic concepts [see Jaynes (1978) for an historical account of development of the method]. The purpose of this chapter is to provide an introductory exposition of the bayesian decision theory; a full discussion of this theory is beyond the scope of this text and can be found in several textbooks (e.g., Winkler, 1972; Lindgren, 1974; Berger, 1985) and articles (see Dukstein and Kisiel, 1971; Davis et al., 1972; Dukstein et al., 1978; Bogardi et al., 1982; Krzysztofowicz, 1983a, 1983b, 1986) that treat this subject extensively.

10.3 GENERAL CONCEPTS OF BAYESIAN UPDATING

The sense of the main idea is best expressed through an example. Suppose that two radically different geologic scenarios have been proposed for leakage based on the available, incomplete data on hand; label the scenarios A and B. One might, for instance, consider that one scenario is a corrosion-driven leakage depending on prevailing subterranean water flow washing against stored containers, whereas the second scenario might consider that thermal expansion and associated fracturing of containers is prevalent due, perhaps, to ongoing chemical interactions in the stored containers.

On the basis of each scenario, one performs a *complete* risk analysis, thereby generating probabilities of leakage $p(L|A)$ and $p(L|B)$, which are conditional on each of scenarios A and B being assumed valid, respectively. One also assigns a probability of each scenario being correct $p(A)$ and $p(B)$, respectively, with the sum being unity $[p(A) + p(B) = 1]$ when no other scenario is possible. Then the probability of leakage $p(L)$, *irrespective of which scenario is correct,* is just

$$p(L) = p(L|A)p(A) + p(L|B)p(B) \tag{10.1}$$

Now suppose that a leak actually *does* occur. One wishes to update the probability of each scenario being correct based on that a posteriori information; i.e., one wishes to calculate $p(A|L)$, the probability of scenario A being correct given that event L occurs, and, correspondingly, $p(B|L)$. Bayes' theorem indicates that

$$p(A|L)p(L) = p(L|A)p(A) \tag{10.2}$$

so that the updated probability of scenario A being correct given that L has occurred is

$$p(A|L) = \frac{p(L|A)p(A)}{p(L|A)p(A) + p(L|B)p(B)} \tag{10.3}$$

With a set of possible scenarios A_i $(i = 1,..., N)$, the generalization of Eq. (10.3) is

$$p(A_i|L) = \frac{p(L|A_i)p(A_i)}{\sum\limits_{j=1}^{N} p(L|A_j)p(A_j)} \tag{10.4}$$

The general argument does not have to be tailored only to leakage probability but is also appropriate for any sort of event L. Thus the argument could be based on the probability of cementation occurring in a formation, on the probability of finding fractures rather than none, or on any other event that is then measured. In each case one starts with a set of different "states" A_i $(i = 1,..., N)$ and assigns the initial probability $p(A_i)$ of the state A_i being the correct one; then one computes the risk assessment of an event L occurring within the framework of state A_i, and then one uses later observations of the occurrence (or nonoccurrence) of event L to update the initial probability of state A_i being correct. The updated probabilities can then be used as initial probabilities and the process repeated, with new a posteriori information being added sequentially to update continuously the state probabilities (see Fig. 10.1). Thus continuous updating of risk assessments of a situation can be made throughout the life of the project—and the life is itself then determined by the economic worth of continuing to develop the opportunity based on all information used in a bayesian update manner.

10.4 NUMERICAL ILLUSTRATION

The ability to use the bayesian method to update probabilities of event occurrence based on new information is one of the most powerful tools available and operates in many settings, not just in the field of toxic burial. Here, an illustration of the procedure is given to provide some familiarity of the many possible uses.

10.4.1 Updating the Likelihood of a State Being Correct

Suppose that, prior to leakage being monitored, there are only two scenarios considered for leakage: corrosion due to water flow and thermal expansion and consequent pitting and fracturing of containers due to chemical interactions in the containers. Denote the two states by A_1 = corrosion and A_2 = thermal expansion. Ahead of leakage being found, take it that, after considerable discussion, probabilities are assigned to the thermal expansion scenario being correct of $p(\text{th})$ = 30%, and correspondingly, $p(\text{corr})$ = 70%. Also take the probability of leakage, if the thermal scenario is correct, as $p(L|\text{th})$ = 70% and $p(L|\text{corr})$ = 90% if the corrosion scenario is correct.

Now suppose that monitoring of buried toxic waste takes place with the result that in the first year of monitoring a leak is registered, in the second year another leak is found, and after a further 3 years a third leak is uncovered by the monitors. At this stage, sufficient concern is raised that the authorities decide to undertake some remedial action. If leakage is due to thermal expansion, then every container must be disinterred, made safe, and reinterred. However, if leakage is due to subterranean water corrosion, then the whole burial site would have to be cleaned and then abandoned, a new site prepared, and also all containers again made safe. Thus the difference in cost and effort of the two remedial actions can be considerable. Clearly, the authorities would like to have a better idea than the a priori estimates of which scenario is now appropriate based on the three leaks observed. The bayesian updating method provides the requisite procedure and, for the three observed leaks, operates as follows.

After Leak 1. Following Eq. (10.3), one can write the probability of the thermal expansion scenario being correct, given that leak L_1 occurred, as

$$p(\text{th}|L_1) = \frac{0.3 \times 0.7}{0.3 \times 0.7 + 0.7 \times 0.9} = 0.25 \qquad (10.5a)$$

and correspondingly, the corrosion scenario probability of being correct is now updated to

$$p(\text{corr}|L_1) = 0.75 \qquad (10.5b)$$

After Leak 2. The updated probabilities from Eq. (10.5) are now used to replace the a priori 30% chance of $p(\text{th})$ and the 70% chance of $p(\text{corr})$ by 25% and 75%, respectively. After the second leak occurs, these 0.25 and 0.75 probabilities are further updated to

$$p(\text{th}|L_2) = \frac{0.25 \times 0.7}{0.25 \times 0.7 + 0.75 \times 0.9} = 0.21 \qquad (10.6a)$$

with, correspondingly,

$$p(\text{corr}|L_2) = 0.79 \qquad (10.6b)$$

After Leak 3. Once more updating, we find that after the third leak has occurred, the updated probabilities are

$$p(\text{th}|L_3) = \frac{0.21 \times 0.7}{0.21 \times 0.7 + 0.79 \times 0.9} = 0.17 \qquad (10.7a)$$

and

$$p(\text{corr}|L_3) = 0.83 \qquad (10.7b)$$

Thus there is an 83% chance that leakage is due to corrosion and only a 17% chance that it is due to thermal expansion. These estimates have increased the original estimate (70%) of leakage due to corrosion to 83% while simultaneously almost halving (from 30 to 17%) the probability that the thermal expansion leakage scenario is correct.

The point to make is that the actual occurrence of events can be used to reevaluate the a priori estimates of conditions.

10.4.2 Updating the Probability of Leakage

Originally, the probability of a leakage $p(L)$ was made up of two estimates, $p(L|\text{th})p(\text{th})$ and $p(L|\text{corr})p(\text{corr})$, which contributed 0.21 and 0.63, respectively, to $p(L)$ for a total leakage probability of 0.84. After the first leak occurs, these two contributions are updated to 0.175 and 0.675 for thermal expansion and corrosion, respectively, leading to an updated $p(L) = 0.850$. After the second leak, the corresponding leakage contributions for thermal expansion and corrosion scenarios are 0.147 and 0.711, respectively, for a total of $p(L) = 0.853$, whereas after the third leak, one has the corresponding values 0.119 and 0.747 for a total $p(L) = 0.876$. Thus information from the three leaks leads not only to an increase in the chance that corrosion due to external water contact against the containers as the most likely culprit (83% chance), but also to an overall increase in the probability of container leakage $p(L)$ *irrespective* of which mechanism is responsible, from an initial probability estimate of 0.84 for a leak to 0.876 after the third leak is registered.

If one had buried $N + 3$ containers initially, then the probability that k of the remaining N will leak is

$$p_N(k) = \frac{N!}{k!(N - k)!}\, p(L)^k\, [1 - p(L)]^{N-k} \tag{10.8}$$

The mean value of k is

$$E(k) = p(L)N \tag{10.9}$$

and the variance $\sigma(k)^2$ is given by

$$\sigma(k)^2 = Np(L)[1 - p(L)] \tag{10.10}$$

The probability that *none* out of the remaining N will leak is

$$P(0) = [1 - p(L)]^N \tag{10.11}$$

so the probability that one or more *will* leak is

$$P(1) = 1 - P(0) = 1 - [1 - p(L)]^N \tag{10.12}$$

Likewise, the probability of two or more containers leaking is

$$P(2) = P(1) - \frac{N!}{1!(n - 1)!}\, p(L)[1 - p(L)]^{N-1} \tag{10.13a}$$

and so the probability of $(k + 1)$ or more containers leaking can be written

$$P(k + 1) = P(k) - \frac{N!}{k!(N - k)!}\, p(L)^k[1 - p(L)]^{N-k} \tag{10.13b}$$

For instance, if 10 containers were interred initially and 3 have leaked so that $p(L)$ has been updated to $p(L) = 0.876$, then of the remaining 7 ($= N$) containers, the probability that none will leak is

$$P(0) = (1 - 0.876)^7 = 0.124^7 = 7 \times 10^{-7}$$

which is so small that the authorities should be seriously concerned about further leaks of the remaining 7 containers.

10.5 BAYESIAN DECISION CRITERIA

10.5.1 Optimal Expected Value Bayes Decision

The objective of this section is to provide an exposition of some more advanced concepts of the bayesian decision theory and to illustrate the use of this theory in risk analyses of environmental projects. This is accomplished with the presentation of a simple example, which has the advantage of providing closed-form solutions that can be reproduced easily by the reader. The procedure in this section follows closely the development by Duckstein et al. (1978) for the case of drainage or not of a plot of land. In their article, Duckstein et al. (1978) have provided a simple, self-contained guide that can be used to analyze hydrologic and economic risk, the worth of additional information, and the sensitivity of the decision making in scientific and economic assumptions. These authors' procedure is applied here in the case of monitoring at a waste disposal site.

Consider again the problem of toxic (or radioactive) waste disposal. In addition to the monitoring network that is mandated by federal or other requirements to operate at a repository site, a decision maker may be considering whether or not to install additional monitoring devices at an annual cost C. Such a situation may arise when components of the repository, which under normal conditions would not have been monitored, are governed by enough geologic uncertainty and present a potential for contamination that consideration is given to extra monitoring of some regions. Cost C represents the total annual cost of the additional monitoring devices and includes the cost of installation and maintenance over T years. Here, uncertainties related to the amortization factor are not considered. The time horizon T corresponds to the lifetime of the additional monitoring network, and the decisions that can be made are $A(0)$ corresponding to the action "no additional monitoring system" and $A(1)$ corresponding to the action "build additional monitoring system." Designate by E an extreme contamination leak event that would not have been detected without the presence of the additional monitoring devices. The remediation cost of a contamination leak is projected to be B. Define the binary random variable by

$$X(i) = \begin{cases} 0 & \text{if } E \text{ does not occur} \\ 1 & \text{if } E \text{ occurs} \end{cases} \tag{10.14}$$

Variable $X(i)$ describes the occurrence or not of a contamination leak in any given time period $i = 1, 2, 3, ..., T$ (years) that can only be detected by the additional monitoring devices. Therefore, the total number N of contamination leak events in the T years of operation of the monitoring system is given by

$$N = \sum_{i=1}^{T} X(i) \tag{10.15}$$

Thus the random variable N fully describes the incidents of contamination leaks in T years that are detected by the supplementary monitoring network, and it is the analysis of this variable that will be performed subsequently.

Assuming that only one contamination leak per year gets detected by the additional monitoring network, the probability of an event E is given by

$$P(X = 1) = p \qquad (10.16)$$

and the probability that no leak occurs by

$$P(X = 0) = 1 - p \qquad (10.17)$$

Here, for reasons of simplicity, it is assumed that the proposed additional system would result in early detection of all contamination leaks that get undetected by the regular monitoring system. Because the variable $X(i)$ is a Bernoulli variable, N in Eq. (10.15) is binomial (Benjamin and Cornell, 1970), and the probability that j contamination leak events will occur in T years of operation of the monitoring network is given by

$$P(N = j) = \binom{T}{j} p^j (1 - p)^{T-j} \qquad (10.18)$$

Here, j takes the values $j = 0, 1, 2,\ldots, T$, and the combination of j out of T depicted at the right-hand side of Eq. (10.18) equals $T!/[\,j!(T - j)!]$ with $j! = 1 \times 2 \times \cdots \times j$ and $0! = 1$.

The financial losses incurred as a result of the different actions or decisions made are encoded in a function that is designated as the loss or objective function. For this study, if action $A(1)$ to build the monitoring system is followed, the financial loss over T years is given by the value of $C \times T$. The values of the loss function can be tabulated in a table of discrete losses such as Table 10.1. These losses clearly depend both on the action followed (i.e., on the action space A) and on the number of contamination events that may occur (i.e., on variable N).

Alternatively, the loss function of this problem can be written in a compact form as

$$L(A, N) = \begin{cases} BN & \text{if } A = A(0) \\ CT & \text{if } A = A(1) \end{cases} \qquad (10.19)$$

The loss function in Eq. (10.19) or Table 10.1 has been assumed to vary linearly with N. Indeed, such dependency will depend on the particular problem at hand. In a later section of this chapter a quadratic form of the loss function is investigated, and comparisons are drawn between the two loss-function models.

The expected value of the loss function with respect to the variable N is termed in the bayesian decision theory as the *goal* or *risk function* (Raiffa and Schlaifer,

TABLE 10.1   Linear Loss Function for Monitoring System Project

Number of contamination events in T years	$A(0)$: Do not build monitoring system	$A(1)$: Build monitoring system
0	0	$C \times T$
1	$1 \times B$	$C \times T$
2	$2 \times B$	$C \times T$
$\vdots$	$\vdots$	$\vdots$
T	$T \times B$	$C \times T$

1961; Duckstein et al., 1978; Berger, 1985). For action $A(0)$, the goal function is given by

$$G[A(0), p] = E^N\{L[A(0), N]\} = E^N(BN) = BTp \qquad (10.20a)$$

where in the last equality the mean value of a binomial variable N, $E(N) = Tp$, was used [see also Eq. (10.9)]. For action $A(1)$, the goal function is given by

$$G[A(1), p] = E^N\{L[A(1), N]\} = E^N(CT) = CT \qquad (10.20b)$$

Equations (10.20a) and (10.20b) indicate what one risks to lose, on average, if actions $A(0)$ and $A(1)$ are taken, respectively. In a standard benefit-risk analysis one would proceed then to evaluate which of the two decisions minimizes the goal function or even consider other measures of uncertainty, as was done in previous chapters. This analysis can be performed easily in this case with the use of Fig. 10.2. Here, the goal functions that correspond to actions $A(0)$ and $A(1)$ are plotted on the left and right y axes, respectively, whereas the probability p is plotted on the x axis. Figure 10.2 indicates (heavy solid line segments) the regions where the optimal action minimizes the goal function. To find the probability value where a change in a decision is made, it suffices to set the two goal functions equal to each other:

$$G[A(0), p] = G[A(1), p] = BTp = CT$$

$$\Rightarrow p = \frac{C}{B} \qquad (10.20c)$$

Thus the ratio C/B provides the break point where a change from the decision $A(0)$, "no additional monitoring system," to $A(1)$, "build additional monitoring system," is recommended.

The particular application of a bayesian analysis in this problem is the recognition that the probabilities p and $(1 - p)$ are themselves uncertain, and hence they need to be considered as random variables. We proceed to discuss the estimation of a subjective prior probability density function (pdf) of the variable p (Duckstein et al., 1978).

It is possible that based on past experience from similar sites, the operator of the repository facility may be able to provide a rough estimate on how often he or

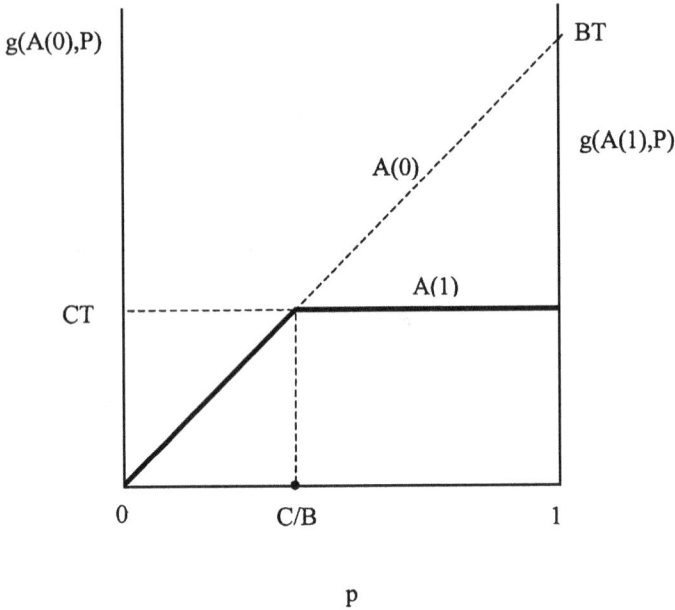

FIGURE 10.2 Optimal decision for monitoring problem.

she expects a contamination leak to occur. Assume that the answer is that he or she expects one serious leak every 4 years. Based on this information, one can construct the prior pdf of p through use of the conjugate prior distribution that for the case of the binomial distribution studied here is the Beta distribution (Benjamin and Cornell, 1970)

$$f_p(p) = \text{Be}(t, r, p) = \frac{(t-1)!}{(r-1)!\,(t-r-1)!}\, p^{r-1}(1-p)^{t-r-1} \quad (10.21)$$

with mean and variance given by

$$m = \frac{r}{t} \quad \text{and} \quad \sigma^2 = \frac{r(t-r)}{t^2(t+1)} \quad (10.22)$$

Discussion of the use of conjugate functions for the development of prior pdf is beyond the scope of this text and can be found in several standard textbooks dealing with the bayesian decision theory (Raiffa and Schlaifer, 1961; Berger, 1985). Using the numbers $r = 1$ and $t = 4$ provided by the operator of the site allows one to evaluate the prior distribution of the probability p as

$$f_p(p) = \frac{3!}{0!2!}\, p^0(1-p)^2 = 3(1-p)^2 \quad (10.23)$$

The Bayes risk R is defined (Duckstein et al., 1978; Berger, 1985) as the expected value of the goal function in Eqs. (10.20a) and (10.20b) with respect to the unknown parameter p. The Bayes risk is then given by

$$R(A) = \begin{cases} R[A(0)] = BT \int_0^1 3p(1-p)^2 \, dp = BT/4 \\ R[A(1)] = CT \end{cases} \tag{10.24}$$

$R[A(0)]$ in Eq. (10.24) can be evaluated by integration or simply by noticing that $R[A(0)] = E^{f(p)}(BTp) = BTE^{f(p)}(p) = BT/4$ by virtue of $m = 1/4$. The optimal Bayes decision is obtained by the action that minimizes the Bayes risk function. For example, if $B = 40$ and $C = 8$, then $A(0)$ returns the value of $R[A(0)] = 10T$, whereas $A(1)$ returns the value of $R[A(1)] = 8T$. Hence, for these values, $A(1)$ should be selected, and additional monitoring devices should be installed. Because the risk function involves a linear and a constant part, respectively, construction of a graph similar to Fig. 10.2 can be done easily, and then the optimal regions can be defined according to

$$\text{If} \quad BT < 4CT, \quad \text{then } A^* = A(0)$$
$$\text{If} \quad BT > 4CT, \quad \text{then } A^* = A(1) \tag{10.25}$$

Here, A^* designates the optimal action for each particular case.

10.5.2   Expected Opportunity Loss

Table 10.1 or Eq. (10.19) depicting the loss function of the problem can be presented as shown in Table 10.2. The first column of the table indicates the two possible states of nature (i.e., the two possible situations that may occur in terms of costs), and the two other columns tabulate the losses incurred under these two possible scenarios. It is straightforward from this table to construct the regret table as shown in Table 10.3.

Table 10.3 tabulates the extra loss that would have been incurred if a particular action were followed given that one had perfect knowledge of the true state of nature. Thus, if it were known that the remediation costs were less than the cost of the additional monitoring system ($BN < CT$), no opportunity would have been lost by following action $A(0)$ (i.e., no regret), but on the other hand, if $A(1)$ was followed, an extra amount ($CT - BN$), not required by the situation, would have been used. Similar arguments apply for the case where the cost $BN > CT$.

One should notice here that N entering Table 10.3 is a random variable, and hence a more appropriate measure of the regret is through the expected value of the loss functions, defined as the goal functions in Eqs. (10.20a) and (10.20b). Using these two equations, one can easily construct the opportunity loss (OL) table as follows shown in Table 10.4.

TABLE 10.2   Loss Table for Two States of Nature

State of nature	$A(0)$: Do not build monitoring system	$A(1)$: Build monitoring system
S1: $BN < CT$	BN	CT
S2: $BN > CT$	BN	CT

TABLE 10.3   Regret for Monitoring Problem

State of nature	$A(0)$: Do not build monitoring network	$A(1)$: Build monitoring network
S1: $BN < CT$	0	$CT - BN$
S2: $BN > CT$	$BN - CT$	0

The opportunity loss provides a measure of the average loss incurred by following a certain action because of imperfect information on the true state of nature. Alternatively, it provides a numerical value of the worth of obtaining more information on the probability of a contamination leak event p.

The values in Table 10.4 can be presented in closed form according to

$$OL[A(0), p] = \begin{cases} 0 & \text{if } 0 \le p \le C/B \\ (Bp - C)T & \text{if } C/B \le p \le 1 \end{cases} \tag{10.26a}$$

$$OL[A(1), p] = \begin{cases} (C - Bp)T & \text{if } 0 \le p \le C/B \\ 0 & \text{if } C/B \le p \le 1 \end{cases} \tag{10.26b}$$

The expected opportunity loss (EOL) can now be evaluated by calculating the expected values of the opportunity losses given in Eqs. (10.26a) and (10.26b) with respect to the prior distribution of the probability of a contamination leak in Eq. (10.23).

$$EOL[A(0)] = 3 \int_{C/B}^{1} (Bp - C)T(1 - p)^2 \, dp$$

$$= 3T \left(\frac{B}{12} - \frac{C}{3} + \frac{C^2}{2B} - \frac{C^3}{3B^2} + \frac{C^4}{12B^3} \right) \tag{10.27a}$$

$$EOL[A(1)] = 3 \int_{0}^{C/B} (C - Bp)T(1 - p)^2 \, dp = 3T \left(\frac{C^2}{2B} - \frac{C^3}{3B^2} + \frac{C^4}{12B^3} \right) \tag{10.27b}$$

TABLE 10.4 Opportunity Loss for Monitoring Problem

State of nature	$A(0)$: Do not build monitoring network	$A(1)$: Build monitoring network
S1: $Bp < C$	0	$(C - Bp)T$
S2: $Bp > C$	$(Bp - C)T$	0

The values of $B = 40$ and $C = 8$ result in

$$\text{EOL}[A(0)] = 4.096T \quad \text{and} \quad \text{EOL}[A(1)] = 2.096T \quad (10.28)$$

Therefore, for this problem and values of the parameters B and C, the worth of obtaining more information about the probability of leak p is $2.096T$, or in other words, this is the amount that on average one would be willing to pay to define the contamination potential. Comparison of the optimal $\text{EOL}[A(1)] = 2.096T$ with the Bayes risk function $R[A(1)] = 8T$ indicates that the expected opportunity loss represents 26.2% of the optimal Bayes risk. For the particular values of the parameters B and C chosen here, the cost of uncertainty appears to be a significant component of the total project costs.

10.5.3 Influence of the Length of the Observation Record on Decision Making

Assume now that the operator's answer is that he or she would expect two contamination leak events in 8 years, or alternatively, in 4 additional years of observation, an extra contamination event was observed. Although the ratio of contamination events per number of years remains the same as that used previously, the numbers $r = 2, t = 8$ provided by the operator of the site have the effect of modifying the prior distribution of the probability p. Thus, from Eqs. (10.21) and (10.22), one obtains

$$f_p(p) = \frac{7!}{5!1!} \, p^1(1 - p)^5 = 42p(1 - p)^5 \quad (10.29)$$

Using Eq. (10.29) and the values $B = 40$ and $C = 8$ yields

$$\text{EOL}[A(0)] = 1680T \int_{C/B = 0.2}^{1} (p - 0.2)p(1 - p)^5 \, dp \approx 3.3T \quad (10.30a)$$

$$\text{EOL}[A(1)] = 1680T \int_{0}^{0.2} (0.2 - p)p(1 - p)^5 \, dp \approx 1.3T \quad (10.30b)$$

Equations (10.30a) and (10.30b) indicate that extending the length of the observation record results in a reduction of the expected opportunity loss for action $A(0)$ from an $\text{EOL}[A(0)]$ of about $4.1T$ to a value of $3.3T$ and for action $A(1)$ from an

EOL[$A(1)$] of about $2.1T$ to a value of approximately $1.3T$. In other words, the larger the observation record, the more one could estimate the true state of nature and hence the less would the regret be for each economic action.

10.5.4 Influence of the Form of the Loss Function on Decision Making

Consider now a loss function of the form

$$L_2(A, N) = \begin{cases} BN^2 & \text{if } A = A(0) \\ CT & \text{if } A = A(1) \end{cases} \tag{10.31}$$

In Eq. (10.31), the cost for adding the supplementary monitoring system has remained the same (CT), but for the decision $A(0)$ not to augment the monitoring network, the consequences have become severe. Thus the losses in Eq. (10.31) increase rapidly the more contamination events remain undetected by the regular monitoring system, effectively penalizing the decision not to install the additional devices. In practice, such a loss function can be used to simulate the situation where an increasing number of contamination events (and associated remediation costs) points to inadequacy of a regular monitoring program (because of poor selection of the mandated devices' locations, underestimation of the uncertainty characterizing the site, etc.).

The goal function for action $A(0)$ is given by

$$G_2[A(0), p] = E^N\{L_2[A(0), N]\} = E^N(BN^2) = BTp(Tp + 1 - p) \tag{10.32a}$$

Here we have used that $\text{var}(N) = E(N^2) - [E(N)]^2$ and that the mean and variance of the Bernoulli distribution are given by $E(N) = Tp$ and $\text{var}(N) = Tp(1 - p)$. For action $A(1)$, the goal function is again given by

$$G_2[A(1), p] = E^N\{L_2[A(1), N]\} = E^N(CT) = CT \tag{10.32b}$$

The probability p^* where a change in action is recommended is obtained by equating the goal functions in Eqs. (10.32a) and (10.32b), yielding the quadratic equation

$$p^{*2}(T - 1) + p^* - \frac{C}{B} = 0 \tag{10.33}$$

which for the values of $B = 40$ and $C = 8$ and a time period of $T = 10$ years returns the value of $p^* = 0.1$.

The Bayes risk for action $A(0)$ is now given by

$$R_2[A(0)] = E^{fp(p)}\{G[A(0)], p\} = E^{fp(p)}[BTp(Tp + 1 - p)]$$

$$= BT^2E^{fp(p)}(p^2) + BTE^{fp(p)}(p) - BTE^{fp(p)}(p^2) \tag{10.34}$$

where the notation $E^{fp(p)}(\cdot)$ stands for the expected value with respect to the prior distribution in Eq. (10.23). For this distribution, the mean and variance are given as $m = 1/4$ and $\sigma^2 = 0.0375$ [Eq. (10.22) for $r = 1$ and $t = 4$]. Substituting the relation $E^{fp(p)}(p^2) = \sigma^2 + m^2 = 0.1$ into Eq. (10.34) yields

$$R_2[A(0)] = 0.1BT(T + 1.5) \tag{10.35a}$$

The Bayes risk for action $A(1)$ is given by

$$R_2[A(1)] = E^{fp(p)}\{G[A(1)], p\} = E^{fp(p)}(CT) = CT \tag{10.35b}$$

For the values of $B = 40$ and $C = 8$, under a quadratic loss function, action $A(0)$ returns a value of $R_2[A(0)] = 4T(T + 1.5)$ and a $R_2[A(1)] = 8T$. The time when the expected losses of both actions are equal to each other can be found by equating Eq. (10.35a) to Eq. (10.35b):

$$0.1BT(T + 1.5) = CT \Rightarrow 4T + 6 = 8 \Rightarrow T = 0.5 \tag{10.36}$$

Hence, for the case where $B = 40$ and $C = 8$, the optimal action is defined according to

$$\text{If} \quad T < 0.5, \quad \text{then} \quad A^* = A(0)$$
$$\text{If} \quad T > 0.5, \quad \text{then} \quad A^* = A(1) \tag{10.37}$$

In general, the optimal action is determined through the following relations:

$$\text{If} \quad 0.1B(T + 1.5) < C, \quad \text{then} \quad A^* = A(0)$$
$$\text{If} \quad 0.1B(T + 1.5) > C, \quad \text{then} \quad A^* = A(1) \tag{10.38}$$

One should notice that Eq. (10.25) determines the optimal action by relating the costs B and C only, whereas, under the quadratic loss model, Eq. (10.38) involves, additionally, the time horizon T of the supplementary monitoring system. Thus, under a linear loss model, action $A(1)$ to install a supplementary system, is to be followed only when the ratio of the costs of remediation to those of the monitoring system exceeds the value of 4. In contrast, under a quadratic loss model, action $A(1)$ is recommended if $B/C > 10/(T + 1.5)$. Table 10.5 indicates the value of the ratio B/C for the two loss models for different time periods. For the linear loss, model B/C remains fixed over the years, whereas for the quadratic loss model, the criterion of how high the contamination costs must be in relation to the cost of the monitoring system in order to install additional devices is relaxed with the length of operation.

The loss function for the case of a quadratic loss function is depicted in Table 10.6. And the regret table of the quadratic loss case is as shown in Table 10.7.

Table 10.7 tabulates the loss that is incurred by each action because information that would have revealed the true state of nature is not available. Thus, if it were known for certain that monitoring would be more expensive than cleaning up a contamination leak, then action $A(1)$ expends $(CT - BN^2)$ more than what is needed to address the situation. On the other hand, if contamination costs turn out to be truly higher than monitoring costs, then action $A(1)$ is using resources efficiently (no regret).

The opportunity loss (OL) table for this case is obtained by taking the expected value of all quantities entering Table 10.7 or directly through the goal functions in Eqs. (10.32a) and (10.32b) (Table 10.8).

Solving Eq. (10.33) or, equivalently, using the inequalities in the first column of Table 10.8 yields a probability $p*$ where a change in action is recommended. Using the probability $p*$ allows presentation of the relations in Table 10.8 according to

TABLE 10.5 Installation of Monitoring Network: Criteria for Different Loss Models

Time period T (years)	Linear loss model: Value of B/C favoring $A(1)$	Quadratic loss model: Value of B/C favoring $A(1)$
1	4	4
2	4	2.9
3	4	2.2
4	4	1.8
5	4	1.5
⋮	⋮	⋮
10	4	0.9

TABLE 10.6 Loss Table for Quadratic Model

State of nature	$A(0)$: Do not build monitoring system	$A(1)$: Build monitoring system
S1: $BN^2 < CT$	BN^2	CT
S2: $BN^2 > CT$	BN^2	CT

TABLE 10.7 Regret Table for Quadratic Model

State of nature	$A(0)$: Do not build monitoring network	$A(1)$: Build monitoring network
S1: $BN^2 < CT$	0	$CT - BN^2$
S2: $BN^2 > CT$	$BN^2 - CT$	0

TABLE 10.8 Opportunity Loss for Quadratic Model

State of nature	A(0): Do not build monitoring network	A(1): Build monitoring network
S1: $Bp(Tp + 1 - p) < C$	0	$[C - Bp(Tp + 1 - p)]T$
S2: $Bp(Tp + 1 - p) > C$	$[Bp(Tp + 1 - p) - C]T$	0

$$OL_2[A(0), p] = \begin{cases} 0 & \text{if } 0 \le p \le p^* \\ [Bp(Tp + 1 - p) - C]\,T & \text{if } p^* \le p \le 1 \end{cases} \qquad (10.39a)$$

$$OL_2[A(1), p] = \begin{cases} [C - Bp(Tp + 1 - p)]\,T & \text{if } 0 \le p \le p^* \\ 0 & \text{if } p^* \le p \le 1 \end{cases} \qquad (10.39b)$$

The expected opportunity loss (EOL) can now be evaluated for the quadratic loss model by calculating the expected values of the opportunity losses given in Eqs. (10.39a) and (10.39b) with respect to the prior distribution of the probability of a contamination leak in Eq. (10.23):

$$EOL_2[A(0)] = 3 \int_{p^*}^{1} [Bp(Tp + 1 - p) - C)]T(1 - p)^2 \, dp \qquad (10.40a)$$

$$EOL_2[A(1)] = 3 \int_{0}^{p^*} [C - Bp(Tp + 1 - p)]T(1 - p)^2 \, dp \qquad (10.40b)$$

The values of $B = 40$, $C = 8$, and $T = 10$ return the value $p^* = 0.1$ [Eq. (10.33)], which, on substitution in Eqs. (10.40a) and (10.40b), yields

$$EOL_2[A(0)] = 30 \int_{p^* = 0.1}^{1} (360p^2 + 40p - 8)(1 - p)^2 \, dp \approx 393 \qquad (10.41a)$$

$$EOL_2[A(1)] = 30 \int_{0}^{p^* = 0.1} (8 - 40p - 360p^2)(1 - p)^2 \, dp \approx 13.37 \qquad (10.41b)$$

For the linear model and a time period of 10 years, the expected opportunity losses were found to be EOL[A(0)] ≈ 40 and EOL[A(1)] ≈ 20.96, respectively

10.6 SUMMARY

The purpose of this chapter has been to illustrate, with particularly simple examples, how one can go about using the occurrence of particular events to update the initial probability estimates. This bayesian aspect of environmental risk assessment is particularly powerful in attempting to improve knowledge of the likely causes of the environmental hazard. In such a way, one can continually update contingency plans for remedial action based on a most likely cause. In addition, one can assess the need to put such plans into action based on the actual events that have occurred and their use to improve the estimates of further disastrous events occurring.

While the example here has been tailored to illustrate simply the Bayes' procedure, it is clear that more than two scenarios can be handled equally well. The dominant theme is to have some a priori way of estimating the probability of a given scenario being correct and then, within the framework of each scenario, to provide an a priori estimate of a particular type of event occurring or not occurring. Once these two aspects of the environmental risk problem have been addressed adequately, it is then a relatively simple matter to update the likelihood of further events occurring and also to update the probability that a given scenario is valid using the bayesian method.

Illustration of this particular goal, and how to achieve it practically, was the objective of this chapter. A more extensive discussion on the bayesian theory can be found in several textbooks and articles, this chapter's aim was to motivate decision-making groups on the use of this powerful method.

CHAPTER 11
MULTIPLE TRANSPORT OF HAZARDOUS MATERIAL: PROBABILITIES OF PROFITABILITY

11.1 INTRODUCTION

Two major concerns suggest that transport of hazardous material can be both worthwhile financially and also a financial catastrophe. One extreme example is transport of a product, such as oil, that has an intrinsic worth. Thus ship tanker transport of crude oil from the Mideast to the shores of America provides a cargo per ship trip that has an intrinsic worth. The transport, however, also involves a risk per trip that a tanker will be wrecked with some massive catastrophic loss for cleanup. The other extreme example is transport of a product, such as radioactive waste, that has no intrinsic value but which must be disposed of by burial because of the deleterious effects of the product on human society. In this case, the transport (by truck, train, or ship) also involves the potential for spillage per trip with, once again, some major cleanup costs.

In either of the two extremes, the issue has to do with safe delivery of product (and associated gains) versus the catastrophic potential (and associated costs). A corporate concern is the worth of undertaking a suite of N such transports with the estimated chance of making a profit versus a loss.

This chapter provides an objective procedure for evaluating the likelihood of making a profit from the N transports based on the probability of each trip being profitable or catastrophically ruinous fiscally.

11.2 STATISTICAL MEASURES FOR INDIVIDUAL TRANSPORT TRIPS

The standard estimates that one can make for worth of an individual transportation operate as follows: Suppose that the product shipped is brought to market (or to the disposal site in the case of hazardous waste) with a gain G to the corporation. Let the costs of transport and delivery per trip be C. Two possibilities exist in the absence of a catastrophic spill: Either the selling price of product at delivery is sufficiently high that a profit G is made, or the product price has declined so much (*viz.* the market spot price for oil) that a particular trip is *not* profitable. To illustrate the extreme-case possibilities, consider that there is a probability p_f for a sale at price G_1 ($<C$) so that one makes a net loss $\Delta C = C - G_1$ on the trip and that there is a probability p_s for a sale at a price G ($>C$) so that one makes a net profit $\Delta G = G - C$ on the trip. In the absence of a catastrophic loss, the success and failure probabilities (p_s, p_f) would sum to unity. However, of greater concern is the small chance p_c of a catastrophic disaster per trip with enormous environmental cleanup costs C_1, much larger than the potential net profit ΔG.

Clearly, one wishes to plan a corporate strategy around all these concerns so as to maximize probable profit and minimize probable loss. In addition, from the results (profit, loss, catastrophic loss) of each transport shipment, the corporation must continuously update the relevant probabilities for the next transport shipment. In addition, after, say, k trips, the corporation also must estimate the probability that the remaining $N - k$ trips will be profitable. Corporate strategy may call for a buyout of a transport contract at a price B if the probable profitability of the $N - k$ trips is not high enough. This factor, too, must be addressed in assessing the probabilities of corporate profit.

11.2.1 Statistics for a Single Event

Consider, first, the expected value E and variance σ^2 of single transportation with probability and value parameters as given in Fig. 11.1, with $p_s + p_f + p_c = 1$. The expected value is

$$E = p_s(\Delta G) - p_f \Delta C - p_c C_1 \equiv p_s G + p_f G_1 - C - p_c(C_1 - C) \quad (11.1)$$

and the variance around the expected value is

$$\sigma^2 = p_s p_f (G - G_1)^2 + p_c[p_s(G - C + C_1)^2 + p_f(G_1 - C + C_1)^2]$$

$$\equiv p_s p_f (\Delta G - \Delta C)^2 + p_c[p_s(\Delta G + C_1)^2 + p_f(\Delta C - C_1)^2] \quad (11.2)$$

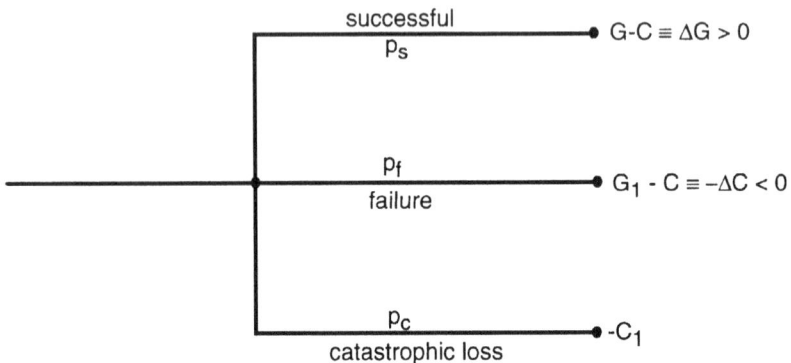

FIGURE 11.1 Schematic decision tree indicating how one includes catastrophic loss in calculation of probability of profit and loss.

The probability $P(V)$ that the single venture will prove profitable with a value greater than or equal to V is then given by

$$P(V) = \pi^{-1/2} \int_{a(V)}^{\infty} \exp(-u^2)\, du \qquad (11.3)$$

where

$$a(V) = \frac{V - E}{2^{1/2}\sigma} \qquad (11.4)$$

so the probability that a single venture will *not* be profitable at all is

$$P_N = 1 - P(0) \qquad (11.5a)$$

where $P(0)$ is the probability of break-even, i.e., no loss but no profit either.

The probability that a loss will occur in excess of an amount A is

$$P_N(A) = 1 - P(-A) \qquad (11.5b)$$

Catastrophic losses C_1 are commonly so high that, despite the fact that their probability of occurrence p_c is usually small compared with p_s or p_f, the product $p_c C_1$ often dominates completely the expected value E of Eq. (11.1). Thus, in the absence of further information, inclusion of the catastrophic probability often can change a project with a positive expected value into one with a negative expected value. The amelioration of this, often large, change in expected value is accomplished with the uncertainty σ on the expected value. Thus, for example, the break-even probability $P(0)$ is

$$P(0) = \pi^{-1/2} \int_{a(0)}^{\infty} \exp(-u^2)\, du \qquad (11.6a)$$

where

$$a(0) = \frac{-E}{2^{1/2}\sigma} \qquad (11.6b)$$

As $C_1 \to \infty$ without bound, all other factors being held fixed, it follows from Eqs. (11.1) and (11.2) that $E_1 \to -p_c C_1$ and $\sigma \to C_1 p_c^{1/2} (1 - p_c)^{1/2}$ so that $a(0) \to \frac{1}{2}[2p_c/(1 - p_c)]^{1/2}$. Because the probability of a catastrophic event p_c is usually extremely small, in most situations it is adequate to replace the limiting form of $a(0)$ by $(p_c/2)^{1/2} \ll 1$. The break-even probability is then approximately

$$P(0) \cong \frac{1}{2} - \left(\frac{p_c}{2\pi}\right)^{1/2} \qquad (11.6c)$$

Thus, at high values of C_1, the standard error and the mean value both increase in proportion to the catastrophic costs. The consequence is that the break-even probability is slightly less than, but not too far from, 50:50 when the chance of occurrence of a catastrophic event is small.

11.2.2 Statistics for Multiple Events

There are three major concerns when one is involved in multiple transport problems. First, ahead of any trips, one must provide some idea of the likely gains, costs, and probabilities, as depicted in Fig. 11.1. Second, after each and every trip one knows the results of that trip (whether it was fiscally successful, whether costs had increased, whether a catastrophic event occurred), as well as the results of all prior trips. This information can be used to update both the expected value and its variance for the next transport trip based on the prior results—in effect, one is performing a bayesian update of prior estimates of success and failure probabilities. The third aspect is then to use the updated probabilities to assess the probability that of the remaining total number of trips, enough will be successful that the corporation will profit from the total venture even if some trips are unprofitable or even catastrophic losses.

The simplest way to provide some assessment of the worth of a total of N trips is to suppose, initially, that each trip is essentially a repeat of previous trips, in the sense that one has the same parameters, and so one treats each trip as independent of all others. Then, because the probability $P(V)$ of making a profit in excess of the amount V is already calculated, the probability P_N that all the trips will *each* make more than the amount V is

$$P_N = P(V)^N \tag{11.7}$$

which amounts to the probability that *all* N of the trips will make a total of $V_N = NV$. A slightly more general way to express this statement is to ask for the probability that the ith trip makes an amount V_i such that the total made from N trips is V_N. Then

$$P_N(V_N) = P_1(V_1)P_2(V_2)\cdots \tag{11.8a}$$

with

$$\sum_{i=1}^{N} V_i = V_N \tag{11.8b}$$

In any event, if C_1 is extremely large (all other parameters being held fixed), note that the probability of all the trips being profitable ($V_N \geq 0$) is

$$P_N(0) = P(0)^N \cong \left(\frac{1}{2}\right)^N \left[1 - \left(\frac{2p_c}{\pi}\right)^{1/2}\right]^N \cong \exp\left\{-N\left[\ln 2 + \left(\frac{2p_c}{\pi}\right)^{1/2}\right]\right\} \tag{11.9}$$

However, it is not necessary that every trip be profitable in order that a net profit is produced after N trips. For example, one could imagine the situation where all except one of the trips lost small amounts of money but the exceptional trip generates more than enough money to offset the combined small losses of all the other trips.

If one is interested in k of the N trips being profitable, then one has

$$P_N(k) = \frac{N!}{k!(N-k)!} P(0)^k [1 - P(0)]^{N-k} \tag{11.10}$$

The probability that *no* trips out of the N will be profitable is

$$P_N(0) = [1 - P(0)]^N \cong \exp[-NP(0)] \tag{11.11}$$

Thus the average number N_{av} of trips that should be made before one has reached a point where the catastrophic chances are likely to overwhelm profitability, even if no catastrophic events have occurred heretofore, is

$$N_{av} \cong \left(\frac{\pi}{2}\right)^{1/2} \frac{(\ln 2)^2}{p_c^{1/2}} \tag{11.12}$$

For a value of p_c of around 1%, one has $N_{av} \cong 21$, whereas at a more typical value of p_c of around 10^{-3}, one has $N_{av} \cong 66$.

Thus, without using information gleaned from each trip concerning its profitability, loss, or catastrophic results, it would seem that one is being less than prudent in estimating the worth of further transport trips.

11.2.3 Bayesian Updating of Probabilities

To include the information available from each transport trip in refining the probabilities, one proceeds as follows: Consider that a corporation is interested in making any profit greater than zero. Initially, the corporation makes the assessments of values p_s, p_f, and p_c for the probabilities of a financially successful trip, a financially unsuccessful trip, or a catastrophic failure. After the first transport trip, one knows for certain whether it was successful, unsuccessful or catastrophic. Thus one can then update the initial assessments of probability based on that first trip information. Thus let S, U, and C denote the states of "successful," "unsuccessful," and "catastrophic failure." The initial probabilities of occurrence of each state are $p(S) = p_s$, $p(U) = p_f$, and $p(C) = p_c$. Suppose, for illustrative purposes, that the first transport is financially unsuccessful but not catastrophic. Then one is interested in $p(S|U_1)$; $p(U|U_1)$, and $p(C|U_1)$, where these conditional probabilities mean, for instance, the probability of a success given that the first trip was fiscally unsuccessful, and so on. These updated probabilities can then be used to calculate an updated expected value and updated variance so that one can then supply an updated value of profitability

$P(0)$ for the next transport trip. In this general way one can update probabilities after each trip.

Suppose, for instance, that transport of oil by ship is undertaken. Two major possibilities could be considered for catastrophic loss: (1) ship breakup at sea during a major storm and (2) shipwreck on near-coastal rocks in either the presence or absence of storm influences. A given ship can undergo only one of these two catastrophes because they are mutually exclusive. Suppose, then, that one initially estimates the relative probability of catastrophic loss due to storm ship breakup at p_{cb} and of near-coastal wreck at $(p_c - p_{cb})$ so that the total catastrophic probability is p_c. A question of concern is how these relative probabilities are to be updated as a consequence of the results from the first trip. Let the catastrophic cleanup costs under the breakup scenario be C_{1b} and C_{1w} under the shipwreck situation.

Then the probability of making a profit $P_b(0)$ under the situation where one has a ship breakup is obtained by substituting p_{cb} for p_c and C_{1b} for C_1 in Fig. 11.1, so one calculates E_{1b} and σ_b and hence $P_b(0)$. Likewise, a similar calculation is performed for $(p_c - p_{cb})$ and C_{1w}, thereby yielding $P_w(0)$ for the wreck scenario.

Then given that the first trip was unsuccessful financially but not a catastrophe, one has the probabilities

$$P(U_1|\text{breakup}) = 1 - P_b(0) \tag{11.13a}$$

and

$$P(U_1|\text{wreck}) = 1 - P_w(0) \tag{11.13b}$$

The bayesian procedure of Chap. 10 then provides the updated probabilities for a ship breakup or near-coastal wreck from

$$P_{b1} \equiv p\,(\text{breakup}|U_1) = p_c \left\{ \frac{[1 - P_b(0)]p_{cb}}{[1 - P_b(0)]p_{cb} + [1 - P_w(0)]\,(p_c - p_{cb})} \right\} \tag{11.14a}$$

and

$$P_{w1} \equiv p\,(\text{wreck}|U_1) = p_c \left\{ \frac{[1 - P_w(0)]\,(p_c - p_{cb})}{[1 - P_w(0)]\,(p_c - p_{cb}) + [1 - P_b(0)]\,p_{cb}} \right\} \tag{11.14b}$$

which give the updated probabilities of shipwreck on near-coastal rocks (P_{w1}) and of ship breakup (P_{b1}) in a major storm. Note that $P_{b1} + P_{w1} \equiv p_c$, so the total chance of catastrophe remains the same.

These updated probabilities can then be used to recalculate $P_b(0)$ and $P_w(0)$ by replacing p_c in Fig. 11.1 with P_{b1} and P_{w1}, respectively, and with the corresponding values of C_{1b} and C_{1w}. Hence the process can be repeated after the results of the second trip are known, with continual updating for each trip to reassess probabilities of fiscal success, fiscal loss, and catastrophic failure for the succeeding trips.

In this way one can continually monitor whether it is a better decision to have ships avoid storms more than near-coastal waters, or vice versa, thereby minimizing the probabilities of catastrophic failure for each trip. There is no updating of the total catastrophic probability p_c at this stage because, without an actual event occurring that involves catastrophic costs, one has no reason to update p_c. Of course, if one initially chooses equal probabilities for p_{cb} and $p_c - p_{cb}$, then the updated probabilities remain equal until the actual occurrence of a catastrophic event.

11.2.4   Residual Trips and Probability Updating

After n trips, r of which are financially successful and $n - r$ of which are financially unsuccessful, the capital accumulation A_n is

$$A_n = r\Delta G - (n - r)\Delta C = r(\Delta G + \Delta C) - n\Delta C \qquad (11.15a)$$

where ΔG is the amount made per successful trip, and ΔC is the amount lost per unsuccessful trip [for simplicity, we have taken each successful (unsuccessful) trip to yield the *same* profit (loss) ΔG (ΔC)]. Note that $A_n \geq 0$ when $(r/n) \geq \Delta C/(\Delta G + \Delta C)$.

If the capital accumulation is positive (and large), then a corporation might choose to be less concerned about catastrophic failure costs C_1 should such occur because once $A_n \gg C_1$, then the corporation can afford to pay the costs of catastrophe out of accumulated capital from preceding trips.

The fraction f_n [$\equiv (r/n)$] of the n prior trips that *must* be successful in order that $A_n \geq C_1$ is

$$f_n \geq \frac{\Delta C}{(\Delta G + \Delta C)} + \frac{C_1}{n(\Delta G + \Delta C)} \qquad (11.15b)$$

Clearly, at large values of n, f_n takes on its smallest value of $\Delta C/(\Delta G + \Delta C)$, which is just the statement that a large enough fraction of all trips must make a profit in order that a net profit be made overall.

The largest value of f_n occurs, of course, on $n = 1$, which is just the requirement that one not have a catastrophic condition arise ahead of accumulating sufficient capital to pay for catastrophic costs. And then one can again go through the bayesian probability updating of the previous chapter to provide improved probability estimates, just as in the preceding subsection.

11.3   BUDGET BUYOUT CONSIDERATIONS

The probable profitability at a total value V of the N trips is $P_N(V)$. Each corporation has a minimal acceptable chance MAC that it sets for profitability of a project at a required value V. If it is estimated that the total project profitability is likely to be less than MAC, then the corporation usually will seek a buyer of the residual component of the project. The buyout price set by the corporation is usually a function of the difference between $P_N(V)$ and MAC. If $P_N(V) > $ MAC, then the corporation retains the project. But if $P_N(V) < $ MAC, then the corporation usually asks the question: How high must V be in order that $P_N(V)$ will again equal MAC? Call this value V_{min}, with $P_N(V_{min}) = $ MAC.

The corporation will then usually offer the project for sale at a buyout value B in excess of V_{min}. In principle, in order to recover costs, the buyout price B is often set so that

$$\text{MAC} - P_N(V) = P_N(B) \tag{11.16}$$

11.4 NUMERICAL ILLUSTRATIONS

To provide an idea of "typical" values, consider three numerical illustrations. First, we look at a simple illustration of multiple transport and accumulated capital before a catastrophe strikes. Then we consider a bayesian update illustration of ship disaster, and finally we consider a buyout situation for a firm that believes the profit margin is too slender to continue involvement in a transport project.

11.4.1 Multiple Transport Before a Catastrophe Occurs

Consider that a firm makes a commitment to transport oil by ship from the Mideast to the United States. If the transport is accomplished successfully, then the firm makes a profit of $1 million per trip, after ship maintenance. However, if there is a major storm during transit, then there is the possibility of ship loss. Historical data indicate that ship and cargo loss occurs for roughly every billion barrels of oil transported. Each ship trip carries approximately 1 million barrels, so the probability of a ship loss is estimated at 10^{-3}. For each thousand successful trips, the accumulated profit is $1 billion, whereas for a catastrophic loss the magnitude of the *Exxon Valdez* disaster in Valdez Bay, Alaska, the cleanup and ancillary costs are estimated at $5 billion. However, if the catastrophic loss occurs in deep ocean waters, far from land, then the loss is the cost of the ship plus the cargo. Thus at $20 per barrel, a million barrel loss amounts to $20 million plus ship cost. Even if the ship costs $20 million, for a total loss of $40 million, such a deep-sea loss is much less than a $5 billion cleanup of environmentally sensitive coastal waters and coastal beaches.

Thus at $1 million profit per trip, for a deep-sea wreck (at a loss of $40 million), one is making the assumption that at least 40 trips can be made profitably before a catastrophic occurrence will happen. Statistical information of a billion barrels shipped between such accidents supports this view. However, for near-coastal calamities, the picture is very different. Here, one would have to make 5000 trips (at a $1 million profit per trip) to cover the costs of a *Valdez*-type disaster, should it occur. However, at a transport of 1 million barrels per trip, it would take only 1000 trips before one has crossed the statistical likelihood of catastrophe occurring. Thus, to guard against such a potential risk, it would be foolhardy of a company not to cover by insurance most of the expected cleanup costs of such an eventuality.

Effectively, one is using historical precedent as a database, together with current costs and profits, to assess the transport risk and to guard against fiscal disaster should such an event occur.

11.4.2   Ship Disaster and Probability Updating

Suppose that the chance of near-shore catastrophe is estimated at $p_c - p_{cb} \cong 10^{-3}$ and of ship breakup in deep water due to storm as $p_{cb} \cong 2 \times 10^{-3}$ so that the total chance of either catastrophe is $p_c \cong 3 \times 10^{-3}$. Let the probability of making a profit under the situation where a deep-water ship breakup occurs be reckoned at $P_b(0) \cong 20\%$ and under the shipwreck in near-coastal waters situation be $P_w(0) \cong 5\%$. The difference in these two estimates reflects the higher cost of environmental cleanup in the case of a near-shore wreck compared with a deep-water loss. The likelihood of a deep-water loss, however, is twice as high as a near-shore wreck for this example. Then, given that the first trip was unsuccessful financially but not a catastrophe, one can estimate the updated probabilities for a ship breakup or near-coastal wreck from Eqs. (11.14a) and (11.14b) as

$$P_{b1} = 1.88 \times 10^{-3} \qquad (11.17a)$$

and

$$P_{w1} = 1.12 \times 10^{-3} \qquad (11.17b)$$

Note that $P_{b1} + P_{w1} = 3 \times 10^{-3}$, as required. Thus the *relative* probability of a breakup versus a near-coastal wreck is $P_{b1}/P_{w1} = 1.68$, whereas the chance of a near-coastal wreck has increased from 10^{-3} to 1.12×10^{-3}, i.e., a 12% increase; correspondingly, the chance of a deep-sea sinking is reduced by 12% from 2×10^{-3} to 1.88×10^{-3}.

As the chance of profit making based on continuing trips is updated after each trip, so the *relative* chances change of near-coastal wreck versus deep-sea loss. However, the total chance p_c of a catastrophe does not change yet because until such a catastrophe does occur, there is no further information available that can be used to update the initial estimate p_c.

11.4.3   Buyout Price and Minimum Acceptable Chance

After operating a hazardous materials transport project for a year, a corporation assesses the cumulative profit made to date at $10 million. Relative to a total investment made of $100 million, the corporation estimates the cumulative net

profit per dollar invested at 10% ($10 million/$100 million). The corporation figures it can do better with its capital than provide a total return of just 10% (perhaps by investing in the stock market or by investing in other ventures such as
titanium dioxide production for paint or coffee bean futures, or whatever). In any
event, the corporate decision makers resolve to sell the remaining part of the contract to deliver hazardous waste. The corporation requires a minimum return on
investments made of 15%. The contract has another 2 years to run. The question
facing the corporation is: What should it set as a buyout price? In assessing the
buyout value, the corporation makes the assumption that the project investment
costs will continue to be $100 million per year over the next 2 years, but it is not
clear to the corporation whether profit margins will continue at the 10% level of
the first year. The corporation economists provide the decision makers with the
assessments that there is a 50% chance of a continuing yearly $10 million profit,
a 90% chance of a profit greater than $1 million, but only a 10% chance of a profit
greater than $20 million. The corporation uses these values to construct a cumu-
lative probability-of-profit curve. The corporation then notes that its MAC value
of 15% return corresponds to a required profit of $V = \$15$ million that has only a
25% chance of being realized. The value $V = \$15$ million corresponds to a gain
per investment dollar of 15%. This profit, however, relative to ongoing investments of $100 million per year, does not correspond to the actual 10% profit real-
ized in the first operational year. The difference amounts to some $5 million/yr ≡
ΔB. Hence, in order to achieve its goal of a MAC of 15% on the remaining 2 years
of the project, the corporation can offer the contract for buyout at a price of $B = 2$
$\times \$15$ million $= \$30$ million. If the corporation wishes to ensure that its internal
MAC value of 15% is met for all *3* years of the contract, then it offers a buyout
price of $2 \times \$15$ million $+ \Delta B ≡ \$35$ million for the contract. Effectively, the corporation is attempting to top up to 15% the 10% return in the first year by offering the remaining 2 years of the contract at a higher price.

   However, because there is only about a 25% chance of reaching the desired
goal of a MAC of 15% (or greater, of course), the corporation may choose to
accept some profit less than its desired goal of 15% in order to have the freedom
to invest its capital of $100 million per year elsewhere at a higher profit potential.
Thus, if the corporation so chooses, it can afford to offer the remaining 2 years of
the contract at a buyout price of less than its desired $35 million in order to achieve
the freedom to invest elsewhere.

11.5 SUMMARY

The purpose of this chapter has been to illustrate how one can assess the probabilities of making a profit from transport of hazardous material in the face of
potential catastrophes when more than one transport trip is involved. The numerical illustrations were designed to reflect as simply as possible the essence of the

arguments. One does, of course, recognize that real case histories are often more complex, as is also clear from some of the developments in preceding chapters for other aspects of environmental risk.

CHAPTER 12
MAXIMIZING PROFIT FOR A TOXIC WASTE SITE MONITORING SYSTEM

12.1 INTRODUCTION

One of the more difficult decisions that needs to be made by a corporation involved in the general business of environmental waste disposal has to do with monitoring of toxic waste in a depository. From a purely monetary perspective, the corporation would like to employ as few monitoring devices as possible because of cost. However, the corporation recognizes that if a toxic leak were to occur and, due to the corporation failure to monitor efficiently, a lawsuit were to follow, then the cost of losing the lawsuit plus the cleanup costs would be financially debilitating. In addition, without adequate monitoring, the corporation would not have a very exact idea of how disastrous the cleanup operation could be. On the other hand, emplacing far too many monitoring devices could be equally financially devastating. While the corporation would then be able to quickly detect toxic leaks, the cost of each monitor plus yearly maintenance could be excessive. Somewhere between these two extremes is where the corporation can operate. The question faced by the corporation is to determine where between the extremes it can operate in order that a profit (or at least no loss) be made. Built into this operational range are also regulatory requirements for monitoring of a toxic site and, most often, a fixed contract price per year for the monitoring component.

In this chapter we demonstrate how one can allow for profit considerations and also for regulatory considerations in the assessment of the best number of monitoring devices to employ. While the general argument depends on a large and often bewildering variety of imposed conditions, it is possible to provide a simple version of a rational procedure that demonstrates how such problems can be addressed. Adding further requirements than those developed here will not change the sense of the argument but will just make the analysis more cumbersome. Accordingly, any such complex government regulations are simplified to allow a sharper perception to be presented of the general procedure.

12.2 SCIENTIFIC CONSIDERATIONS

Consider, for simplicity, a toxic waste disposal site of total surface area A_T. Let the toxic waste initially have a uniform concentration q_0 per unit area, and let toxicity decline with time in an exponential manner. Then, at time t after initial storage, the toxicity is $q(t) = q_0 \exp(-bt)$, where b is the decline rate. Let n monitoring devices be emplaced equally spaced such that each device monitors a surface area A centered on the device and such that the devices monitor throughout the thickness of the site. (It is possible to redo all the calculations that follow using monitoring over local *volume* elements, but such calculations just provide technical frills without adding any particular new insight.) The total area monitored is then nA, so the total area unmonitored is $A_T - nA$. Clearly, a limit on the number of monitoring devices is $n \leq A_T/A$ or one then monitors more than the total depository area.

Because the areal toxicity declines exponentially, the total toxic amount monitored at time t after storage is

$$q_0(t) = q_0 \exp(-bt)\, nA \qquad (12.1)$$

Thus the total toxicity that is *not* monitored at time t is

$$\Delta q(t) = q_0 \exp(-bt)\, (A_T - nA) \qquad (12.2)$$

Often, regulations on monitoring requirements can be simplified to versions of two basic rules.

Requirement 1. There shall be enough monitoring devices emplaced that a prescribed fraction f of the total depository toxicity shall be measured always.

This requirement translates, at time t, into

$$q_0 \exp(-bt)\, nA \geq fq_0 \exp(-bt)\, A_T \qquad (12.3a)$$

which can be written as

$$nA \geq fA_T \qquad (12.3b)$$

Thus a first requirement on the number of monitoring devices in this case is

$$A_T \geq nA \geq fA_T \qquad (12.4)$$

A second requirement often has to do with the level of toxicity considered to be hazardous.

Requirement 2. Monitoring can be discontinued when the total toxicity in the depository falls below a prescribed value M; otherwise, monitoring will be continued.

For our illustration, this requirement is that monitoring must continue so long as

$$q_0 \exp(-bt)\, A_T \geq M \qquad (12.5)$$

Equation (12.5) provides a time scale t_* at which one can stop monitoring given by

$$bt_* = \ln (q_0 A_T / M) \tag{12.6}$$

12.3 FINANCIAL CONSIDERATIONS

Take it that each monitor is purchased at the beginning of depository fill at a cost of $\$C_0$, for a total monitor purchase (in year $t = 0$ dollars) of $\$C_0 n$. Also take it that the monitoring costs per year are proportional to a fixed basis cost B and a cost per monitor m at year $t = 0$. If inflation is at a rate r per year, then these ongoing costs have to be increased each year. Thus at time t the costs of monitoring in fixed year $t = 0$ dollars are

$$C_M = (B + mm) \exp (rt) \tag{12.7}$$

Thus the total monitoring cost by time t is

$$C_{\text{total}} (t) = nC_o + \int_0^t (B + mn) \exp (rt) \, dt \tag{12.8}$$

To this cost one also must add the development cost D of the toxic site in year $t = 0$ dollars for a total outlay by time t of

$$\text{Total} = C_{\text{total}}(t) + D \tag{12.9}$$

Consider that the contract offered by the regulatory authorities for toxic site monitoring provides a *fixed number* dollar value per year V_0 for the monitoring program. Because of inflation, the actual worth of this value at year t is

$$V(t) = V_0 \exp (-rt) \tag{12.10}$$

Thus the total gains that can be made by year t are

$$G(t) = \int_0^t V(t) \, dt - D - C_{\text{total}}(t) \tag{12.11}$$

which can be written as

$$G(t) = V_0 r^{-1} [1 - \exp (-rt)] - (D + C_0 n) - (B + mm) r^{-1} [\exp (rt) - 1] \tag{12.12}$$

Notice that on $t = 0$, one has a negative value

$$G(0) = - (D + C_0 n) \tag{12.13}$$

and at large values of t, $G(t)$ is again negative. If there is to be any profit at any time, then the maximum value of $G(t)$ must, at the least, be positive. The maximum of $G(t)$ occurs at time

$$t_{max} = (2r)^{-1} \ln \frac{V_0}{B + mn} \tag{12.14a}$$

and t_{max} is positive as long as $V_0 > B + mn$, and then at $t = t_{max}$, one has

$$G(t_{max}) = V_0 r^{-1} (1 - \varepsilon^{1/2})^2 - (D + C_0 n) \tag{12.14b}$$

where $\varepsilon = (B + mn)/V_0 \leq 1$, which is just the requirement that the rate of gains minus costs at year $t = 0$ should be positive or there is no hope of ever making a profit. Thus, in order to have any hope of profit at any time, one must have

$$1 > 1 - \left(\frac{B + mn}{V_0} \right)^{1/2} \geq \frac{r^{1/2} (D + C_0 n)^{1/2}}{V_0^{1/2}} \equiv \gamma^{1/2} \tag{12.15}$$

When the inner component of Eq. (12.15) is satisfied, then the outer component requires $V_0 > r(D + C_0 n)$, which is just the requirement that the total amount from the contract payment must, at the least, exceed the site-development costs plus the costs of monitoring equipment or again there is no hope of ever making a profit.

When the preceding inequalities are satisfied, there is a range of times around $t = t_{max}$ at which $G(t)$ can be positive. Indeed, one can write $G(t)$ in the form

$$G(t) = V_0 r^{-1} \exp(-rt) [(1 + \varepsilon - \gamma) x - 1 - \varepsilon x^2] \tag{12.16}$$

with $x = \exp(rt)$.

Clearly, $G(t)$ is zero at the two times t_- and t_+ given through

$$t_- = r^{-1} \ln ((2\varepsilon)^{-1} \{1 + \varepsilon - \gamma - [(1 + \varepsilon - \gamma)^2 - 4\varepsilon]^{1/2}\}) > 0 \tag{12.17a}$$

and

$$t_+ = r^{-1} \ln ((2\varepsilon)^{-1} \{1 + \varepsilon - \gamma + [(1 + \varepsilon - \gamma)^2 - 4\varepsilon]^{1/2}\}) > t_- \tag{12.17b}$$

with

$$t_+ > t_{max} > t_- > 0$$

Thus one should retain the monitoring contract until at least $t = t_-$ or there is no profit because $G(t) <$ for earlier times. Equally, however, one should *not* retain the contract beyond time $t = t_+$ or again $G(t) < 0$. Ideally, one should relinquish the contract at time $t = t_{max}$ when one has made the largest possible profit given by Eq. (12.15b). Now, all these parameter values depend on the number n of monitoring devices in place. And this number is limited by Eq. (12.4) rewritten as

$$n_+ \equiv \frac{A_T}{A} \geq n \geq \frac{fA_t}{A} \equiv n_- \tag{12.18}$$

Equally, monitoring must continue until time $t = t_*$, given by Eq. (12.6). Thus, ideally, one would like to adjust parameters so that $t_{max} = t_*$. If such is not possible, then one would like to arrange that $t_* > t_-$ so that some profit can be made and that $t_+ > t_*$ so that, again, one does not run into negative profit regimes. In general, then, one requires that the inequality

$$t_+ > t_* > t_- \tag{12.19}$$

is satisfied for profit making.

Recognizing that t_+ and t_- depend on n, through the parameters ε and γ in Eq. (12.17), one can then use Eq. (12.19) to find minimum and maximum values for n at which a profit can be made. These values can then be used in Eq. (12.18) to see if they satisfy that override requirement. If so, then the contract can be accepted, for one can make some profit; if not, then the contract should be refused because there is no way any profit can ever be made.

Alternatively, note from Eq. (12.16) that, on $t = t_*$, $G(t_*)$ is positive so long as

$$(1 - \gamma) x_* > 1 + \varepsilon x_* (x_* - 1) \tag{12.20}$$

where $x_* = \exp(rt_*)$. Inserting $\gamma = r(D + C_0 n)/V_0$ and $\varepsilon = (B + mn)/V_0$ into Eq. (12.21) yields an expression for the number n of monitoring stations in order that a profit be generated at time $t = t_*$ in the form

$$n \leq t_*^{-1} [rC_0 + m (t_* - 1)]^{-1} [(x_* - 1) (V_0 - x_* B) - x_* rD] \equiv n_* \tag{12.21}$$

which requires

$$V_0 \geq x_* \frac{B + rD}{x_* - 1} \tag{12.22}$$

in order to be satisfied at all. But once assurance can be provided that Eq. (12.22) is satisfied, then one has merely to check whether the inequality $n \leq n_*$ is consistent with the override requirement $A_T/A > n \geq fA_T/A$ to ensure one has a profit. Thus one has to check that

$$n_* \geq \frac{fA_T}{A} \tag{12.23a}$$

and then any n in the range

$$\min \{n_*, A_T/A\} \geq n \geq fA_T/A \tag{12.23b}$$

will return a profit at time $t = t_*$.

Inequality (12.22) also can be written as a constraint on x_* as

$$x_- \leq x_* \leq x_+ \tag{12.24a}$$

with

$$x_\pm = (2B)^{-1}\{(B + V_0 + rD) \pm [(B + V_0 + rD)^2 - 4BV_0]^{1/2}\} \tag{12.24b}$$

If x_* does *not* lie in the range given by Eq. (12.24a), then there is *no* possibility of profit.

12.4 NUMERICAL ILLUSTRATION

To illustrate how these calculations can be made, consider a toxic waste depository of 1 km^2 ($\equiv A_T$). Let each monitoring instrument be capable of accurately measuring toxic waste leakage in an area of 100 m^2 ($\equiv A$) around the instrument. Then $A_T/A = 10^4$. Let each monitoring device cost $100 ($\equiv C_0$) at time $t = 0$, and let the monitoring cost per year per device be $10 ($\equiv m$) at $t = 0$. The fixed basis cost per year for monitoring is taken at $10,000 ($\equiv B$). Let inflation be calculated at 3% per year ($r \equiv 0.03$), and also let the site-development cost be 10^6 ($\equiv D$). Let the government contract call for payment to the contracting firm of $10 million ($\equiv V_0$) per year. The government requires that 30% ($\equiv f$) of the toxic waste potential leakage area be measured by the monitoring devices. The decay lifetime of the toxic material is estimated at 100 years ($\equiv b^{-1}$), and it is required that the toxicity be monitored for a time t_* until the total level has dropped to e^{-1} of its initial value, that is, $q_0 A_T/M \equiv e$. Then from Eq. (12.6) one has $bt_* = 1$ so that a monitoring time of 100 years ($\equiv t_*$) is required from the contractor.

The override requirement on the number of monitors to install is $A_T/A \geq n \geq fA_T/A$, which can be written as

$$10^4 \geq n \geq 0.3 \times 10^4 \tag{12.25}$$

The values $x_\pm$ of Eq. (12.24b) are given by $x_+ = 110.09$ and $x_- = 0.91$. Now $x_* \equiv \exp(rt_*) = \exp(0.03 \times 100) = \exp(3) = 20.08$. Thus x_* *does* lie between x_- and x_+ for the numbers given. Hence there is a chance of profit. Then, from Eq. (12.21) one has

$$n_* = 3.86 \times 10^4 \tag{12.26}$$

Thus any $n < n_*$ will allow a profit to be made. Remembering that the override requirement is as given by Eq. (12.26), we see that this situation *does* allow profit.

In particular, inserting the relevant parameters into the profit equation (12.17), one has

$$G\,(t_*) = \$6.44 \times 10^3\,(n_* - n) \tag{12.27}$$

With the requirement that the minimum allowed number of monitors n is 3×10^3, it follows that the maximum profit (in year $t = 0$ dollars) that can be achieved over the monitoring time of 100 years under the restrictions imposed is $G_{max} = \$2.18 \times 10^8$, for an average of \$2.18 million per year. If the restriction on monitored area is tightened so that one *must* monitor the total area of the depository, then $n = 10^4$. In this case the cumulative profit (in year $t = 0$ dollars) decreases to $G_{max} = \$1.73 \times 10^8$, or \$1.73 million per year.

The point of this example is to show that one can allow for variations in contract requirements and monitoring conditions and that one can, indeed, estimate what it takes to achieve profitability in a monitor project but still remain within the scientific demands imposed by the regimenting authorities.

12.5 SUMMARY

While one can question the simplifying assumptions made in this chapter, such as homogeneity of the depository, exponential decay with time of the toxicity, etc., and while one can question the monitoring requirements imposed as illustrative devices, nevertheless, the logic pattern used remains the same under all conditions. One must first evaluate the constraints on the number of monitoring devices in order to satisfy scientific and regulatory requirements and determine the relationship between time and number of monitoring devices so that a profit can be made and when one can expect such a profit. Then one merely has to check that those requirements are not at odds with the regulatory monitoring requirements. In such a way it is a relatively easy matter to assess when a profit is likely and still stay within the regulatory requirements.

Perhaps one of the major concerns with such calculations is that they deal with long-term future estimates (for the particular numerical example, a 100-year monitoring program is required). It is doubtful that parameters estimated today will remain valid over the monitoring lifetime required (e.g., 3% inflation for 100 years is surely not accurate given the last 100 years' history of inflation rate fluctuations). Thus the direct use of calculations made here for specific parameter values can be called into question. To handle this uncertainty, one can proceed as in previous chapters: Assign ranges to each parameter, compute a suite of Monte Carlo results for multiple choices of specific parameter values chosen from the ranges, and then provide cumulative probability curves of profitability likelihood. In this way one covers the eventualities of a large group of choices and one also determines which parameter uncertainties are causing the greatest contributions to the probability of profit making. This particular aspect is not covered here because it depends on the regulatory aspects of each contract and on the specific financial aspects of each corporation.

CHAPTER 13
OPTION PAYMENTS FOR FUTURE INFORMATION

13.1 INTRODUCTION

Part of the problem faced by a firm involved in the business of hazardous and/or toxic material transport and burial is the unknown future. For instance, suppose the scientific knowledge at a given time advocates toxic chemical material burial and that this advice is followed. Later, advances in scientific knowledge (or more site-specific information) indicate that burial at that site was a poor option because the release of vapors from the burial site (or groundwater contamination) is toxic to neighboring communities. The corporation is then charged with the task of environmental cleanup of the site and compensation to the affected people. The cost of such remedial action can be bankrupting to the corporation and is also a future time bomb in the sense that the corporation has no idea at the time of internment of the material that future information will lead to a toxic connection. The dramatic impact of changing conditions is illustrated by the following example of remediation action at a wood processing company's site (Rotbart, 1997). An environmental consultant was hired who developed a remediation plan and solicited bids from environmental service contractors, eventually selecting a contractor who estimated the project cost at $10 million. Remediation work was initiated, but after a short period these efforts disturbed the local community with vapor clouds, which led to abandonment of the applied technology, replacement of the environmental consultant, and use of a different cleanup technology for an additional $20 million. Although in this case the burden of unanticipated costs was carried by the client company, this example amply illustrates the financial repercussions of uncertainties from site conditions and the inefficiency of technological solutions in environmental projects.

Three major options are usually available to an environmental company. First, the environmental firm can attempt to include in a contract with the regulatory authorities a limited-liability clause for a specific period of time beyond which all liability reverts to the regulatory authority. Second, it may purchase some form of insurance to provide protection against claims that arise from (1) errors, omissions, or negligence by the consulting firm (professional liability), (2) physical activities such as construction, sampling, or remedial activities (contractor's pollution liability), (3) incomplete contamination cleanup of remediated properties (postremediation coverage), or (4) third-party suits for property damage and bodily injury from disposal sites used by the insured (off-site disposal coverage) (Herrmann, 1995; Dixon, 1996; Dunn, 1997; Fletcher and Paleologos, 2000). The insurance scenario has been considered in Chap. 5.

Third, and of interest here, is the alternative of an option scenario. Consider the case of a consulting firm that is involved in the transport and burial of hazardous material or the cleanup of contaminated properties. Instead of paying an insurance premium every year against the remote chance of leakage or incomplete cleanup during the liability period, an environmental company may elect to pay a one-time amount for the option to limit its maximum liability to a fixed amount. Should no unpredictable event occur, then the company has expended only the amount paid for the original option. Should a major unanticipated environmental liability arise, however, the company has a fixed maximum liability charge plus the option cost. The remaining liability costs then have to be carried by the agency with which the environmental company negotiated the option, which may be a partner corporation, an insurance company, or a regulatory authority.

Environmental insurance and option payments have been used to facilitate merger and acquisition transactions because they translate liability responsibilities into finite, fixed costs of the transaction (Milligan, 1998). In many instances, buyers of industrial property require long-term guarantees that any potentially undiscovered contamination at the site will be paid by the seller. In addition, it is common for buyers to request indemnity agreements and guarantees against liability claims arising from historic contamination of the site. Such conditions are usually unacceptable to a selling corporation because it will have to carry on its balance sheet an unlimited cleanup cost overrun item and a high amount of contingent environmental liability for third-party claims that will severely limit the corporation's prospects in future financial transactions (Taylor, 1998).

This chapter is concerned with the use of option payments in decision-making situations in the environmental industry. Two procedures are discussed and illustrated numerically for evaluation of the maximum amount of options in the case of projects involving burial of hazardous wastes.

13.2   MAXIMUM OPTION PAYMENTS

Consider a hazardous material disposal site project that, if successful, would return a net profit G to the environmental corporation. By successful, it is meant that no leakage occurs during the period prior to which liability reverts to the regulatory authorities. If unsuccessful (i.e., if leakage occurs), then let the liability costs of site remediation be C. Let the probability of success be estimated at p_s, and so the probability of a leak is given by $p_f = 1 - p_s$. Estimates of such probabilities can be obtained by expert opinion, historical records of similar projects that the corporation has been involved in, or records of such projects from the whole environmental industry.

If the corporation does not take out an option, then it is responsible for the total remediation costs. In this case, as shown in Fig. 13.1a, the expected value E_0 of the project to the corporation is

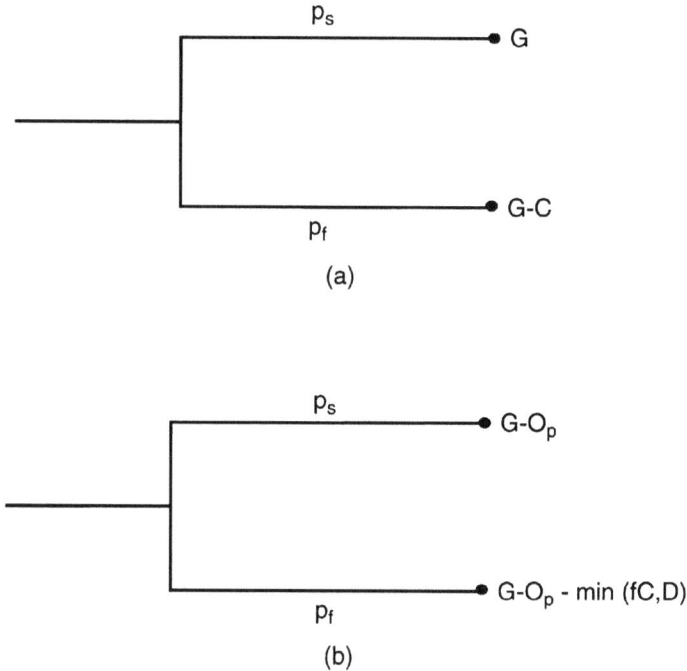

FIGURE 13.1 Decision-tree diagram when (a) no option payment has been included and (b) when option has been paid.

$$E_0 = G - p_f C \tag{13.1a}$$

with a variance σ_0^2 given by

$$\sigma_0^2 = p_s p_f C^2 \tag{13.1b}$$

If the corporation does pay an option amount O_p, then it limits liability during the period prior to reversion to regulatory authorities (or to the insurance agency that has underwritten the option contract) to a fraction f of the total remediation costs C, up to a maximum amount D. For any amount above D, the corporation is not liable, but instead, the option holder is responsible for the excess cost. Thus the decision-tree diagram is now changed, as shown in Fig. 13.1b. Effectively, the corporation lowers its net profit, if no leakage occurs, from G to $G - O_p$ in order to option against the fiscally damaging effects of unlimited liability for remediation should leakage occur.

   In this case, the expected value E_1 in Fig. 13.1b of the project to the corporation is

$$E_1 = G - O_p - p_f \min(fC, D) \tag{13.2a}$$

with the variance σ_1^2 given by

$$\sigma_1^2 = p_s p_f \, [\min(fC, D)\,]^2 \tag{13.2b}$$

There are several ways to assess the option amount to be paid for protection against unlimited liability, depending on corporate philosophy. Here, we consider two of the more common approaches.

13.2.1 Mean Value Assessment of Option Costs

The simplest, but not necessarily the best, method to determine the option amount to be paid to an insurance agency is to require that the expected project worth with optioning be greater than without:

$$E_1 \geq E_0 \tag{13.3a}$$

which requires that

$$O_p \leq p_f C[1 - \min(f, D/C)] \tag{13.3b}$$

Thus two factors determine the amount of option to be paid. First is the fraction f of total remediation cost negotiated; second is the maximum acceptable liability D relative to the remediation costs C. The most difficult task, of course, is to a priori estimate the potential remediation costs (and/or other liabilities) in the event of leakage; should they prove to be truly excessive, then the ratio D/C will be smaller than the negotiated fraction f. In this case, the maximum option amount that should be paid is limited to

$$O_p \leq p_f(C - D) \tag{13.4a}$$

Equation (13.4a) can be rearranged to provide the option amount as a fraction m of the total profit G as

$$m \leq p_f\left(\frac{C}{G} - \frac{D}{G}\right) \tag{13.4b}$$

It is clear from Eq. (13.4b) that the fraction m is controlled, primarily, by two factors: (1) the probability of failure and (2) the ratio of remediation (or liability) costs to the profit amount—both of which are highly unpredictable. Indeed, there have been several well-documented cases where remediation costs turned out to be astronomical compared with anticipated profits (e.g., accidents during the transport of oil and surface water and groundwater contamination resulting from mining activities). Another approach is to try to provide a limit for O_p (or m) based on the requirement that E_1 is positive, which yields

$$O_p \le G - p_f \min(fC, D) \tag{13.5a}$$

With unknown remediation costs, a corporation often will choose a maximum payment of less than the worst liability so that

$$m \le 1 - p_f \frac{D}{G} \tag{13.5b}$$

provides an absolute maximum that should be paid under any conditions if the expected value E_1 is to remain positive. Expression (13.5b) is of little practical value because for catastrophic events ($p_f \ll 1$) it translates to $m \approx 1$ (or $O_p \approx G$), which does not allow any profit to the corporation.

13.2.2   Probabilistic Assessment of Option Costs

For a project with a mean value E and a variance σ^2, the cumulative probability P_+ of making a profit greater than or equal to zero is provided by the equivalent gaussian expression

$$P_+ = (2\pi)^{-1/2} \int_{-q}^{\infty} \exp\left(\frac{-u^2}{2}\right) du \tag{13.6}$$

where $q = E/\sigma$. If q is large and positive, then $P_+ \to 1$ as $q \to \infty$, whereas if q is large and negative then $P_+ \to 0$ as $q \to -\infty$. Thus, for positive expected values, P_+ is always greater than 50%. The degree to which P_+ then exceeds 50% is determined by the ratio of E/σ. If there is great uncertainty about the expected value such that $E/\sigma \ll 1$, then P_+ is only slightly above 50%, whereas if $E/\sigma \gg 1$, then P_+ is close to 100%. These facts can be used to assess the amount of option payment as follows: For the project depicted in Fig. 13.1b, note that [Eq. (13.2)] although E_1 depends on the option amount, the variance σ_1^2 is independent of that amount.

Hence, in general, for any option amount one can write Eq. (13.6) in the form

$$P_+ = (2\pi)^{-1/2} \int_{-Q}^{\infty} \exp\left(\frac{-u^2}{2}\right) du \tag{13.7a}$$

where

$$Q = \frac{G - O_p - p_f \min(fC, D)}{\min(fC, D)\sqrt{p_s p_f}} \tag{13.7b}$$

As the option cost O_p is varied, then so too the cumulative probability of making a profit varies. Corporations often require that no matter what costs are entailed, a project should have a minimum probability of realizing a profit [in the oil industry this is referred to as the *minimum acceptable chance MAC* of making a profit (Lerche and MacKay, 1999)]. Each corporation has its own internal risk tolerance and sets a MAC value accordingly. Thus one equates Eq. (13.7a) with the corporate-determined MAC value and then finds Q and the maximum option amount that allows a profit probability of MAC or greater to be achieved. For example, if MAC = 0.5 (50% chance of some profit) is set, then

$$O_p \le G - p_f \min(fC, D) \equiv O_{50} \tag{13.8a}$$

whereas if MAC = 0.84 (84% chance of profit) is set, then

$$O_p \le G - \sqrt{p_s p_f}(1 + \sqrt{p_f/p_s})\min(fC, D) = O_{84} \tag{13.8b}$$

Note that $O_{84} < O_{50}$, so the higher the probability of profit a corporation insists on having, then the smaller is the maximum option amount the corporation is prepared to pay to avoid a catastrophic event. Alternatively, if the option amount that a corporation is prepared to pay is fixed, then for a mandated 50% probability of making a profit the potential gains must satisfy

$$G \ge O_p + p_f \min(fC, D) \equiv G_{50} \tag{13.9a}$$

whereas if the mandated probability of profit bearing is 84%, then the gains must be higher at

$$G \ge O_p + \sqrt{p_s p_f}(1 + \sqrt{p_f/p_s})\min(fC, D) = G_{84} > G_{50} \tag{13.9b}$$

13.3 NUMERICAL ILLUSTRATIONS

Consider an environmental project for which a corporation estimates total gains of $7 million. The corporation estimates that there is a 99% probability that there will be no leakage of hazardous material during its management of the project, prior to liability reverting to the regulatory authorities. However, if leakage were to occur, environmental remediation costs are estimated to range up to a maximum of $100 million. The corporation is interested in evaluating the utility of signing an agreement with an insurance company according to which the corporation will pay 10% of the actual remediation costs up to a maximum of $5 million.

In this case the parameters of the problem are $p_s = 0.99$, $p_f = 0.01$, $C \le \$1000 \times 10^6$, $D = \$5 \times 10^6$, $G = \$7 \times 10^6$, and $f = 0.1$. What is unknown is the actual cost of remediation. The insurance company has asked for a premium of $6 million to cover the corporate liability limit, and it is prepared to absorb all remediation costs over $5 million. Note that when the corporate liability limit of $5 million is charged, then at 10% the total remediation costs will have already reached $50 million. The insurance company would then be responsible for the remaining $45 million (and above) in costs. However, if the corporation had agreed to the option amount of $6 million, then the insurance company can use this amount to offset its total charge. Effectively, then, for both direct and indirect costs, the corporation would pay a total of $11 million maximum, whereas the insurance company would be paying for costs of $39 million and above. To assess whether the $6 million option is reasonable, the corporation first performs mean value calculations.

13.3.1 Mean Value Assessment of Option Costs

Inserting the parameter values into Eqs. (13.1a) and (13.1b), one has (in $ million)

$$E_0 = 7 - \frac{C}{100} \tag{13.10a}$$

and

$$\sigma_0 = \frac{0.995C}{10} \tag{13.10b}$$

Equations (13.2a) and (13.2b) allow calculation of the expected value and standard error of the worth of the project taking into account the option amount

$$E_1 = 7 - O_p - 0.01 \min(0.1C, 5) \tag{13.11a}$$

$$\sigma_1 = 0.995 \min(0.01C, 0.5) \tag{13.11b}$$

For actual remediation costs less than $700 million the expected value $E_0 \geq 0$; otherwise, $E_0 < 0$. Inspection of Eq. (13.11a) shows that for fractional remediation in excess of $50 million the corporation liability cap becomes effective so that, for this situation, one has (in $ million)

$$E_1 = 7 - O_p - 0.05 \tag{13.12}$$

In this case, in order that $E_1 \geq 0$, the option amount should be less than $6.95 million. If the actual remediation costs range between $50 and $0 million, then the option payment rises linearly from $6.95 to $7 million for $E_1 \geq 0$. Thus, on the basis of requiring a positive expected value, an option payment of $6 million is a good alternative for the environmental corporation.

13.3.2 Probabilistic Assessment of Option Costs

For a minimum acceptable chance MAC of 0.5, the maximum option payment (in $ million) can be calculated from Eq. (13.8a) as

$$O_p \le 7 - 0.01 \min(0.1C, 5) \equiv O_{50} \qquad (13.13a)$$

whereas for a mandated MAC of 84%, the maximum option (in $ million) [Eq. (13.8b)]

$$O_p \le 7 - 1.095 \min(0.01C, 0.5) \equiv O_{84} \qquad (13.13b)$$

For anticipated cleanup costs of less than $50 million, Eq. (13.13a) provides a maximum option cost (in $ million) of

$$O_{50} = 7 - 10^{-3} \times C \qquad (13.14a)$$

whereas for cleanup costs in excess of $50 million, the corresponding option (in $ million) is

$$O_{50} = 6.95 \qquad (13.14b)$$

Equally, if the corporation insists on an 84% chance of a profitable venture, then (in $ million)

$$O_{84} = 7 - (1.095 \times C/100) \qquad (13.15a)$$

if the remediation costs are less than $50 million, whereas it is

$$O_{84} = 7 - 0.5475 = 6.4525 \qquad (13.15b)$$

for remediation costs over $50 million. Thus the option offer from an insurance company of $6 million to limited liability appears acceptable.

Alternatively, if the corporation has no negotiating power in determination of the amount of option payment, then it can ask how large should the gains be in order to satisfy the corporate-mandated risk tolerance. At $O_p = \$6$ million, a MAC of 50% then requires gains in excess of (in $ million)

$$G_{50} = 6 + 0.01 \min(0.1C, 5) \qquad (13.16a)$$

whereas for a MAC of 84% one would require gains in excess of (in $ million)

$$G_{84} = 6 + 1.095 \min(0.01C, 0.5) \qquad (13.16b)$$

13.4 SUMMARY

Environmental projects are burdened by the prospect of liability claims that may arise long after activities at a site have been completed and the impact of low-probability and high-cost events. This situation is accentuated by the incomplete scientific information that is available at most sites and the limited technological solutions that exist to address, in a cost-effective way, remediation tasks. This condition has led many environmental companies to explore alternatives for risk reduction through the use of innovative insurance policies, partial involvement in projects as members of a consortium, etc. Payment of a one-time amount for the option to limit a corporation's maximum liability to a fixed amount has been used in many cases to facilitate financial transactions and merger and acquisitions and to set well-defined limits to future risks. Should no unpredictable catastrophic event occur, then the environmental company has expended only the amount paid for the original option. However, should a major unanticipated environmental liability arise, the company has a fixed maximum liability charge plus the option cost. The remaining liability costs then have to be carried by the agency with which the environmental company negotiated the option, which may be a partner corporation, an insurance company, or a regulatory authority.

While the option example presented in this chapter was chosen specifically for simplicity, the general method of analysis presented here is applicable to more complex situations in the environmental arena. The most challenging aspect of this approach is assessing what future unknown remediation or liability costs might be or at least bracketing the range of these costs for a project in order for a corporation to be able to evaluate what amount of option it is willing to pay. In this way a corporation can set limits today on its future potential liabilities and option out of paying. Of course, in complex situations, a corporation will have to evaluate all the risk-reducing alternatives in order to assess which one provides the optimal coverage for the least expenditure in order to maximize corporate profit.

CHAPTER 14

THE WORTH OF RESOLVING UNCERTAINTY FOR ENVIRONMENTAL PROJECTS

Aspects concerning the risk of an environmental project often are not included in projecting the value or benefit expected of the project. Three measures, involving the variance around the expected value, the volatility, and the cumulative probability of making a profit, have been discussed in previous chapters as a way to better characterize risk. This approach allows one to distinguish between projects that may have the same expected value and otherwise would be considered equal in worth. In addition, the acquisition (at some cost) of further data and the conduct of further studies in attempts to resolve better the uncertainties of a project are both shown to depend on two factors: (1) whether the information is to be acquired in order to produce desired changes in the success and failure chances of the project (which such newly acquired data may not end up doing in a substantial manner that justifies the additional cost), and (2) even if the data do as required, one cannot use the difference between expected values of the project in the absence and presence of the new data to provide an estimate of value added to the project assessment. Such a determination can arise only when measures of risk and uncertainty are used in addition to expected value because expected value, on its own, does not contain the risk information required. Numerical examples are presented to illustrate these points.

14.1 INTRODUCTION

One of the more popular and simple devices for assessing the worth of becoming involved in an environmental project in advance of actually doing the project is the decision tree. Basically, one takes the best scientific information available for a project and computes two major components: (1) the probability that one will indeed make a profit (and equally, therefore, the probability that one will not make a profit), and (2) the cost of the project and also the likely gains to be made if indeed one does go ahead with the project. It is well known that both the probability and the gains and costs are themselves uncertain (Lerche and MacKay, 1999) ahead of actually undertaking the project. Only after the project is completed does one have the knowledge that provides ground truth information.

Environmental projects often have a direct measure of worth expressed in terms of human health risk reduction or environmental remediation success. Both these measures of worth can be converted to cash values. In the case of human health risk reduction, one can equate the decreased lawsuit possibility or the lessened medical costs to a cash equivalent; in the case of a successful environmental remediation, one can use the land so cleaned for more profitable pursuits than otherwise, such as selling it to a housing or commercial development, instead of the land not returning profit. In this way the direct profitability is tied to improved use and a decreased lawsuit potential from purchasers of the land. The process of assigning cash-equivalent values is fascinating and complex in its own right but is not considered here because the dominant thrust of this chapter is toward assessing the added worth from new data collection. It is taken that the cash conversion of human health risk reduction and/or environmental risks has already been performed.

In efforts to improve on an uncertain situation, often a corporation will commit to collecting more data in an effort to accomplish two goals: (1) to improve the probability of a successful (i.e., profitable) operation of a project from the value that is available without the additional data, and (2) to reassess the scale of the likely environmental problem so that a better perception of the assumed profit becomes available. However, the new information is not obtained without additional cost. The question of interest to a corporation is: Does the extra cost of the supplementary data bring to the project either such an increase in potential worth (as well as in the probability of success) as to exceed the cash outflow for acquisition of new data and their interpretation, or are the probabilities of being (or not being) successful sufficiently well resolved with the existing data that one does not seek further involvement? In other words, the value of the added data can be to enhance the profitability of a project or to sharpen up knowledge sufficiently that a decision to be involved or not can be made more accurately. Answers to the preceding question are critical to corporate decision-making teams that consider involvement in multiple projects with variable probabilities of success and related profits (or costs) and can lead to actions that range from full involvement or abandonment of an environmental project to only partial commitment in those projects where the uncertainty in the outcome exceeds the risk tolerance of the corporation.

The simpler procedure for making such an estimate of value-added information has been decision-tree methods. At its simplest level, a decision tree, such as that depicted in Fig. 14.1, has two channels: a chance of success, labeled p_s, and a chance of failure, labeled p_f. If no other channels are available, then $p_s + p_f = 1$. For the success channel, one has potential gains G and a cost for the environmental project of C, whereas for the failure channel, one has just the cost of the project C. The way one customarily assesses the worth of the project is to calculate the expected value (EV) of the two branches together as

$$EV = p_s G - C \qquad (14.1)$$

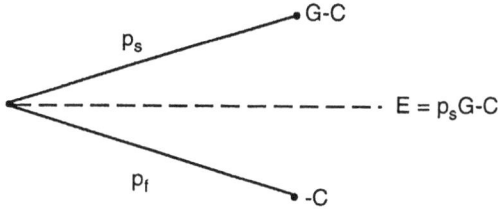

FIGURE 14.1 Sketch of a two-channel decision-tree diagram for possible outcomes of an environmental project.

If EV is positive, then it is usually assumed that, on average, the project is worth undertaking. The problem that arises is that an average value assignment fails to distinguish three major factors: (1) there is not the possibility of doing an infinite number of such environmental projects, which is where the notion of the expected value came from; instead one has just one specific project to consider; (2) the EV so calculated is not a possible outcome of the project; when one undertakes the project, the outcome is either a gain at the value $G - C$ or a loss at $-C$; and (3) even if one were to use the EV as a measure of worth of the project, there is no categorization of the uncertainty of this worth. Thus a project with $p_s = 0.1$ and gains of \$250 million at a cost of \$10 million would return an EV of \$15 million, but a project with a success probability of 0.2, gains of only \$150 million, and a cost of \$15 million would return precisely the same EV and, if this were to be the only criteria used, would be ranked as coequal with the first project. Thus one has no idea of the risk of each project or, indeed, of the relative risk between projects with the same EV values. It is often taken in the oil industry that it is sufficient to calculate the EV of a project in the absence of data collection and also to recalculate the EV under assumptions about how the new data will resolve uncertainties better as measured through the success and failure probabilities. Should the new EV be estimated higher than the old EV, then it is often taken that acquisition of the data will improve matters, and therefore, the acquisition is worthwhile. Note that there are two assumptions inherent in this procedure: (1) that one knows, ahead of acquiring the data, precisely how the success probability will be altered by the data collection, and (2) that it is an adequate procedure to use just the difference in the EV values to asses the worth of data collection relative to its cost. Neither of the assumptions is valid; indeed, they can be so misleading as to cause major upheavals in environmental project programs. And this method still does not provide the capability of distinguishing between two projects that return the same EV values both before and after data collection when one uses EV as the only statistic of relevance for a decision.

For example, even recently, McMahon et al. (1999) considered the worth of seismic acquisition and of further studies on the value of an oil exploration project based solely on the difference between the EV of the project with and without the extra data collection. These authors advocate going ahead with the seismic data costs and collection based on the EV difference and on assessments of what the differences would be in the probability of success. Their particular illustration (McMahon et al., 1999, their appendix figure) is reproduced here in modified form in Fig. 14.2. Section 14.3 provides a simple procedure that uses additional statistical measures to evaluate the worth of additional data for the situation depicted in Fig. 14.2.

The purpose of this chapter is to show that there are more useful statistical measures of worth than just the EV and that introduction of such measures permits a more accurate assessment to be obtained of the risk of projects with the same EV. Such measures also allow one to obtain a better representation of project success probability as well as the worth of collecting new data at some cost. Such understanding is not possible based on the EV as the sole measure of worth.

14.2 REPRESENTATIONS OF RISK AND UNCERTAINTY

In addition to the EV of a project as defined earlier, three other measures can help to characterize the risk and chances of success. Bearing in mind that the EV is *not* one of the two outcomes possible with the decision tree of Fig. 14.1, the most relevant quantity to obtain is the uncertainty on EV. For the two-branch decision tree of Fig. 14.1, the uncertainty is defined in terms of the standard error of EV. The second moment E_2 of the two-branch system is given by

$$E_2 = p_s(G - C)^2 + p_f C^2 \tag{14.2}$$

so the variance σ^2 is then given in general by

$$\sigma^2 = E_2 - EV^2 \tag{14.3}$$

which corresponds to an uncertainty (standard error) on EV of $\pm\sigma$. For the two-branch decision-tree of Fig. 14.1 one then obtains

$$\sigma^2 = p_s p_f G^2 \tag{14.4}$$

Use of the standard error tells one how uncertain the EV is; a small value of σ relative to EV means that the expected value is well determined; a large value of σ relative to EV means that there is considerable uncertainty on EV. Thus the inclusion of σ in the process of analyzing the worth of projects enables one to identify which projects are riskier than others even if they have identical EV values. For instance, in the example given earlier, where one had two projects with the same EV of \$15 million, one has $\sigma = \$64$ million for the project with potential gains of \$150 million, whereas one has $\sigma = \$75$ million for the project with potential gains of \$250 million. While both have the same EV, the first project is less risky in terms of uncertainty on EV than the lower probability of success but higher

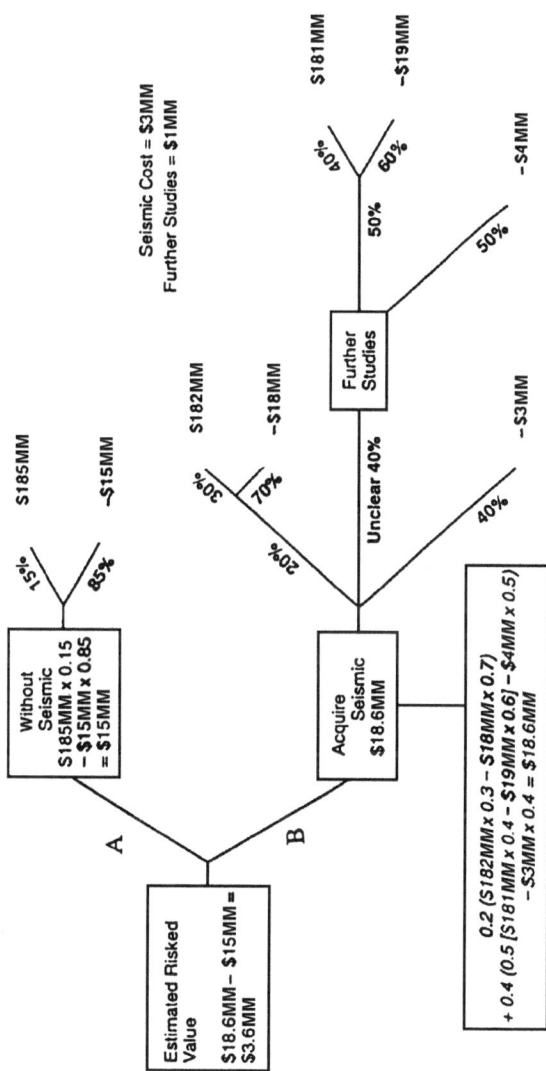

FIGURE 14.2 Decision-tree diagram on the value added to a project by a new seismic survey and modeling studies. [Modified from McMahon et al. (1999).] ($MM = $million)

potential gains project. Hence the use of standard error provides a quantitative measure lifting the ambiguity that would otherwise exist on projects with the same EV as well as providing a measure of the accuracy of EV as a parameter capable of quantifying a project's worth.

As a device for assessing the quality of the risk of a project, a combination of standard error and EV can be used, called the volatility v, defined by

$$v = \frac{\sigma}{|EV|} \tag{14.5}$$

The volatility provides a measure of the accuracy of the EV value. A volatility much less than unity v ($\ll 1$) indicates that there is but little uncertainty in EV, and so it can be used as an accurate representation of the worth of a project. However, a volatility much larger than unity ($v \gg 1$) indicates that there is considerable uncertainty, and so the use of EV as a representative measure of the worth of a project is severely compromised. The basic information relayed in the case of high volatility is that the EV value is so far removed from either the success branch worth or the failure branch worth that there is really very little relevance to EV as a characteristic measure of a project's value. For instance, in the two cases presented earlier, the first project has a volatility of 4, whereas the second has a volatility of 5, so both projects are fairly high risk relative to using EV as the project worth measure, and the second project is 25% riskier than the first (volatility ratio of 5/4) relative to using volatility as the measure of risk.

Based on only the EV and the standard error information σ for a project, one can write the equivalent gaussian probability $P(W)$ that the project should return a worth greater than or equal to a specified value W as

$$P(W) = (2\pi\sigma^2)^{-1/2} \int_W^\infty \exp\left[\frac{-(x - EV)^2}{2\sigma^2} \right] dx \tag{14.6}$$

which, with the substitution $x = EV + 2^{1/2}\sigma u$, can be rewritten in the form

$$P(W) = \pi^{-1/2} \int_B^\infty \exp(-u^2)\, du \tag{14.7}$$

where $B = (W - EV)/(2^{1/2}\sigma)$. If one is interested in the probability of making a profit of EV or greater, then $W = EV$ (and so $B = 0$) so that $P(EV) = 50\%$, as it should. One of the conventional uses of this cumulative probability measure of worth is to estimate the probability of making any profit at all, i.e., $W = 0$, corresponding to $B = -1/(2^{1/2}v)$. Thus there is a relatively tight intertwining of EV, standard error σ, volatility v, and probability of profit measured by Eq. (14.7). Each of the three measures uses different combinations of EV and standard error to provide different perspectives on the worth of going ahead with involvement in a project and of assessing the risk of the project. In particular, it is clear that such measures provide a much sharper picture of what is occurring with a project than does the EV alone, which can be extremely ambiguous, does not include the riskiness of the project, and often does not bear much resemblance to the possible outcomes of a project—a major deficit that is compensated for using the other three measures.

14.3 APPLICATION TO THE QUESTION: IS VALUE ADDED?

The application of the preceding measures of risk is of considerable use when evaluating the worth of acquiring more data, or of doing further studies, in order to refine the estimate of the worth of a project. This type of concern is generally referred to as the *value of added information*. In particular, in this section we work through an application of these risk procedures using Fig. 14.2, modified from McMahon et al. (1999), to evaluate the worth of obtaining additional seismic data for an oil exploration project.

Consider, then, the upper decision tree of Fig. 14.2, labeled *A*. Branch *A* does not involve any additional collection of seismic data, and at a probability of success of 15%, a profit of $185 million can be returned by the project, whereas a cost of $15 million will be incurred with a probability of failure at 85%. Figure 14.2 indicates that the expected value for this branch is $EV_A = \$15$ million, and the simple calculation of this value involves the multiplication of profit (or cost) by the probability of success (or failure). Using the expressions of Section 14.2, it is then a simple matter to show that $\sigma_A = \$71.4$ million, so the volatility of the upper section of the decision tree is $\nu_A = 4.76$, considerably in excess of unity. Thus EV_A is extremely poorly determined, and indeed, the corresponding probability of making any profit at all is only $P(0)_A = 58.4\%$, making the project one of high risk and marginally likely to be worthwhile.

Consider then the lower decision tree of Fig. 14.2, labeled *B*. Initially, an amount of $3 million can be spent for a new seismic survey, the results of which are estimated to produce three possible outcomes. With a probability of 20%, the new survey is expected to delineate the probabilities of success and failure to 30% and 70%, respectively, with a profit now of $182 million ($185 million minus the cost of the survey, $3 million) or a loss of $18 million (the $15 million cost from the previous case of branch *A* plus the cost of the survey). The new survey also may not resolve the situation unequivocally, with a probability of such an eventuality estimated at 40%, or simply may validate the initial assumptions of branch *A* in terms of probabilities and costs but at an expense of $3 million. It is assumed that further studies at an extra amount of $1 million are capable of improving the resolution of the project and in an unambiguous manner either validate the initial assumptions, with a probability of 50% of such an event occurring, at a total cost of $4 million now or clearly redefine the probabilities of success or failure at 40% and 60%, respectively, but with an increased cost that reduces the profitability of the project. These altered probability and cost parameters of the project that includes supplementary seismic data are now shown on branch B of Fig. 14.2. The question that a corporation is facing is whether such an expenditure of money is worthwhile. A total of $4 million more is being spent than would have been the case if the corporation had just used the upper branch A of the decision tree to decide on involvement in the project. While the lower branch B of the decision tree is considerably more complex than branch A, the statistical measures to extract are just EV_B and σ_B. EV_B was calculated by McMahon et al. (1999), and it is easy to show that the value of the expected value of branch B is $EV_B = \$18.6$ million.

The difference in the EV values for branches A and B in Fig. 14.2 is $3.6 million, but it is not an accurate measure of the value added without including the attendant uncertainty, which is significant. Thus, in the case of branch B, it can readily be shown using Fig. 14.2 that $\sigma_B = \$66.02$ million, so there is still considerable uncertainty on EV_B because the volatility is now $v_B = 3.55$. While reduced compared with the value of 4.76 for branch A, the volatility is still large compared with unity, and so the project is still extremely risky. The relative improvement in risk, as measured by the ratio of volatilities (3.55/4.76), is some 25%, so it is true that the project uncertainty has been reduced by acquisition of the survey and the cost of the further studies. However, the improvement has not converted the project into one in which there is a high chance of profitability. Indeed, the cumulative probability of turning any profit at all for branch B is $P(0)_B = 61.2\%$—only a 2.8% change in the probability of profit estimate from branch A that was obtained in the absence of spending $4 million on the new seismic survey and the further studies. And the chance of being profitable is still not very far above the 50:50 break-even probability value of EV or greater. In branch A one has only an extra 8.4% chance of making any profit at all, and branch B only improves matters to an extra 11.2% above the 50% chance of getting a return at EV or greater. There is a considerable expenditure of extra money ($4 million) for virtually no change in the conditions or resolution of the project. And this statement can be quantified easily because the variance on the value-added information can be written as

$$\sigma_{VA}^2 = \sigma_A^2 + \sigma_B^2 \tag{14.8}$$

which, for the values $\sigma_A^2 = 71.4^2$ and $\sigma_B^2 = 66.02^2$, yields the uncertainty measure on the mean value added (VA = $3.6 million) of $\sigma_{VA} = \$97.24$ million, reflecting directly the large uncertainties of the EV values as representative measures of branches A and B. Correspondingly, the volatility of the value-added mean estimate is $\sigma_{VA}/VA = 27.01$, which is extremely large compared with unity, strongly suggesting that there is no meaning to the $3.6 million difference between branches A and B because the uncertainty is so large on both the EV values.

One also can compute the cumulative probability that one should spend anything at all on acquisition of new data and further studies. The calculation basically uses the same cumulative probability method as given in the preceding section; one finds then that the probability one should not spend any extra money is 48.5%. And the probability that a value of $3.6 million or greater will be brought to the project is only 50%. In short, the expression of mean value added is not very trustworthy because of its large uncertainty. At one standard error uncertainty, the value added can range from a positive value of $(3.6 + 97.24) million = $100.84 million all the way to $(3.6 − 97.24) million = −$93.64 million. This range is so large compared with the mean value of branch A that one really has no indication that acquiring a new seismic survey as well as doing further studies is either adding to or detracting from the project worth obtained in the absence of spending the $4 million extra for such new information. In short, when uncertainties are allowed for, the acquisition of new seismic data and further studies do not better resolve the project. They should not be undertaken.

It must be stressed here that uncertainty also exists on the project parameters for success probabilities, as well as on the potential gains that could accrue (due to both future economic uncertainties and scientific uncertainties in the estimates of probabilities and costs of the project). Furthermore, one also must take into account that the predicted improvements in the probability values on branch B are conditional on the resolution of the seismic survey, i.e., the estimated probabilities of 20%, 40%, and 40% of the seismic survey unequivocally clarifying the situation, of not being able to provide resolution, and of validating the initial assumptions, respectively, all of which are highly questionable. The same argument applies for the capability of further studies to resolve the situation. It would seem that the improvement of 2.8% in probability of profit is buried in the "noise" caused by the uncertainty of the parameters. The project risk has not been changed substantially. No survey or further studies should be undertaken; the extra cost of $4 million far outweighs the potential improvement estimated to occur and does little to nothing to lower the risk of the project to any significant degree.

14.4 SUMMARY

The purpose of this chapter has been to show that using just the expected value (EV) of a project can be misleading if one is interested in evaluating the worth of an environmental project and its risk together with the likelihood the project will prove profitable. The quintessential reason is that the EV does not represent one of the possible outcomes of the project, and because one does not have the luxury of undertaking an infinite number of trials (for which the EV would indeed represent the average of all outcomes), then one must include some measure of the uncertainty around the EV. Failure to do so means that one has a very poor idea how risky a project is, nor does one have any idea of its profit potential. Neither of these factors is included in the EV, which is not one of the possible outcomes for a single project and so does not represent fairly how a single project should be analyzed. This point was made using an illustration of two projects with the same EV but very different risk factors, as measured by the standard error of the EV as well as by the volatility.

Based on a modified example from McMahon et al. (1999) that seeks to evaluate the worth of additional information to an oil exploration project through a seismic survey and further studies, it was shown that such an assessment cannot be addressed using only the EV for the extra value. In particular, the changes in the estimated probabilities of success and failure are *only* estimates, made prior to the acquisition of new data and further studies. One does not know to what extent the information obtained would really change parameters to the desired values. And even then, one has to work through not just a difference in estimated EV values in the presence and absence of the new information, but one must, more properly, take into account the uncertainty of the EV values. In this way one correctly estimates volatility and profit probability for the project incorporating some measure of the uncertainty of EV. Further, there is uncertainty on all parameters entering such project assessments from both future economic concerns of uncertainty and scientific uncertainties concerning the environmental parameters of the project. The simple procedure presented in this chapter can be applied as an initial evaluation of the worth of obtaining information on a project, with more advanced techniques available in the bayesian decision literature discussed in Chap. 10.

CHAPTER 15
SCIENTIFIC UNCERTAINTY IN ENVIRONMENTAL PROBLEMS: MODELS AND DATA

This chapter addresses the issue of scientific uncertainty in the quantitative description of physical processes that are of interest in environmental problems. The objective of this chapter is to illustrate the point that, in addition to the economic uncertainties that have been dealt with in this book, there exists significant uncertainty in the scientific component of environmental projects. These scientific uncertainties usually arise from the incomplete quantification in a model of the complex interaction of physical, chemical, and biologic processes that characterize most environmental problems, the small number of data that are collected to elucidate these processes, and the measurement errors that usually characterize data sets. In this respect, this chapter can be studied independently of the material of this text that deals with economic risk analysis, its purpose being to serve as a reminder of the complexity of issues entering environmental projects, full treatment of which requires an integrated analysis of both the scientific and economic uncertainties.

In order to illustrate the issue of scientific uncertainty, three major problems in groundwater flow determinations are discussed. First, a quantitative method is provided to show how data can be used to help differentiate between competing model representations of groundwater flow, and a quantitative definition is provided for such resolution. An example is given from steady-state one-dimensional flow to illustrate the method using a continuous, perfectly measured head data field. Second, the method is generalized to finitely sampled but perfect data. Third, the effects of data uncertainty are examined, and a procedure is provided that allows determination of model parameters (and their uncertainties) from uncertain data fields. An example is given to illustrate when collection of additional data improves predictive capabilities and when improved model resolution is needed first before further data collection is embarked on.

The importance of the quantitative procedure presented here is that it allows one to recognize when sufficient data resolution has been achieved that models can be distinguished one from another, and the optimal model representation of a groundwater flow system can be established. Alternatively, the procedure provides information on when a model is not adequately describing the physical system so that better representation of the hydrogeologic flow regime is necessary.

15.1 INTRODUCTION

The purpose of this chapter is to demonstrate how one can use observations of hydraulic head to differentiate proposed hydrodynamic flow models in attempting to unravel subsurface flow conditions. The sense of the argument is most simply given using a one-dimensional steady-state depiction of hydrodynamic flow, although the basic foundation of the argument is quite general.

Imagine, then, that one had a set of measurements of hydraulic head $h(x)$ for various subsurface depths x ranging from the surface ($x = 0$) to some depth L. In the simplest case (discussed in Sec. 15.2), one can consider that $h(x)$ is measured continuously with increasing depth; in more realistic situations (discussed in Sec. 15.3), $h(x)$ is taken to be measured at a finite set of depths only. Consider that different steady-state flow models have been proposed to account for the observations. Clearly, not all such models can be equally valid (although all could be equally invalid). The basic question posed here is: How can one use the observations and the models in a quantitative sense to help differentiate between more appropriate versus less appropriate models, and how does one define quantitatively *appropriate?* The ability to provide an answer to this question would help enormously in attempts to address the physical subsurface flow regime and to provide an assessment of the hydraulic conductivity most relevant to a given situation (Yeh, 1986).

Further, it is most often the case that measurements of hydraulic head have some associated error, either statistical or systematic. Thus the estimates of relevant model behaviors also must incorporate this uncertainty (Delhomme, 1979). Questions of some significance are: When different models are involved, how does one assign this observational error between the models? And how does one provide a quantitative measure of when more measurement precision and accuracy are required to make further headway with unraveling the more appropriate from the less appropriate models?

Answers to these questions are important in attempts to evaluate subsurface flow conditions. Determination of such conditions is of primary importance in environmental remediation projects because if an incorrect model is used, it is quite possible that one would do more harm than good in initiating remedial action. Thus the ability to provide rational, objective, reproducible quantitative measures is of fundamental significance.

Throughout the body of this chapter we consider the situation of just two competing models and carry through the detailed analysis, including illustrative examples, for such a situation. The general method involving an arbitrary number of competing models is relegated to appendices. The basic reason for this relegation is that the points to be made are most easily portrayed with two competing models. The general analysis is no more difficult, but is more tedious and tends to obscure the essential points of the depiction without adding any new significant insights.

Three basic sections constitute the body of this chapter. Specifically, in Section 15.2 it is taken that the hydraulic head is measured continuously and precisely (no error), and we then show how one can provide a quantitative assessment of the more appropriate of two competing models for describing the subsurface flow. Section 15.3 relaxes the assumption of continuous measurement of hydraulic head and deals with discrete measurements made at a finite number of depths but still retaining the assumption of error-free measurements. Section 15.4 relaxes the assumption of error-free measurements and includes statistical error so that one can assess the worth of improving the models in relation to the worth of improving measurement error.

In this way a progressive trend of patterns of behavior is provided that enables one to determine quantitatively where one should place further emphasis and, more importantly, why one should do so. Illustrative examples are provided so that one can follow through, in detail, how the various calculations are sensitive to the differences between proposed models in relation to available measurements.

15.2 CONTINUOUS MEASUREMENTS OF HYDRAULIC HEAD

Suppose that measurements of hydraulic head $h(x)$ have been made continuously in $0 \leq x \leq L$. Two different models of hydraulic head dependence on x have been proposed, labeled $h_1(x)$ and $h_2(x)$, respectively. It is required to determine which, if either, is a more appropriate model to satisfy the continuous observations. The procedure for such a determination operates in two parts as follows.

15.2.1 Weighting Factors

Because two different models have been proposed, it is possible that the geologic conditions pertaining to hydraulic conductivity are themselves uncertain. Thus it may be that there is some truth to both models and that a linear combination of both is a better representation of the observed measured head than would either model taken on its own. Assign constant weighting factors w_1 and w_2, respectively, to the models with the requirement that

$$w_1 + w_2 = 1 \qquad (15.1)$$

The combined representation of hydraulic head is then just $w_1 h_1(x) + w_2 h_2(x)$. If this representation is to most accurately portray the observed head measurements throughout the interval $0 \leq x \leq L$, it then follows that one should seek to minimize the quadratic functional χ^2 given by

$$\chi^2 = \int_0^L [h(x) - w_1 h_1(x) - w_2 h_2(x)]^2 \, dx - 2\lambda(1 - w_1 - w_2) \qquad (15.2)$$

with respect to changes in w_1 and w_2 for *prescribed* behaviors of the two models $h_1(x)$ and $h_2(x)$. The factor λ in Eq. (15.2) is a Lagrange undetermined multiplier to be determined so that Eq. (15.1) is also satisfied simultaneously with the minimization of Eq. (15.2).

Differentiating Eq. (15.2) with respect to w_1 and setting the derivative to zero yield

$$\frac{1}{2}\frac{\partial \chi^2}{\partial w_1} = \int_0^L [h(x) - w_1 h_1(x) - w_2 h_2(x)]\, h_1(x)\, dx - \lambda = 0 \qquad (15.3a)$$

whereas similar differentiation of Eq. (15.2) with respect to w_2 yields

$$\frac{1}{2}\frac{\partial \chi^2}{\partial w_2} = \int_0^L [h(x) - w_1 h_1(x) - w_2 h_2(x)] h_2(x)\, dx - \lambda = 0 \qquad (15.3b)$$

Equations (15.1), (15.3a), and (15.3b) represent three linear equations in the three unknowns w_1, w_2, and λ. Direct solution yields

$$w_1 = \frac{1}{D}\int_0^L [h_2(x) - h(x)][h_2(x) - h_1(x)]\, dx \qquad (15.4a)$$

and

$$w_2 = \frac{1}{D}\int_0^L [h_1(x) - h(x)]\,[h_1(x) - h_2(x)]\, dx \qquad (15.4b)$$

together with

$$\lambda D = \left[\int_0^L h_1(x)h_2(x)\, dx\right]^2 - \left[\int_0^L h_1^2(x)\, dx\right]\left[\int_0^L h_2^2(x)\, dx\right]$$

$$+ \sum_{i=1}^{2}\left[\int_0^L h(x)\, h_i(x)\, dx\right]\left\{\int_0^L h_j(x)\left[h_j(x) - h_i(x)\right]dx\right\} \qquad (15.4c)$$

where in Eq. (15.4c) $j = 2$ when $i = 1$ and $j = 1$ when $i = 2$ and

$$D = \int_0^L [h_1(x) - h_2(x)]^2\, dx \qquad (15.4d)$$

Note from Eqs. (15.4a) and (15.4b) that if $h_2 \to h$, then $w_1 \to 0$ and $w_2 \to 1$. Equally, if $h_1 \to h$, then $w_2 \to 0$ and $w_1 \to 1$. Thus the weighting factors represent the quantitative degree to which each proposed model is supported by the data available.

15.2.2 Parameter Determination

With the values w_1 and w_2 given by Eqs. (15.4a) and (15.4b), it follows that $w_1 + w_2 = 1$ always. Therefore, one now considers the two quadratic functionals

$$Y_1^2 = \int_0^L [h(x) - h_1(x)]^2 \, dx \tag{15.5a}$$

$$Y_2^2 = \int_0^L [h(x) - h_2(x)]^2 \, dx \tag{15.5b}$$

The reason for considering these two functionals is that each proposed model contains parameters that are, in general, to be determined. For instance, if it is taken that the hydraulic conductivity has a linear increase or decrease with depth, then the scaling length over which the conductivity doubles is the relevant parameters to be determined by matching that proposed model to the continuous hydrodynamic head data. Similarly, for each proposed model there will be a suite of parameters to be determined. Let the vector $p_1(p_2)$ denote the set of all such parameters in the first (second) model. Then one can write

$$Y_1^2(p_1) = \int_0^L [h(x) - h_1(x, p_1)]^2 \, dx \tag{15.6}$$

where the dependence of the model hydrodynamic head on p_1 has been explicitly displayed. Clearly, a best least squares fit of model head h_1 is achieved when, for all components of the vector p_1, there is a minimum of $Y_1^2(p_1)$. For the jth component of the vector p_1, that is, p_{1j}, one then has

$$\int_0^L [h(x) - h_1(x, p_1)] \frac{\partial h_1(x, p_1)}{\partial p_{1j}} \, dx = 0 \tag{15.7}$$

For each component of p_1 a similar equation is available. Thus one finds the best set of parameters for each proposed model that allows closest satisfaction of the measured head data. However, because there is no fundamentally compelling reason for the data to conform to the assumptions of a preconceived model, in general a model does not fit the head data $h(x)$ perfectly. Equipped with the best available parameters for each model, one calculates w_1 and w_2 as above to determine the most appropriate model combination to use to best satisfy the continuous head data in $0 \leq x \leq L$.

15.2.3 Illustrative Example

Two Models. Consider two one-dimensional steady-state models as follows: Model 1 assumes saturated groundwater flow with a steady input flux q and a constant hydraulic conductivity K_1 between $x = 0$ and $x = L$. On $x = 0$, the head is h_0. Then, from Darcy's law,

$$q = -K \frac{dh}{dx} \qquad (15.8)$$

the modeled behavior of hydraulic head is

$$h(x) = h_0 - \frac{q}{K_1} x \qquad (15.9)$$

On $x = L$, a head of $h(L)$ then corresponds to a flux of

$$\frac{q}{K_1} = \frac{h_0 - h(L)}{L} \qquad (15.10)$$

so that one can equally well write Eq. (15.9) in the form

$$h(x) = h_0 + \left[h(L) - h_0 \right] \frac{x}{L} \equiv h_1(x) \qquad (15.11)$$

For model 2 it is taken that the hydraulic conductivity varies with x in the form

$$K(x) = \frac{K_2}{1 + \varepsilon x/L} \qquad (15.12)$$

where K_2 and ε are constants. For the same input flux q and the same head conditions, the corresponding model head behavior is

$$h(x) = h_0 - x \frac{q}{K_2} \left(1 + \frac{\varepsilon x}{2L} \right) \equiv h_2(x) \qquad (15.13a)$$

where

$$\frac{q}{K_2} = \frac{h_0 - h(L)}{L(1 + \varepsilon/2)} \qquad (15.13b)$$

Note that it is not possible to maintain both the same head $h(L)$ at $x = L$ and simultaneously the same flux for models 1 and 2 if K_1 and K_2 are chosen to be the same. Alternatively, if one chooses K_2 in Eq. (15.13b) to be $(1 + \varepsilon/2)$ larger than the corresponding K_1 in model 1, then the same flux and same head at $x = L$ can be preserved in both models. Between $x = 0$ and $x = L$, however, the hydrodynamic head varies differently with increasing values of x in both models. Indeed, with $K_2 = K_1(1 + \varepsilon/2)$, one has from Eq. (15.13a)

$$h_2(x) = h_0 + x \frac{[h(L) - h_0][1 + \varepsilon x/(2L)]}{L(1 + \varepsilon/2)} \qquad (15.14)$$

showing directly that $h_2 \neq h_1$ except on $x = 0$ and $x = L$.

Data. Consider now a set of continuous measurements of steady-state hydrodynamic head in $0 \leq x \leq L$. Take the measurements to provide the representation

$$h(x) = h_0 + [h(L) - h_0]x \frac{\exp[-b(x - L]}{L} \qquad (15.15)$$

where b is a known constant. Notice that in this case neither model 1 nor model 2 can fit the continuous head data precisely at all x, although the head data at $x = 0$ and $x = L$ are satisfied by both models. The best that can be achieved is to adjust parameters in the models to come as close as possible to satisfying the observed continuous head data. In model 1 there are no free parameters once the head at $x = 0$ and $x = L$ is adjusted to conform to the observed values. In model 2 there is just the one free parameter, ε, which can be adjusted to minimize discord between the observations [as represented through Eq. (15.15)] and the model [as represented through Eq. (15.14)].

However, before carrying out this operation, there is a point worth mentioning. It could be argued that there is no need to go through the exercise of comparing models because one could just differentiate Eq. (15.15) to evaluate the corresponding hydraulic conductivity K_{eff} (Dagan, 1989; Gelhar, 1993; Paleologos et al., 1996; Dagan and Neuman, 1997) from

$$q = -K_{eff} \frac{dh}{dx} \qquad (15.16)$$

yielding

$$K_{eff} = -qL[h(L) - h_0]^{-1} \frac{\exp b(x - L)}{1 - bx} \qquad (15.17)$$

Such an argument is formally correct *provided* one makes the assumption that the observations are indeed from a one-dimensional saturated groundwater system. However, when one remembers that the observations represent measurements only at one location, albeit in $0 \leq x \leq L$ for the third dimension, there is no particularly good reason that one should be allowed to use Eq. (15.16) to recover an effective hydraulic conductivity. In particular, suppose that the observations yield a value b such that $bL > 1$. Then the effective conductivity in Eq. (15.17) will go from positive infinity to negative infinity as x crosses $1/b$—which it does in the range! In this case, one can be absolutely sure that the one-dimensional measurements of head at a location represent flow from lateral directions as well as from vertical directions.

The point is that the assumed model representations of flow behavior do not have to be honored by the real data. What one is attempting to do is to ascertain how well models can be adjusted to best conform to observed data. The simple models presented earlier, and the simple representation of a continuous data field, best illustrate this point without the obfuscating technical details that attend a more general investigation (but see the appendices).

Parameter Determination. Consider first the model parameter determination. The only model with a free parameter is model 2, where ε is available. The least squares control function is chosen to be a weighted quadratic of the mismatch between observed and modeled heads in the form

$$Y^2 = \int_0^L \left(1 + \frac{\varepsilon}{2}\right)^2 [h(x) - h_2(x)]^2 \, dx \tag{15.18}$$

Substituting Eqs. (15.14) and (15.15) into Eq. (15.18) yields

$$Y^2 = \left[\frac{h(L) - h_0}{L}\right]^2 \int_0^L x^2 \left\{\left(1 + \frac{\varepsilon}{2}\right) \exp\left[-b(x - L)\right] - \left(1 + \frac{\varepsilon x}{2L}\right)\right\}^2 \, dx \tag{15.19}$$

Minimizing Eq. (15.19) with respect to ε produces an equation for ε in the form

$$\frac{\varepsilon}{2} \int_0^L u^2 \left\{\exp\left[-B(u - 1)\right] - u\right\}^2 \, du = \int_0^L u^2 \left\{\exp\left[-B(u - 1)\right] - u\right\}$$

$$\times \left\{1 - \exp\left[-B(u - 1)\right]\right\} \, du \tag{15.20}$$

where $u = x/L$ and $B = bL$. Hence, for a given value of B, the integrals in Eq. (15.20) can be evaluated to provide an expression for ε for model 2.

Weighting Factors. With models 1 and 2 prescribed through Eqs. (15.11) and (15.14), respectively, and with the best-fit parameter ε for model 2 given by Eq. (15.20) in terms of integrals over the continuous data field of Eq. (15.15), it follows that one can now evaluate directly the relative weights w_1 and w_2 with which the two models provide a cumulative best fit to the continuous data. In this way one obtains that combination of models 1 and 2 which most accurately satisfies the data field. Then, with $B = bL$, one can evaluate the integrals in Eq. (15.20) (Gradshteyn and Ryzhik, 1980; Macsyma, 1996) to obtain

$$\left(1 + \frac{\varepsilon}{2}\right)\left(\frac{e^{2B}}{4B^3} - \frac{12e^B}{B^4} + \frac{12}{B^4} + \frac{47}{4B^3} + \frac{11}{B^2} + \frac{3}{2B} + \frac{1}{5}\right)$$

$$= \frac{2e^B}{B^4}(B - 3) - \frac{1}{20} + \frac{1}{B^4}(B^2 + 4B + 6) \tag{15.21}$$

and also calculate the weight w_1 from Eq. (15.4a) as

$$\frac{1}{60} w_1 \varepsilon = \frac{-1}{12} - \frac{\varepsilon}{40} - \left(1 + \frac{\varepsilon}{2}\right) B^{-4} [(6 - 2B) \exp(B) - 6 - 4B - B^2] \quad (15.22a)$$

Note that as $B \to 0$, then

$$\frac{\varepsilon}{2} \to -B\left(1 + \frac{B}{4}\right) \quad (15.22b)$$

and that as $B \to \infty$, then

$$\frac{\varepsilon}{2} = -1 + 8 \exp(-B) \quad (15.22c)$$

In order to avoid even the possibility that one would end up in the nonphysical regime ($\varepsilon < -1$) corresponding to the reversal of model permeability in the region $0 \leq x \leq L$, for numerical illustration we take throughout $B = 1/4$. This choice is made to ensure that the model has at least a chance of honoring some of the data. In the simple case where $B = 1/4$, one has $\varepsilon = -1.973$. The corresponding weight factors are $w_1 = 0.982$ and $w_2 = 0.018$. The implication is that the best combination of the two models to fit the data is 98.2% of the constant conductivity model and 1.8% of the parabolic head model 2 and with the best scale parameter for model 2 at $\varepsilon = -1.973$. This combination gives the representation of

$$w_1 h_1 + w_2 h_2 = h_0 + [h(L) - h_0] \frac{x}{L} \left(2.315 - \frac{1.315x}{L}\right) \quad (15.23a)$$

whereas the continuous observations give [Eq. (15.15) for $B = bL = 1$ and $u = x/L$]

$$h = h_0 + \left[h(L) - h_0\right] \frac{x}{L} \exp\left(1 - \frac{x}{L}\right) \quad (15.23b)$$

It may appear counterintuitive that a linear model (model 1) has a higher weighting coefficient than the parabolic model (model 2) in attempts to fit data with the x-dependent structure $\{(x/L) \exp[-B(x/L - 1)]\}$, but the reason is relatively clear. The structure of the combined model fit $w_1 h_1 + w_2 h_2$ can be written as

$$\frac{w_1 h_1 + w_2 h_2 - h_0}{h(L) - h_0} = \left(w_1 + \frac{w_2}{1 + \varepsilon/2}\right) \frac{x}{L} + w_2 \left(\frac{\varepsilon/2}{1 + \varepsilon/2}\right) \left(\frac{x}{L}\right)^2 \quad (15.24)$$

The linear term in (x/L) has a coefficient of 2.315, whereas the quadratic term coefficient is -1.315 [see also Eq. (15.23a)]. Direct comparison of Eq. (15.23b) with Eq. (15.23a) indicates that the combined model attempts to approximate

$$\frac{x}{L}\, e^{(1\,-\,x/L)} \approx \frac{x}{L}\left[2.315 - 1.315\left(\frac{x}{L}\right)\right] \tag{15.25}$$

Thus the constant factor 2.315 is doing the best it can to match the exponential at $x = 0$, whereas the factor -1.315 in the quadratic term is arranged to balance the fact that at $x = L$, both the left- and right-hand sides of Eq. (15.25) are in agreement at unity. However, the least squares control function cannot simultaneously arrange w_2h_2 at $x = L$ to dominate over the term w_1h_1 because w_2 must handle *both* the linear term in (x/L) *and* the quadratic factor. Because w_1h_1 is itself linear in (x/L), the quadratic functional then arranges that the *total* linear term comes as close as it can to satisfying $(x/L)\exp(1 - x/L)$ while at the same time adjusting the quadratic term to handle the behavior near $x = L$. This compromise is then best served by assigning high weight w_1 onto the linear term in (x/L) and a smaller balancing form to the $(x/L)^2$ term of $w_1h_1 + w_2h_2$. While perhaps initially counterintuitive, once the result is seen, the reasons for its appearance are clear and intuitive in hindsight.

Thus the combined model is not a very good fit to the observations, but it is the best available under the constraints of the two model assumptions. Further, the weights indicate that considerably more understanding of the data field is obtained with the parabolic head model, thus suggesting that more such spatially variable models should be used to obtain a greater degree of consistency between models and data. It is this point that the illustrative example speaks to most strongly.

15.2.4 Spatial Changes and Partitions

The preceding arguments apply to a continuous data field and models in $0 \le x \le L$. However, it can (and does) happen that groundwater flow may take place through a medium in which major changes are suspected to occur in hydraulic conductivity. A simple example is that in $0 \le x \le a$ it is estimated that the hydraulic conductivity is almost constant so that model 1 is appropriate, whereas in $a \le x \le L$ it is thought that the conductivity is more closely described by model 2. The problem addressed here is to obtain the best value of the transition distance a so that the misfit is minimized between head observations $h(x)$ and the linearly combined weighting of the two models.

In this case one assumes that there are two constant values for each weighting factor, w_1 being appropriate in $0 \le x \le a$ and W_1 being relevant in $a \le x \le L$, with $w_2 = 1 - w_1$ and $W_2 = 1 - W_1$, respectively. The least squares control function is now made up of two pieces:

$$\chi^2 = \int_0^a [h(x) - w_1h_1(x)(1 - w_1)h_2(x)]^2\, dx$$
$$+ \int_a^L [h(x) - W_1h_1(x)(1 - W_1)h_2(x)]^2\, dx \tag{15.26}$$

Some care has to be exercised in this case because the free constant parameters in each model have to be determined consistent with the inherent assumption of a shift from model 1 dominance to model 2 dominance as x crosses a. For instance, if it is *assumed* that model 2 alone were to be correct, then the parameter ε would be best determined by Eq. (15.5b) but with the integral limits as $x = a$ and $x = L$, respectively, rather than $x = 0$ to $x = L$. More generally, one merely differentiates Eq. (15.26) with respect to w_1, W_1, ε, and a in order to obtain four equations at the extrema that can be used as before to determine all four unknowns once the continuous head data field $h(x)$ is given. An alternative is available for a minimum search. Note that Eq. (15.26) is quadratic in the two unknown weights w_1 and W_1 but is, in general, highly nonlinear in the remaining parameters. Thus differentiation with respect to the weights will produce linear equations—just as for Eq. (15.4)—which then enable w_1 and W_1 to be expressed in terms of integrals over the head data field $h(x)$ and the two models $h_1(x)$ and $h_2(x)$. When these general variations are substituted into Eq. (15.26), one then has a behavior for χ^2 that is dependent on the remaining two parameters ε and a. The nonlinear search procedure given in Appendix 15B can then be employed to determine the best values. While not difficult to implement, such an analysis is tedious analytically and is best examined numerically.

The point of presenting an outline of the analysis here for the two simple models is to demonstrate that the same general sense of argument can be used to find the best partitioning for assumed model behaviors and given data. In this case, *best* is interpreted so that the global mismatch of data and weighted models throughout the total domain of observations is minimized. In this way one can determine when the weighting factors shift over from dominance of one model to dominance of the other; this shift then provides an indication of the depth at which there are major changes in the physical hydraulic conductivity.

15.3 DISCRETE MEASUREMENTS OF HYDRAULIC HEAD

While Sec. 15.2 provides the general argument for assessing model dominance relative to observations, it makes the assumption that a continuous field of hydraulic head measurements is available. In practice, however, it is usually the case that observations are made only at a finite number of depths in the range $x = 0$ to $x = L$. For $k = 1,\ldots, N$, let these depths be labeled $x(k)$ and the corresponding head values $h(k)$. In this case one would like to ensure that the assumed model behaviors are weighted so that there is minimum mismatch at the observation positions. In the two-model case, the quantity to minimize is the discrete analog to the preceding section, i.e.,

$$\chi^2 = \sum_{k=1}^{N} [h(k) - w_1 h_1(k) - (1 - w_1) h_2(k)]^2 \qquad (15.27)$$

Then one obtains

$$w_1 = \frac{\sum_{k=1}^{N} [h_2(k) - h(k)] [(h_2(k) - h_1(k)]}{\sum_{k=1}^{N} [h_1(k) - h_2(k)]^2} \tag{15.28}$$

Equally, determination of the free parameters is then accomplished by replacing the integrals of the preceding section by summations over the data points observed. The major difference is that now the determination of w_1 is sensitive to two factors: (1) the number of measurement points and (2) the locations and distribution of the measurement points in $0 \le x \le L$. Accordingly, a sensitivity survey must be done to determine the stability of the values with respect to number and distribution of points—a task best suited to numerical investigation.

15.4 STOCHASTIC DETERMINATIONS

15.4.1 General Considerations

The preceding two sections yield precise values for the weighting coefficients and for the free parameters because of the intrinsic assumption that the measured hydraulic head values are statistically sharp. In reality, the measurements have some uncertainty so that, even if the functional forms of the models are known precisely, both the weighting factors and the free parameters also must carry this uncertainty; i.e., they are not statistically sharp. For instance, in model 2, the relevant functional form of concern is the factor $[1 + (\varepsilon x/2L)]/(1 + \varepsilon/2)$. While the functional form is precise (because it represents an assumed model dependence), the value of ε cannot be precise if there is uncertainty in the measurements of head values. The same argument applies to the weight factors. The general question is: What does uncertainty in the measured values do to the determination of weight factors and free parameters? Clearly, an answer to this question allows one to assess whether more (and less uncertain) measurements are needed to improve the determination of the "best" weighting factors and also whether the free parameters are sufficiently well known that no further improvements are needed; an operational definition of such accuracy statements also must be given, of course.

The simplest way to exhibit the procedure is as follows: Imagine that a continuous sequence of hydraulic head measurements has been made for a saturated groundwater problem. If the sequence is repeatedly measured, then, at each depth x, eventually one will produce an ensemble average head value $\langle h(x)\rangle$ and statistical fluctuations $\delta h(x)$ with zero mean $\langle \delta h(x)\rangle = 0$ such that any particular sequence can be written $h(x) = \langle h(x)\rangle + \delta h(x)$ (Bakr et al., 1978; Glezen and

Lerche, 1985). Throughout this section angular brackets will be reserved for ensemble average quantities. In the two-model case described previously, it follows that the free parameter ε also must have an ensemble average constant value $<\varepsilon>$ and a fluctuating component $\delta\varepsilon$ with zero mean value $<\delta\varepsilon> = 0$. Equally, the weight coefficient w_1 also must be composed of an ensemble average value $<w_1>$ and a fluctuating component δw_1 with $<\delta w_1> = 0$.

15.4.2 Parameter Determination

For each realization of the measured head one would again write the quadratic functional

$$\chi^2 = \int_0^L \{h(x) - w_1 [h_1(x) - h_2(x, \varepsilon)] - h_2(x, \varepsilon)\}^2 \, dx \qquad (15.29)$$

and one would again write the functional form determining the parameter ε as

$$Y_2^2 = \int_0^L \left[h(x) - h_2(x, \varepsilon) \right]^2 dx \qquad (15.30)$$

However, because $h(x)$ is partially deterministic and partially stochastic through its statistical uncertainty, a slightly different procedure is now followed than previously.

Because $h_2(x, \varepsilon)$ depends on the value of ε, h_2 also must have an ensemble component $<h_2>$ and a fluctuating component δh_2 that has zero average value $<\delta h_2> = 0$. If $|\delta\varepsilon| \ll <\varepsilon>$, then one can write

$$h_2(x, \varepsilon) \equiv h_2(x, <\varepsilon> + \delta\varepsilon)$$

$$= h_2(x, <\varepsilon>) + \delta\varepsilon \, \frac{\partial h_2(x, <\varepsilon>)}{\partial <\varepsilon>} + \frac{1}{2} \, \delta\varepsilon^2 \, \frac{\partial^2 h_2(x, <\varepsilon>)}{\partial <\varepsilon>^2} \cdots \quad (15.31)$$

Note that in Eq. (15.31) the term $h_2(x, <\varepsilon>)$ and its derivatives with respect to $<\varepsilon>$ are deterministic; hence

$$<h_2(x, \varepsilon)> = h_2(x, <\varepsilon>) + \frac{1}{2} <\delta\varepsilon^2> \frac{\partial^2 h_2(x, <\varepsilon>)}{\partial <\varepsilon>^2} + \cdots \qquad (15.32)$$

Subtracting Eq. (15.32) from Eq. (15.31) yields

$$\delta h_2(x, \varepsilon) = \delta\varepsilon \, \frac{\partial h_2(x, <\varepsilon>)}{\partial <\varepsilon>} + \cdots \qquad (15.33)$$

which to quadratic order in $\delta\varepsilon$ gives

$$<\delta h_2^2> = <\delta\varepsilon^2> \left(\frac{\partial h_2(x, <\varepsilon>)}{\partial <\varepsilon>} \right)^2 \qquad (15.34)$$

From Eq. (15.32), to quadratic order in $\delta\varepsilon$, one obtains

$$<h_2>^2 = h_2(x, <\varepsilon>)^2 + <\delta\varepsilon^2>h_2(x, <\varepsilon>) \frac{\partial^2 h_2(x, <\varepsilon>)}{\partial <\varepsilon>^2} \quad (15.35)$$

A measure of the statistical sharpness of the model hydraulic head is the volatility v defined by (Lerche and MacKay, 1999)

$$v = \frac{<\delta h_2^2>}{<h_2>^2} \quad (15.36)$$

If the volatility is small compared with unity, then the fluctuations in the model are small; conversely, a volatility much larger than unity implies a significant uncertainty. Volatilities also can be defined for the parameter determination and/or for the weight factors as

$$v(w_1) = \frac{<\delta w_1^2>}{<w_1>^2} \quad (15.37a)$$

$$v(w_2) = \frac{<\delta w_2^2>}{<w_2>^2} \quad (15.37b)$$

$$v(\varepsilon) = \frac{<\delta\varepsilon^2>}{<\varepsilon>^2} \quad (15.37c)$$

Note that since $w_1 + w_2 = 1$, it suffices to calculate $v(w_1)$ because $v(w_2)$ can be written as

$$v(w_2) = \frac{<\delta w_1^2>}{(1 - <w_1>)^2} \quad (15.37d)$$

To calculate the relative measures of statistical sharpness, one proceeds as follows: Split the quadratic functional χ^2 into an ensemble piece $<\chi^2>$ and a fluctuating component $\delta\chi$ that has zero average value $<\delta\chi> = 0$. Then one extremizes the components with respect to variations in $<w_1>$, δw_1, $<\varepsilon>$, and $\delta\varepsilon$. The mathematics is tedious but not difficult and so for this reason is omitted here. The result is that if one considers fluctuations to quadratic order only, then one can write that $\delta\varepsilon$ is given by

$$\delta\varepsilon \int [<h(x)> - h_2(x, <\varepsilon>)] \frac{\partial h_2(x, <\varepsilon>)}{\partial <\varepsilon>} dx = \int [<h(x)> - h_2(x, <\varepsilon>)] \delta h(x) dx \quad (15.38)$$

which shows directly that $\delta\varepsilon$ is linear in the measured head fluctuations δh. When the substitution of $\delta\varepsilon$ is made into the average component $<\chi^2>$, what results is a functional form that involves two factors: the mean and statistical fluctuations in the head measurements and the model 2 behavior but involving only the mean value of the free parameter $<\varepsilon>$. The reason is that one has used Eq. (15.35) to eliminate the fluctuating component of $\delta\varepsilon$ in favor of expressions involving fluctuations in the head measurements δh modified by model 2 behavior but involving only the mean component $<\varepsilon>$. The resulting expression for the mean quadratic functional is then as follows:

$$<\chi^2> = \int \left\{ \left[<h(x)> - h_2(x, <\varepsilon>) - \frac{1}{2} <\delta\varepsilon^2> \frac{\partial^2 h_2(x, <\varepsilon>)}{\partial <\varepsilon>^2} \right]^2 + <\delta h(x)^2> \right.$$

$$\left. + <\delta\varepsilon^2> \left[\frac{\partial h_2(x, <\varepsilon>)}{\partial <\varepsilon>} \right]^2 - 2 \left(\frac{\partial h_2(x, <\varepsilon>)}{\partial <\varepsilon>} \right) <\delta h(x)\, \delta\varepsilon> \right\} dx \qquad (15.39)$$

Note that Eq. (15.36) is a function solely of the one unknown, $<\varepsilon>$, so once the functional form of the fluctuation correlations in the head $<\delta h(x)\delta h(y)>$ are given, then one can replace $<\delta h\delta\varepsilon>$ in Eq. (15.36) using the form of Eq. (15.35). Equation (15.36) then truly involves only measured quantities as well as the functional form for model 2, together with the one parameter $<\varepsilon>$ to be adjusted until $<\chi^2>$ takes on its minimum value. Then one merely substitutes this value for the best-fitting $<\varepsilon>$ into Eq. (15.35), and so one can compute $<\delta\varepsilon^2>$ directly. If it turns out that the volatility in $<\varepsilon>$ is small compared with unity, then no sharper statistical set of head measurements is needed to lower the volatility further. However, that a parameter is well determined with low volatility does not automatically imply that the weighting factors, including their uncertainties, are equally well determined. The reason is that one weighting coefficient could be extremely small compared with unity so that a small change implies a large volatility. While the coefficient is small, and so the particular model involved would then not contribute much to the overall fit of models to data, the point is that such a calculation must be worked through in order to determine which, if either, weighting coefficient falls into such a situation. We consider this aspect next.

15.4.3 Weighting Factors

Once the free parameter, both its mean and fluctuating components, is determined, one can then consider determination of the corresponding components of the weighting factors. The simplest way to investigate the effects of fluctuations in measured head is to note that, for each realization, the minimum of χ^2 yields the equation for w_1 for that realization in the form

$$\int [h - w_1(h_1 - h_2) - h_2] \left(h_1 - h_2 \right) dx = 0 \qquad (15.40)$$

However, Eq. (15.37) must be valid for each and every realization and thus for the mean and fluctuating components too. Hence one writes $h = <h> + \delta h$ and $w_1 = <w_1> + \delta w_1$ in Eq. (15.37) and then takes the ensemble average value of the equation. The result is

$$\int [<(h_1 - h_2)\,(h - h_2)>\,dx = <w_1>\int <(h_1 - h_2)^2>\,dx$$
$$+ 2\int <(h_1 - h_2)>\,<\delta w_1\,(h_1 - h_2)>\,dx \quad (15.41)$$

Subtraction of Eq. (15.38) from Eq. (15.37) yields an equation describing the fluctuating part of w_1 in terms of the fluctuations in head measurements and the fluctuations in model 2 behavior as

$$\delta w_1 \int (h_1 - <h_2>)^2\,dx = \int (h_1 - <h_2>)(\delta h - \delta h_2)\,dx + 2 <w_1> \int \delta h_2 (h_1 - <h_2>)\,dx$$
$$(15.42)$$

Now note that in Eq. (15.41) one is interested in the combination of factors $<\delta w_1 \delta h_2>$. By multiplying Eq. (15.42) by δh_2 and then taking the ensemble average of the result, one obtains directly the required expression that, when substituted into Eq. (15.41), enables one to write an expression for $<w_1>$ alone. The result involves the mean and fluctuating components of the measured head together with the mean and fluctuating components of model 2 through its dependence on the parameter $<\varepsilon>$ and the fluctuating part $\delta\varepsilon$. One obtains the equation

$$<w_1> \left\{ \int < (h_1 - h_2)^2 >\,dx \right.$$

$$+ \frac{4\left\{ \int dy <h_1(y) - h_2(y)> \left[\int dx <h_1(x) - h_2(x)> <\delta h_2(x)\delta h_2(y)> \right] \right\}}{\int (h_1 - <h_2>)^2\,dx} \right\}$$

$$= \int < (h_1 - h_2)(h - h_2) >\,dx$$

$$- \frac{2\{\int dy[h_1(y) - <h_2(y)>] \int dx[h_1(x) - <h_2(x)>][<\delta h(x)\delta h_2(y)> - <\delta h_2(x)\delta h_2(y)>]\}}{\int (h_1 - <h_2>)^2\,dx}$$

$$(15.43)$$

Equation (15.43) provides an explicit expression for the ensemble value of the weighting factor $<w_1>$ that, when used in Eq. (15.42), enables one to compute δw_1 and, correspondingly, $<\delta w_1^2>$. Hence one determines the volatility $v(w_1) = <\delta w_1^2>/<w_1>^2$. Thus all the information is provided to see if the fluctuations in head measurements are causing significant shifts in parameter values for particular model depictions or in the weighting coefficients of the relative contributions of each model to both the mean and fluctuating head data.

15.4.4 Numerical Illustration

Parameter Determination. Consider the same two models as previously. For head measurements, take it that the mean hydraulic head $<h(x)>$ is given precisely by the hydraulic head of Eq. (15.15). For the fluctuating component $\delta h(x)$, take it to represent a fixed degree of uncertainty in the mean hydraulic head. Also assume that the fluctuations in head measurement are uncorrelated from location to location with depth. Then one can write

$$<\delta h(x)\delta h(y)> = f^2 L \delta(x - y) \tag{15.44}$$

where f is the root mean square head uncertainty.

For simplicity of illustration, it is easiest to consider volatility only to quadratic order in fluctuating quantities. This simplification means that $<w_1>$ and $<\varepsilon>$ are given to the required accuracy by the calculations of Section 15.2 as though there were no fluctuations at all in the measured head.

From Eq. (15.35) it then follows that

$$<\delta\varepsilon^2> = \frac{4LF^2(1 + \eta)^4 \int u^2 \{\exp[-B(u - 1)] - (1 + \eta u)/(1 + \eta)\}^2 \, du}{(\int u^2(1 - u)\{\exp[-B(u - 1)] - (1 + \eta u)/(1 + \eta)\} \, du)^2} \tag{15.45}$$

where $\eta = <\varepsilon>/2$, $B = bL$ as before, and $F = f/[h(L) - h_0]$ measures the fractional head uncertainties across the system. Also, the variable u in Eq. (15.45) is just x/L and so ranges from $u = 0$ to $u = 1$ in the integrals. For the numerical value of $B = 1$ used previously, one has $\eta = -0.861$ and $<\varepsilon> = -1.722$ so that one can evaluate the integrals in closed form in Eq. (15.45) and obtain the ratio

$$\left(\frac{<\delta\varepsilon^2>}{<\varepsilon>^2}\right)^{1/2} = 2.83F \tag{15.46}$$

Thus, if the parameter is to be determined to, say, 1% accuracy, then the root mean square fluctuations in relative pressure head measurements F must be held to better than 0.5% accuracy for the preceding example.

Weighting Factor Determinations   It is not, however, only the parameter determination that is of importance; the weighting factors play a critical role. It may be that a parameter is poorly determined in a particular model, but it also may be that both the mean value of the associated weighting factor and its uncertainty are so small that the particular model hardly contributes to the weighted sum of models attempting to satisfy the head data and the fluctuations in head data. In such a case, better parameter determination is not relevant. However, if the particular model is of high relative weight, then accurate parameter determination is necessary.

For the assumption of uncorrelated head measurement error, as given through Eq. (15.44), it is not difficult to show (but it is extremely tedious!) that

$$\frac{<\delta w_1^2>}{<w_1>^2} = \frac{<\delta \varepsilon^2>}{\eta^2} \tag{15.47}$$

and that

$$\frac{<\delta w_2^2>}{<w_2>^2} = \frac{<\delta \varepsilon^2> <w_1>^2}{[\eta(1 - <w_1>)]^2} \tag{15.48}$$

For the numerical values reported above, one then has

$$\left(\frac{(<\delta w_1^2>}{<w_1>^2)} \right)^{1/2} = 5.67F \tag{15.49a}$$

$$\left(\frac{(<\delta w_2^2>}{<w_2>^2)} \right)^{1/2} = 0.054F \tag{15.49b}$$

Thus there is considerably more uncertainty on the weighting factor for model 1 than for model 2 for this numerical illustration, as was to be expected because model 1 only contributes, on average, about 1% to the composite sum, whereas model 2 contributes the remaining 99%. Hence, in order to have an accurate contribution, to, say, better than 1% in a relative sense of model 1, it follows that the relative head uncertainty measurement F should be more accurate than about 0.2%. However, to obtain the same 1% relative accuracy for the model 2 contribution, one needs only a relative head measurement accuracy of better than about 20%—principally because model 2 is already dominating the linear combination of the two models attempting to fit as best they can the total continuous head data.

15.5 AVERAGING MEASURES FOR HYDRODYNAMIC PROBLEMS

This short section shows that different choices made for averaging of information in hydrodynamic flow problems can have significant implications for interpretations of the system. Using a simple illustration, it is shown that care has to be exercised in obtaining physically meaningful results and that, depending on the model assumptions and the data available, there may not be acceptable models. It is also shown that there may be more than one model behavior that is acceptable, again depending on the data. The results have implications for the hydrodynamic upscaling problem for flow in permeable media, for ensemble averaging methods, and for parameter determination for deterministic models of permeable flow.

In many problems concerning hydrodynamic flow in permeable media, some form of averaging is usually performed on modeled processes, often using Darcy's flow equation as a basic starting point for system description. Three fundamental sorts of averaging are customarily in place: averaging over "small" domains in order to describe the hydrodynamic flow on coarser scales (often referred to as the *upscaling problem*) (Rubin and Gomez-Hernandez, 1990; Desbarats, 1992), averaging to describe an ensemble of possible situations because of uncertainty or variation in system parameters and/or boundary conditions and/or measurement uncertainty (Gomez-Hernandez and Gorelick, 1989); and averaging to describe how well a particular model matches a set of data, often with model parameter determination obtained by minimizing specific functionals describing the global mismatch of data and model behaviors (Clifton and Neumann, 1982; Carrera and Neumann, 1986a, 1986b; Desbarats and Dimitrakopoulos, 1990).

The purpose of this section is to illustrate, by specific example, that the choice of the measure of mismatch can have profound implications for inferences pertinent to the hydrodynamic system.

15.5.1 Specific Hydrodynamic Flow Model

Consider a one-dimensional model of steady-state flow in which the head is h_0 on $x = 0$ and $h(L)$ on $x = L$. Take the hydraulic conductivity $K(x)$ to vary with x in the form

$$K(x) = K_0\left(1 + \frac{2\eta x}{L}\right)^{-1} \tag{15.50}$$

where η is a dimensionless scaling constant and K_0 is also constant.

From Darcy's law describing the relation between flow q and head $h(x)$ in the form

$$q = -K(x)\frac{dh}{dx} \tag{15.51}$$

note that Eq. (15.51) requires that the head $h(x)$ be either monotonically increasing or decreasing with increasing x. The general solution to Eq. (15.51) is

$$h(x) = h_0 - (Lq)K_0^{-1}\frac{x}{L}\left(1 + \eta\frac{x}{L}\right) \tag{15.52}$$

which automatically satisfies the boundary condition $h(x) = h_0$ on $x = 0$. On $x = L$, the boundary condition $h(x) = h(L)$ requires

$$\frac{Lq(1 + \eta)}{K_0} = -[h(L) - h_0] \tag{15.53}$$

so the solution can be written in the more transparent form

$$h(x) = h_0 + \left[\frac{h(L) - h_0}{1 + \eta} \right] \frac{x}{L} \left(1 + \frac{\eta x}{L} \right) \qquad (15.54)$$

Now suppose that this model of head variation is used in an attempt to satisfy a set of continuous, statistically sharp head data $H(x)$ in $0 < x < L$ and where the data are taken in the form

$$H(x) = h_0 + \left[h(L) - h_0 \right] f\left(\frac{x}{L} \right) \qquad (15.55)$$

where $f(x/L)$ is dimensionless with the values $f(0) = 0$ and $f(1) = 1$. Note that the head data $H(x)$ are measured with depth x at one spatial location. The data do *not* have to conform to the assumed one-dimensional model behavior. The aim is to determine how well such models, under their basic assumptions, satisfy the head data.

In particular, one is interested in the value of the parameter η that provides the best fit of the model $h(x)$ to the data field $H(x)$. Customarily, such problems are handled by some form of global measure of mismatch between $h(x)$ and $H(x)$, which is then extremized to obtain the best value of η, i.e., the value that minimizes the measure chosen.

For instance, one quadratic measure is

$$\Psi_0^2 = \int dx \left[f(\frac{x}{L}) - \frac{(x/L)(1 + \eta x/L)}{1 + \eta} \right]^2 \qquad (15.56)$$

with the integral range from $0 < x < L$. However, an equally acceptable quadratic measure is

$$\Psi_1^2 = \int dx \left[(1 + \eta) f\left(\frac{x}{L} \right) - \left(\frac{x}{L} \right)\left(1 + \eta \frac{x}{L} \right) \right]^2 \qquad (15.57)$$

And, when minimized with respect to η, Ψ_0^2 and Ψ_1^2 return *different* values for η. In fact, the generalization

$$\Psi_p^2 = \int dx \left[\left(1 + \eta \right)^{p+1} f\left(\frac{x}{L} \right) - \left(\frac{x}{L} \right)\left(1 + \eta \frac{x}{L} \right)\left(1 + \eta \right)^p \right]^2 \qquad (15.58)$$

when minimized with respect to η, returns the relation between the power p and the best value of η in the form

$$p = -(1 + \eta)\frac{I_{\alpha\beta} + \eta I_{\beta\beta}}{I_{\alpha\alpha} + 2\eta I_{\alpha\beta} + \eta^2 I_{\beta\beta}} \qquad (15.59)$$

where $\alpha(u) = \dfrac{x}{L} \equiv u$, $\beta(u) = (x/L)^2 \equiv u^2$, and

$$I_{ab} = \int du [f(u) - a(u)][f(u) - b(u)] \qquad (15.60)$$

with the integral in Eq. (15.60) taken over the range $0 < u < 1$.

Note that if p is chosen to be zero (corresponding to Ψ_1^2), then the solutions to Eq. (15.59) are

$$\eta = -1 \qquad (15.61a)$$

or

$$\eta = -\frac{I_{\alpha\beta}}{I_{\beta\beta}} \qquad (15.61b)$$

whereas if $p = -1$ (corresponding to $\Psi_0{}^2$), then Eq. (15.59) yields

$$\eta_0 = \frac{I_{\alpha\beta} - I_{\alpha\alpha}}{I_{\beta\beta} - I_{\alpha\alpha}} \qquad (15.62)$$

Also note that on $\eta = 0$, Eq. (15.59) returns

$$p_0 = \frac{-I_{\alpha\beta}}{I_{\alpha\alpha}} \qquad (15.63)$$

For values of p other than $p = 0$ and $p = -1$, Eq. (15.59) provides two values of η. A sketch of the curve p versus η is given in Fig. 15.1. Thus, by choosing different measures of global mismatch (represented here by the power p for the simple purposes of illustration), one obtains different best values for η.

There is a physical requirement, however. The value of η may *not* be smaller than -1 because then the model head $h(x)$ would reverse direction at the point $x^* = L/|\eta|$, which would lie in the domain $0 < x < L$. On physical grounds, such a reversal is forbidden because Darcy's law requires that $h(x)$ be monotonically increasing or decreasing in the range $0 < x < L$. This unphysical region ($\eta \leq -1$) is marked by vertical dashed line at $\eta = -1$ on Fig. 15.1.

In fact, a second, and much stronger, requirement exists. The value of η *must* exceed $-1/2$. This requirement follows because the model hydraulic conductivity

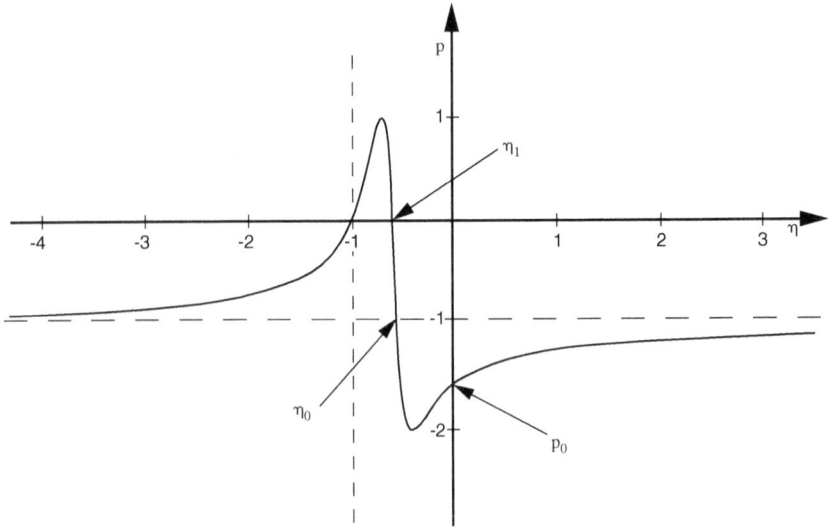

FIGURE 15.1 Sketch of the general shape of the p versus h curve represented by Eq. (15.59). The critical value of $\eta = -1$ is marked by a vertical dashed line. The two values ($p = 0$ and $p = -1$) at which Eq. (15.56) provides linear equations for η (with $\eta > -1$) are marked as η_1 (for $p = 0$) and η_0 (for $p = -1$), respectively. The value marked p_0 corresponds to $\eta = 0$.

is taken as proportional to $(1 + 2\eta x/L)^{-1}$. If η were to be smaller than $-1/2$, then the model conductivity could not be positive definite everywhere on $0 < x < L$, which it must be physically. Figure 15.2 provides a sketch of the behaviors allowed depending on whether η_0 and/or η_1 are themselves greater or less than $-1/2$. In Fig. 15.2*a*, neither η_1 nor η_0 can represent physically acceptable minima; in Fig. 15.2*b*, η_1 is not acceptable but η_0 is, whereas in Fig. 15.2*c*, both η_1 and η_0 are physically acceptable.

The first point to make is that depending on the global measure of mismatch used, there can be many "best" solutions. The second point is that the best solution classes have to be constrained by the requirements of the problem. In the example given, the requirement of a monotone increasing (or decreasing) model head with increasing x demands $\eta > -1$, whereas the requirement of a positive model hydraulic conductivity everywhere demands enforcement of the stronger constraint $\eta > -1/2$.

However, once these constraints are satisfied, an infinite number of solutions is still possible for the parameter η, each depending on the value used for the parameter p. Indeed, inspection of Fig. 15.1 or Fig. 15.2 shows that there are always two solutions possible for η for each choice of p.

While it could be argued that the choice of the weighting factor $(1 + \eta)^{1+p}$ of the mismatch between observed and modeled head is special, and while it also could be argued that a different model dependence of hydraulic conductivity could

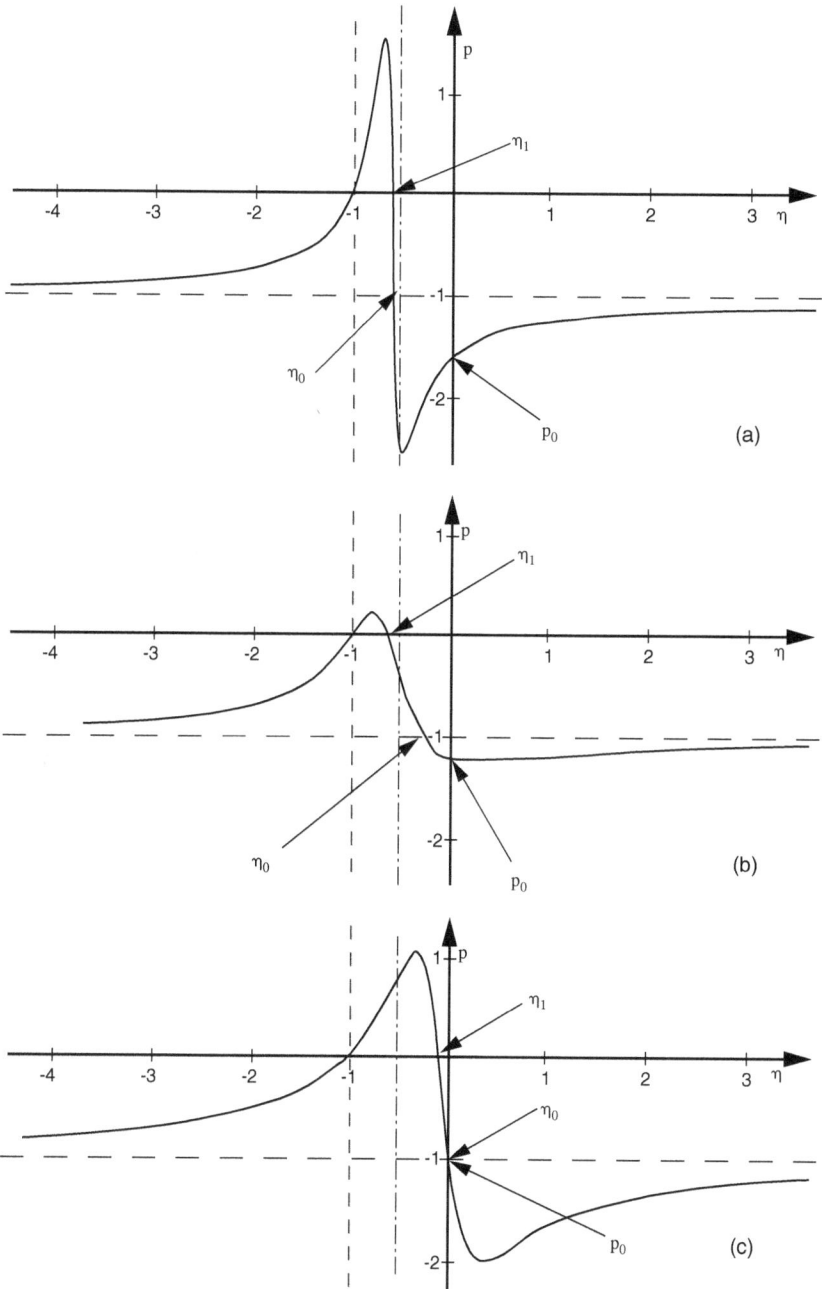

FIGURE 15.2   Sketch of the three behaviors for physical solutions of Eq. (15.59) when the extra constraint $\eta > -1/2$ is added (represented by the vertical dot-dash line at $\eta = -1/2$), as a consequence of requiring that the model hydraulic conductivity be positive everywhere in $0 < x < L$.

be chosen, these points are not relevant to the argument. The point is that the model choices are not fundamentally forbidden. Thus, for instance, if the spatial dependence of the measured head data $f(x/L)$ is such that η_0 and η_1 are both smaller than $-1/2$, then one cannot use measures of global mismatch that are of the form of Eqs. (15.53) and (15.54), for both will yield unphysical values for η. Instead, with a weighting factor $(1 + \eta)^{p + 1}$ one must then arrange to have a value of p such that the curve of p versus η from Eq. (15.56) can never intersect the point $p = p^*$ at $\eta = -1/2$, where

$$ p* = \frac{-(2I_{\alpha\beta} - I_{\beta\beta})}{4I_{\alpha\alpha} - 4I_{\alpha\beta} + I_{\beta\beta}} \qquad (15.64) $$

or else one cannot obtain physically meaningful results. And there will always be two "best" values of the parameter η for each chosen p value.

Other choices of weighting factors with similarly free parameters such as p also would have to ensure that the corresponding η values exceed $-1/2$.

15.5.2 Discussion

What can be learned from the simple illustration reported here has implications for general hydrodynamic flow problems. First, note that it is not sufficient, in general, just to choose arbitrary measures of mismatch for modeled head versus observations. The simple example given shows that any such measure must be constrained in order that parameters determined make physical sense. In more complex, three-dimensional, heterogeneous, partially saturated, time-dependent, anisotropic hydrodynamic problems, it may not be as clear as it is in the simple example considered here to ascertain that all physical conditions have been satisfied. Indeed, it may not even be possible to identify all such conditions.

Second, the weighting one attaches to a global measure of mismatch is, apparently, not only nonunique but also, for a given choice with free parameters, constrained by conditions not incorporated directly in the measure of mismatch and also may lead to multiple solutions for physical parameter values. There seems not to be available any specific requirement on what choice of weighting is most appropriate nor, indeed, any objective way of deciding the issue. This particular point is most vexing because the implication is that different authors will (and do) choose different weightings so that intercomparisons of results from different works are often difficult, if not impossible.

In addition, the question of how and what one averages in an upscaling hydrodynamic flow to correctly representing the finer-scaled information is a major concern. Also note that the simple illustration dealt only with a spatially independent weighting factor. The complications that can ensue when a spatially dependent weighting factor is used, which also may depend on the physical parameters being sought and which also may contain its own free parameters akin to the power p, are clearly problems of considerable concern in their own right.

Third, measures of global mismatch for, say, hydrodynamic head do not contain all the salient information of the system being modeled—otherwise, there would be no need to impose extraneously the two requirements $\eta > -1$ (modeled head monotonic) and $\eta > -1/2$ (hydraulic conductivity positive) in the simple illustration. Thus another question of serious concern in general hydrodynamic flow problems is: When can one be sure that minimization measures used do contain all the information of the system so that any results are guaranteed to make physical sense?

Fourth, for a minimization measure that returns a best parameter not within a physically acceptable range, one inference is that the model chosen is a poor approximation to the observations and one is advised to seek more relevant model behaviors—or, of course, a more appropriate measure of mismatch, or both.

Finally, while the example presented is extremely simple, it is its very simplicity that sharply illuminates the problems. The complications that arise when complex deterministic (and even more complex stochastic) models of flow are under consideration are, quite frankly, uninviting in this regard. The complexities would merely cloud the salient points, which are better brought out by the simple example used here.

It would seem, however, that the points raised need to be considered in virtually all hydrodynamic flow problems in a manner more detailed than appears to have been the case so far.

15.6 SUMMARY

The purpose of this chapter has been to indicate how one can use observations of hydrodynamic head to help distinguish between competing models of groundwater flow, including uncertainty in the data contributing to uncertainty in parameter determination for individual models, as well as including continuous or discrete models. To be sure, the specific example used of steady-state one-dimensional model comparisons against a prescribed data field was rather simplistic, but the essential elements of the procedure to be used to help disentangle the worth of each model are most easily portrayed with such a simplistic situation.

Other situations follow the same line of logical development, albeit with more technical details to be incorporated, which, if they had been included here, would have defeated the purpose of providing a simple exposition.

While the method of analysis has been carried through analytically here, more generally, resort to numerical procedures is necessary. For instance, the models chosen incorporated only parameters of fractional weighting (and its uncertainty) together with one "free" parameter in just one of the two models. More complex models contain more parameters so that determination is best handled numerically. Equally, the number of data measurements and their uncertainties are best handled numerically too.

In addition, while the stochastic assessment has been given here using analytical procedures and also the simplest measurement uncertainty statistics of uncorrelated error, it becomes clear from the complexity in this simple case that more general investigations with more complex models and more general correlated statistics for the head fluctuations are best handled by Monte Carlo numerical procedures. Despite these limitations, the essence of the argument is accurately portrayed by the simple situations considered here.

In essence, what one has is a quantitative procedure that allows a prescription to be given of which of several competing models or combinations best account for a given set of head data and determines the fractional contribution of each model to the total. In this way one can decide where further emphasis needs to be placed to improve matters and, indeed, whether such improvement is necessary.

One also can determine when it is necessary to shift from one group of models to another grouping with a given data set in order to portray more accurately changes in subsurface hydrogeologic conditions, and significantly, one also can determine the best depth at which to effect such a transfer.

The importance of allowing for uncertainties in the measured data and their influence on parameter determinations in particular models, as well as the influence on the uncertainty of each model in contributing to a total to account as best as possible for the observations through the weighting factors and their uncertainties, are clearly of relevance. The procedure for developing quantitative measures to address this aspect also was most easily portrayed with the simple model prescriptions given here.

APPENDIX 15A: GENERAL DEVELOPMENT OF PROCEDURES FOR N MODELS

While the body of this chapter dealt with just two simple models in order to minimize complexity, it is possible to generalize the arguments given to any number N of models. Consider that $h_i(x)$ represents the modeled hydrodynamic head for the ith model. Let $h(x)$ be the measured head data field in $0 \leq x \leq L$. Analyze the problem into three parts as in the body of the chapter: continuous, discrete, and stochastic measurements.

15A.1 Continuous, Perfect Measurements

In this case the optimal linear combination of N models that satisfies the data field is given by minimizing a quadratic functional χ^2 with respect to weighting factors w_i ($i = 1,..., N$) with

$$\chi^2 = \int_0^L \left[h(x) - \sum_{i=1}^{N} w_i h_i(x) \right]^2 dx - 2\lambda \left(1 - \sum_{i=1}^{N} w_i \right) \qquad (15.65)$$

where λ is a Lagrange undetermined multiplier. The minimum of Eq. (15.65) with respect to w_j occurs when

$$\int_0^L \left[h(x) - \sum_{i=1}^N w_i h_i(x) \right] h_j(x) \, dx = \lambda \qquad j = 1, 2,..., N \qquad (15.66)$$

Let

$$A_j = \int_0^L h(x) h_j(x) \, dx \qquad (15.67a)$$

and

$$S_{ij} = \int_0^L h_i(x) h_j(x) \, dx = S_{ji} \qquad (15.67b)$$

Equation (15.66) can then be written as

$$\sum_{i=1}^N w_i S_{ij} + \lambda = A_j \qquad j = 1, 2,..., N \qquad (15.68a)$$

together with the requirement

$$\sum_{i=1}^N w_i = 1 \qquad (15.68b)$$

Equations (15.68a) and (15.68b) represent $N + 1$ equations in the $N + 1$ unknowns, λ, and w_i ($i = 1, 2,..., N$) and reminds one of the formulation for ordinary kriging (Isaaks and Srivastava, 1989). In general, the system of equations (15.68a) and (15.68b) must be positive definite to yield one and only one solution for λ and w_i. Technical execution rests on the ease with which one can calculate A_j and S_{ij}.

15A.2 Discrete, Perfect Data

If the data field $h(x)$ is measured only at M discrete depth points $x_1, x_2,..., x_M$, yielding measured head values $h(x_j)$ ($j = 1, 2,..., M$), then the discrete analogue to Eq. (15.65) is given by

$$\chi^2 = \sum_{j=1}^M [h(x_j) - \sum_{i=1}^N w_i h_i(x_j)]^2 - 2\lambda \left(1 - \sum_{i=1}^N w_i \right) \qquad (15.69)$$

Differentiation of Eq. (15.69) with respect to all w_i leads to the minimization equation

$$\sum_{i=1}^{N} w_i S_{ik} + \lambda = A_k \qquad \text{for } k = 1, 2, \dots, N \qquad (15.70)$$

where

$$A_k = \sum_{j=1}^{M} h(x_j) h_k(x_j) \qquad (15.71a)$$

$$S_{ik} = \sum_{j=1}^{M} h_i(x_j) h_k(x_j) = S_{ki} \qquad (15.71b)$$

This combination of equations is precisely the discrete analogue of Eqs. (15.68a), (15.67a), and (15.67b). The method of solution is identical, although the values for each w_i now depend on how many data points are used and how the measurement points are distributed spatially.

15A.3   Stochastic Measurements

When the measurements of hydraulic head have a random statistical component, then one proceeds effectively as in the body of the text. There are two situations to consider. Either the models of head behavior are taken to have predetermined parameters so that only the weighting coefficients have random components associated with the random components of measured head values, or the parameters for each model are adjusted to obtain the best fit of each model alone to the data field (including uncertainties), and then these best-fit models are used to determine weighting coefficients. Consider each case separately.

Predetermined Models.   In this case no components of any model are adjustable, only the weighting coefficients. From Eq. (15.71) one then has that S_{ik} is statistically sharp. Write, as in the body of the text, $h(x) = <h> + \delta h$. Correspondingly, in Eq. (15.71) one has $A_k = <A_k> + \delta A_k$. Also write $w_i = <w_i> + \delta w_i$, and $\lambda = <\lambda> + \delta\lambda$. Then

$$\sum_{i=1}^{N} <w_i> S_{ik} = <\lambda> + <A_k> \qquad (15.72a)$$

and

$$\sum <w_i> = 1 \qquad (15.72b)$$

together with

$$\sum \delta w_i S_{ik} = \delta \lambda + \delta A_k \tag{15.72c}$$

and

$$\sum \delta w_i = 0 \tag{15.72d}$$

The procedure for solution in this case is identical to that presented previously in this appendix. The only difference in technical execution is the shift to the sum of statistical fluctuations in weighting coefficients, which is zero in Eq. (15.72d) from unity in Eq. (15.72b) for the mean coefficients. The calculation of the associated volatilities $<\delta w_i^2>/<w_i>^2$ proceeds as described in the body of the text.

Adjustable-Parameter Models. When the parameters of each model can be adjusted to best satisfy the data field for that model, then the resolution is more complex. First, the mean values of the model parameters and their uncertainties must be determined; second, the implication is that each model function will then have an ensemble average component as well as a random component. Thus S_{ik} will no longer be statistically sharp, and the fluctuating component of A_k will no longer be due solely to the corresponding head measurement fluctuation $\delta h(x)$ but also will have a component due to the fluctuations in $h_k(x)$.

In the preceding section one had

$$A_k = \int h(x) h_k(x)\, dx = \int <h(x) h_k(x)>\, dx + \int \delta h(x) h_k(x)\, dx \equiv <A_k> + \delta A_k \tag{15.73}$$

However, when $h_k(x)$ is no longer statistically sharp and so also has to be written in the form $h_k = <h_k> + \delta h_k$, it then follows that

$$<A_k> = \int [<h(x)> <h_k(x)> + <\delta h(x) \delta h_k(x)>]\, dx \tag{15.74a}$$

and that

$$\delta A_k = \int [<h(x)> \delta h_k(x) + \delta h(x) <h_k(x)> + \delta h(x) \delta h_k(x) - <\delta h(x) \delta h_k(x)>]\, dx \tag{15.74b}$$

so that considerably more terms have to be evaluated. Likewise, S_{ik} now has an ensemble average part plus a random component with

$$<S_{ik}> = \int [<h_i(x)> <h_k(x)> + <\delta h_i(x) \delta h_k(x)>]\, dx \tag{15.75a}$$

and

$$\delta S_{ik} = \int [\delta h_i(x) <h_k(x)> + \delta h_k(x) <h_i(x)> + \delta h_k(x)\delta h_i(x) - <\delta h_k(x)\delta h_i(x)>]\, dx$$

$$(15.75b)$$

Hence solution of Eq. (15.72) is now more difficult because one has

$$\sum <w_i> <S_{ik}> + \sum <\delta w_i \delta S_{ik}> = <\lambda> + <A_k> \qquad (15.76a)$$

and

$$\sum <w_i> = 1 \qquad (15.76b)$$

together with

$$\sum \left(\delta w_i <S_{ik}> + <w_i>\delta S_{ik} \right) = \delta\lambda + \delta A_k - \sum \left(\delta w_i \delta S_{ik} - <\delta w_i \delta S_{ik}> \right) \quad (15.76c)$$

and

$$\sum \delta w_i = 0 \qquad (15.76d)$$

Notice that now the fluctuating component of the weighting factors is involved in the equation for the mean components and that the ensemble average components are involved in the fluctuating-component equations. As shown in the body of the text, even in the case of just two models with one weighting coefficient and one adjustable parameter, together with stochastic components, the development and resolution of the coupled equations to determine $<w_i>$ and $<\delta w_i^2>$ were tediously involved although not fundamentally difficult.

In the case of multiple models of hydrodynamic flow with many weighting coefficients and many parameters, the analytical approach becomes extremely involved, and numerical Monte Carlo procedures are to be preferred (Gomez-Hernandez and Gorelock, 1989; Desbarats, 1992; Deutsch and Journel, 1992). Thus one would choose a value for $\delta h(x)$ around the mean flow head, and then one would solve that case to obtain the best adjustable-parameter values for each of the models separately (see Appendix 15B for this procedure). Then one would directly calculate the associated weighting coefficients. The procedure is then repeated for a Monte Carlo suite of choices for $\delta h(x)$. In this way, one constructs the mean and root mean square fluctuating components of w_i, together with the mean parameter values and their uncertainties. While completely general, such Monte Carlo schemes are extremely computer-intensive when many parameters in many models have to be determined. Nevertheless, the effort spent is definitely worthwhile relative to the need to ascertain if better and more precise measurements are needed and also to ascertain which direction of functional model dependence is indicated by the weighting coefficients.

APPENDIX 15B: INDIVIDUAL PARAMETER DETERMINATION PROCEDURE

For a given model functional behavior of hydraulic head, there can be many adjustable parameters, each with a different dimension. For instance, if a model of hydraulic conductivity were to be chosen in the form of $K(x) = K_0[1 + \alpha^2 \sin^2 (\pi x/L)]^{-b}$, then the parameters α and b are dimensionless, whereas the parameters L and K_0 have dimensions of length and length over time, respectively.

In addition, while the relative weighting coefficients are determined directly from a functional that is quadratic in the weighting coefficients, thus leading to linear equations for the weighting coefficients, the same is not true of parameters entering each model behavior where highly nonlinear parameter behaviors are the rule.

The purpose of this appendix is to address all these problems simultaneously. The procedure operates as follows: Denote by the vector p all parameters entering a particular model $h_k(x, p)$. Then construct the positive control function

$$C(p) = \int_0^L [h(x) - h_k(x, p)]^2 \, dx \tag{15.77}$$

Suppose for the moment that some initial estimate has been chosen for the vector p, and let this value be p_0. Then construct the dimensionless quantity

$$U = \frac{C(p)}{C(p_0)} \tag{15.78}$$

Before one can proceed with the nonlinear method for obtaining the best parameter values, one further transformation is needed. The vector p is a conglomerate of scalar parameters, all of which have different dimensions. It is best to normalize this concern out of the problem as follows: Suppose that a component of p is constrained to lie between a minimum p_{min} and a maximum p_{max}. Then write

$$a = \frac{p - p_{min}}{p_{max} - p_{min}} \tag{15.79}$$

Regard the vector a as being fundamental, with p derived from a through application of Eq. (15.79). Then a is dimensionless, and the range of any component is $0 \le a \le 1$. Thus an initial choice p_0 automatically generates an initial choice for a, a_0. Consider then the following nonlinear iteration scheme for each scalar component a_j of the vector a, with the iteration scheme to be run through N times:

$$a_j(n) = \sin^2 \theta_j(n) \tag{15.80a}$$

$$\theta_j(n + 1) = \theta_j(n) \exp\left(-\tanh\left\{\frac{\delta_j \alpha_j \partial C[a(n)]}{\partial a_j(n)}\right\}\right) \tag{15.80b}$$

$$a_j(n + 1) = \sin^2 \theta_j(n + 1) \tag{15.80c}$$

where

$$\alpha_j = \left|\frac{\partial C(a_0)}{\partial a_{j0}}\right|^{-1} \ln[1 + (Na_{j0})^{-1}] \tag{15.80d}$$

$$\Gamma_j(n) = \frac{|a_j(n + 1) - a_j(n)|}{a_j(n)} + \beta^2 \tag{15.80e}$$

$$\delta_j = \frac{\Gamma_j(n)}{\sum_{k=1}^{J}[J^{-1}\Gamma_k(n)]} \tag{15.80f}$$

and where the initial choice of the component a_j is written a_{j0}. The derivative is calculated numerically from

$$\frac{\partial C[a(n)]}{\partial a_j(n)} = \frac{C[a_1(n),\ldots, a_j(n)(1 + \beta),\ldots] - C[a_1(n),\ldots,a_j(n),\ldots]}{\beta a_j(n)}$$

$$\tag{15.81}$$

where the default value of β is 0.1.

The preceding nonlinear iteration scheme has several interesting characteristics. First, it guarantees that if a_j is positive at any iteration, then it stays forever positive, as required. Second, it guarantees that $C(a)$ will always be smaller after each iteration if the derivatives are calculated exactly; thus $C(a)$ always heads to a minimum. Third, it is a relatively simpler matter, once a minimum is achieved, to "freeze" all except one parameter at the values corresponding to the minimum point and then perform a linear search between the preset minimum and maximum given above for each parameter component. Fourth, with respect to the choice of initial estimates for the parameters and their ranges of minima and maxima, it now becomes clear how to ensure that they will be close to the appropriate range needed. Note that if a component of a tends toward zero or unity, then a better place to look for a minimum is when the lower, or upper, limit on the range of the corresponding component of the p vector is moved to lower, or higher, values, respectively. Thus the procedure also guarantees correction of a poor initial choice of range for a given parameter. Fifth, as a tactical maneuver, a more rapid approach to a minimum can be achieved if N nonlinear iterations are done, and then the final values of the parameters are used to replace the initial estimates, and a further N iterations done. This tactic is far superior to performing $2N$ iterations around the initial estimate.

APPENDIX 15C: MEASURES OF MODEL AND OBSERVATIONAL MISMATCH

In the body of the text, the quadratic functional used to provide a global measure of mismatch between observations, represented through [Eq. (15.15)] $h(x) = h_0 + [(h(L) - h_0)/L]x \exp[-b(x - L)]$, and model behavior, represented through (Eq. 15.14) $h_2(x) = h_0 + [(h(L) - h_0)/L]x[1+\varepsilon x/(2L)]/(2L)]/(1 + \varepsilon/2)$, was written in the form

$$Y^2 = \int_0^L \left(1 + \frac{\varepsilon}{2}\right)^2 [h(x) - h_2(x)]^2 \, dx \qquad (15.82)$$

And then Y^2 was minimized with respect to ε in order to find the best value of the parameter ε leading to greatest consistency between observations and data, as represented through $h(x)$ and $h_2(x)$, respectively.

However, this measure of mismatch is not the only quadratic functional measure that can be used. As an example consider

$$Y'^2 = \int_0^L [h(x) - h_2(x)]^2 \, dx \qquad (15.83)$$

which, when minimized with respect to ε, also provides (for $\varepsilon \neq -2$) a linear equation for the best choice of ε in the form

$$\frac{\varepsilon}{2} \int_0^1 u^2 (1 - u)(e^{-B(u - 1)} - u) \, du = \int_0^1 u^2 (u - 1)(e^{-B(u - 1)} - 1) \, du \qquad (15.84a)$$

or

$$\frac{\varepsilon}{2} = -\frac{D + 1/12}{D + 1/20} \quad \text{with } D = \int_0^1 u^2(u - 1)e^{-B(u - 1)} \, du \qquad (15.84b)$$

where $u = x/L$ and $B = bL$. This value of ε is *not* the same as the value used in the text from the minimization of Eq. (15.82). For instance, on $B = \frac{1}{4}$, as used in the body of the text, Eq. (15.84b) returns the value $\varepsilon = -1.133$, different from the value of $\varepsilon = -1.973$ from Eq. (15.82). The corresponding weight factors [Eq.

(15.22)] are now drastically different at $w_1 = 0.046\%$ and $w_2 = 99.954\%$. This combination gives the representation of

$$w_1 h_1 + w_2 h_2 = h_0 + [h(L) - h_0] \frac{x}{L} \left(2.306 - 1.306 \frac{x}{L} \right) \qquad (15.84c)$$

which is very similar to that of Eq. (15.23a), which corresponds to $\varepsilon = -1.973$.

Indeed, if one were to replace $(1 + \varepsilon/2)$ in Eq. (15.82) by $(1 + \varepsilon/2)^p$, where p is an arbitrary power, then each such measure of mismatch between observations and models yields a different best value for the one parameter ε, dependent on p. In fact, weighting $h(x)$ and $h_2(x)$ with any arbitrary, but prescribed, function $f(\varepsilon)$ and then performing the differentiation to find the best value of ε yields different values, each dependent on the specific form of $f(\varepsilon)$.

The issues to be addressed here are severalfold. First, is there a choice of $f(\varepsilon)$ that will produce, in some sense, a lowest value for a quadratic functional mismatch between observations and models? Second, why is there more than one solution possible? Third, what are the implications for the physical problem in the text and, more generally, for any problem involving measures of mismatch between measurements and models? Answers to these questions are important because the ability to be specific about the worth of parameters to an understanding of model representations can be compromised unless clear unequivocal resolutions can be provided.

The answer to the first question is clearly negative because one could always choose $f(\varepsilon)$ to be the reciprocal of the quadratic functional used, in which case the corresponding new functional would be constant so that any value of ε would be appropriate. Any departure of $f(\varepsilon)$ from the preceding choice is permissible, of course, and each and every such choice will yield a different value for ε. Thus there is no corresponding "best" choice.

This problem of measure of misfit is related to the general problem of extracting a signal from a set of data (Percival and Walden, 1993). Consider a data set D. Let the data be constituted from a signal S and a residual "noise" component N. Then one writes

$$D = S + N \qquad (15.85)$$

The aim is to extract the signal. In order to do so, one must prescribe either the properties of the signal or of the noise (or both) that one *desires* to have present, and then one devises technical procedures using the prescribed properties to extract the desired signal. In addition, the noise must have properties that are not contained in the signal, and vice versa, or one could not distinguish noise from signal. Then one uses some form of projection operator (Bracewell, 1965) P so that

$$PD = S \qquad PN = 0 \qquad (15.86)$$

and then

$$(D - PD) = N \qquad (15.87)$$

The general classes of all possible ways to define signals and associated noise are legion, starting with the Gauss least squares mismatch procedure.

The operator measures given by Eqs. (15.82) and (15.83) have two obvious attributes. First, the model signal $h_2(x)$ is uniquely defined in its structure in x, except for the value of the constant parameter ε. Second, the measure version [Eq. (15.82)] operates under the premise that if $\varepsilon/2$ tends toward -1, then one *desires* to retain a finite measure. The price that is paid here for this desire is that the model $h_2(x)$ [modulated by the factor $(1 + \varepsilon/2)$] remains finite as $\varepsilon/2 \to -1$, but the involvement of the data field is minimized because the weighting $(1 + \varepsilon/2)h(x)$ then reduces the contribution to zero. On the other hand, the measure value [Eq. (15.83)] *desires* to keep the data field representation finite at all parameter values; the price that is paid for this desire is that the model tends to infinity as $\varepsilon/2$ tends toward -1. The quadratic measure [Eq. (15.84)] is then not bounded at all finite values of ε. Either desire is an appropriate measure because each represents the measure definition of signal properties versus data misfit. This same measure definition problem of signal properties prevails with any data set and any model where different filtering devices organize different components of the data to produce desired signals (or desired noise properties). Viewed in this light, the reason for different values for a "best fit" value of ε with different measures of mismatch is clear. The determination depends on the measure of mismatch used.

The general problem of comparison of results in the field of hydrology now becomes significant. In particular, such measure definitions play critical rules for upscaling of results in heterogeneous media, for the measures used to define smaller-scale "block" properties, for stochastic versus deterministic results, and for a variety of allied problems. It is the properties that one chooses to assign that determine the measures of mismatch to be used. Different authors have used different measures, making direct (and often even indirect) comparisons of results difficult, if not impossible.

The simple representation provided by Eqs. (15.82) and (15.83) allows a clean, sharply focused illustration of this basic pattern of effects.

CHAPTER 16
HUMAN, WATER, CHEMICAL, BIOLOGIC, AND RADIOACTIVE RISKS

While the main aim of this book has been to evaluate the financial risk aspects of environmental projects from a corporate perspective, there are a variety of other risks for which it is difficult to assign a monetary value. These risks broadly include those from naturally occurring process and anthropogenic impacts on the environment. In some cases, such environmental risks have an immediate, direct, and harmful effect on humanity or on the associated biosphere that humans use. In other cases, the consequences can be more indirect, subtle, long-lasting, and more deleterious.

Naturally occurring processes continually alter the Earth's atmosphere, topography, and biomass loads and their distributions around the world. When these processes adversely affect the environment relative to the perceived needs of humanity, they are considered environmental problems. Remediation can be performed, but preventative measures seem difficult to provide.

Anthropogenic processes are those produced by human activities in exploiting and modifying the environment. In many instances, such activities may exert a load on the environment the consequences of which are difficult to evaluate in advance. Thus it is not clear in many cases precisely how toxic storage, spillage, or leakage will later affect the environmental situation, and often it is many years later before a connection can be established between toxicity and human health. Indeed, at the time of storage it may be that no toxicity is known, and only much later can it be shown that some unsuspected agent in the stored material is indeed harmful to humanity.

The problems a corporation faces from such storage, spillage, or leakage concerns have to do not only with the financial aspects of site-remediation costs but also with the influence on humanity. This chapter provides a brief overview of such noncash risks related to natural and anthropogenic processes on the environment and concerns itself with the interrelated nature of storage, spillage, and leakage on humankind. Here the emphasis is on a broad overview of major classes of environmental problems the scale of which exceeds a standard risk-analysis procedure and which requires an integrated effort from society as a whole. Remediation methods are considered as part of the entrepreneurial activities of humanity, whereas preventative measures are viewed from the self-preservation perspective of a society. Finally, some basic rules of engagement are given when one is handling environmental problems that enable a logical, ordered approach to be taken subject to the expediencies of cost/benefit/health and intrinsic need. Arguments also can be given to evaluate quantitatively the relevant conditions for prioritizing remediation and avoidance procedures.

16.1 INTRODUCTION

Humanity needs several conditions to be present at the domain of the Earth in order to survive. In order of precedence, determined as the absence of a factor leading to rapid death, these factors are air, water, food, shelter, clothing, and energy. Regarded as the sine qua non for environmental conditions to maximize the well-being of humankind, these factors are precisely those which human activities sometimes do a poor job to maximize. In many respects, we *are* our own environmental problem.

Within this broad classification of humanity's needs, it is relevant to examine how natural and anthropogenic effects affect the survival and well-being of humanity and how we can learn to control, modulate, and/or remediate what we determine to be deleterious effects for societies. One problem that is pervasive is who decides, and how one decides, precisely what constitutes a deleterious effect and whether it persists on a short or a long time scale as far as humanity's involvement is concerned.

16.2 NATURAL ENVIRONMENTAL PROBLEMS

As far as can be determined, evolution of the air, water, biomass, and land masses of the Earth has been continuous throughout the total history of the planet. Until the advent of humankind, the response of living creatures to such natural events was essentially passive: adaptation, mutation, and species extinction. The appearance and spread of human beings across the face of the globe, plus the ever ongoing demands of a place to live, places to grow food, and modification of the local environment by replacement of the local plant material by cities, roads, houses, industry, and farms, has led to more aggressive actions by the natural processes of nature on the desired habitat of humanity—a so-called natural environmental problem.

Basically, the natural processes of both rapid and slow geologic evolution continue, and humankind can perform only remediation after the fact but cannot preordain or control to any significant extent the natural processes themselves. Perhaps the most important processes to list that influence the environment where humankind lives are (1) climatic variations (e.g., El Niño), (2) floods, (3) droughts, (4) earthquakes, (5) landslides and avalanches (snow, mud, and submarine turbidities), (6) forest fires, (7) volcanic explosions, (8) food pests, (9) meteoritic impact, (10) hurricanes, typhoons, monsoons, and tornadoes, (11) sea-level fluctuations, and (12) ice floes and ice sheets.

Some of these processes influence humankind immediately (such as earth-quakes, hurricanes, and floods), whereas others have a longer-term impact over years or centuries (such as droughts, the effects of which may be ongoing over tens to hundreds of years), and a large meteoritic impact (as in the case of the Cretaceous-Tertiary boundary event and the associated extinction of the dinosaurs) would have an immediate and likely nonrecoverable influence on the survivability of humankind as a species. Thus natural environmental events range across the full spectrum of time and space scales and range from local, recover-able events to irrecoverable long- or short-term events.

While one can do little to control the natural processes, the problem humanity faces is to provide remedial action to those unfortunate enough to be affected by any such event. Remedial action in such cases involves saving survivors, recoup-ing the influenced area if possible, population redeployment if the impact is non-recoupable (such as in drought-stricken areas), and emergency medical, shelter, and food and water supplies. While such remediation actions help humanity, they do nothing to ameliorate the basic natural processes. And humanity has a long-term record of reoccupying those land areas which have been subjected to ongo-ing or periodic natural environmental processes. For instance, Kobi in Japan, Istanbul in Turkey, and San Francisco in the United States are all sited on extremely earthquake-prone regions, as known from historical records of earth-quake occurrence. Yet, after each such earthquake event, rebuilding has continued on precisely the same areas again and again. Presumably, the reasons for this apparent illogical activity are that there are more compelling reasons for reoccu-pation than for desertion (e.g., excellent port facilities, strategic control of a narrow strait, excellent farm land, no more land available to a constrained pop-ulation).

There seems little that one can do to safeguard against such natural processes; we are faced with the eventuality of having to learn to live with natural environ-mental processes and their impacts on humankind. Remediation action after each such event seems all that we are able to do. The situation is extremely different, however, for anthropogenic environmental events, and these are considered next.

16.3 ANTHROPOGENIC ENVIRONMENTAL PROBLEMS

As remarked in the Introduction, there are several basic requirements humankind needs in order to survive: air, water, food, shelter, clothing, energy, and a place to live. In addition, the collective activities of human beings result in industrial devel-opment, mining, modulation of land uses, and transportation requirements. Humankind is extremely efficient at taking the basic requirements for survival and converting them through collective activities into improved living conditions. Unfortunately, at the same time, some of these activities produce wastes and create environmental problems that may undermine the conditions for human survival in the long term. The magnitude of some environmental problems today has been shown to exceed, in some instances, the resources of a single organization, and the impact has been shown to affect so many aspects of the human and natural conditions that no simple risk analysis or other procedure can address it. For such problems, society as a whole is required to mobilize its resources and attempt resolution.

We first provide some illustrations of these sorts of environmental problems and then consider both preventative and remedial courses of action. We then look at the political and social implications for prevention controls and remediation of anthropogenic environmental problems.

16.3.1 Examples of Anthropogenic Environmental Problems

Air. The fundamental requirement for existence is air to breathe. And yet in many instances pollution of the air, as the result of anthropogenic activities, undermines this absolutely fundamental requirement. Automobile exhaust pollution in major cities of the world (Los Angeles, Bangkok, São Paulo, Beijing) is rapidly making such locations less than viable sites for human survival without major adverse impacts on human health and longevity. From an industrial perspective, humankind seems to emit ever-increasing amounts of SO_2 and CO_2 into the atmosphere, leading to acid rain (from sulfur dioxide converting to sulfuric acid) that destroys forests and pollutes good farm land, whereas the industrially produced carbon dioxide enters the atmosphere and so provides one of the major components contributing to the greenhouse effect on a global scale. The release of massive amounts of fluorocarbons into the atmosphere has the major negative effect of destroying the ozone layer and so of allowing more solar ultraviolet light to penetrate to the Earth's surface, leading to major skin cancer problems, to say the least.

Water. After air, water is the next fundamental factor needed for human survival, both for internal consumption for physiologic maintenance and for food supply control, be it for watering land-based plants and animals or for freshwater and saltwater farming of fish, crustaceans, seaweeds, etc. And yet, here again, pollution of the lakes, rivers, and oceans of the world has increased dramatically during the past century in a variety of manners—less than conducive to the use of water bodies as sustainable prerequisites for human existence. The main sorts of primary contaminants are chemical, biologic, physical, and nuclear, which may be accidentally or intentionally introduced into the fluid environment. Thus disposal of untreated sewage and agricultural and industrial waste in the oceans continues to be the practice of most developed and developing countries. In the United States, several types of waste, such as solid waste, construction and demolition debris, industrial waste, dredge spoils, and radioactive waste, had been disposed in the oceans in the past (Council of Environmental Quality, 1970). Currently, federal law prohibits ocean dumping in the United States of chemical, radiologic, and biologic warfare agents and high-level radioactive waste, and the regulation of other types of waste is monitored by several federal agencies.

Primary chemical contamination of the fluid environment is as varied as the chemicals produced by the industrial activities of humankind, ranging from heavy metals to arsenic (from gold mining mainly), to oil (from pipeline breakage and tanker ship disasters), to paper pulp residue (as in the southern part of Lake Baikal, Russia), to mixes of residual chemicals used in the paint and lacquer industries. This list can be extended almost endlessly depending on mining activities and industrial activities around the world.

Primary biologic contamination of the fluid environment can occur from a variety of sources: human and animal waste, hospital detritus, biologic weapons residue (such as anthrax stored on the Aral Sea Island), decayed organic material from residual foods and unused animal parts, and so on. The impact of this primary biologic contamination on pure water supplies can be enormous both in terms of human toxic contamination and also in terms of secondary concentration of biologic contamination through the food chain. Such an effect is, of course, also a major problem with chemical contaminants.

Physical contamination arises when, for example, used building materials are indiscriminately dumped in a water supply. The presence of asbestos, or of lead piping, or of the fine detritus can pollute water to such an extent that fish die, oysters become toxic, and the water has so much physical detritus in suspension that it becomes undrinkable. Major impacts on oceanic and coastal life include (1) death of marine organisms by toxic pollutants, (2) cultural eutrophication (the excessive bloom of algae leading to oxygen depletion in shallow waters), (3) habitat changes that affect entire marine ecosystems, and (4) depletion of dissolved oxygen because of the demand for the decomposition of organic wastes (Council of Environmental Quality, 1970). Other aspects of physical contamination can arise by performing different combinations of operations on water supplies: either disposing of different mixes and types of physical contaminants into a given water supply or change the water flow conditions intentionally (dams, levees, canals, etc.). Thus heavy management of many rivers (such as the Colorado River) has led to increased salinity at parts of the rivers that render their waters unsuitable for drinking purposes or even irrigation (Graf, 1985). Both these impacts can have deleterious results for water quality.

Two other major forms of water pollution are provided by humankind: nuclear and oil. In the case of nuclear pollution, one has merely to look at the Kara Sea region, full of rusting hulks of Soviet nuclear submarines with fuel-laden reactors still onboard, as well as nuclear cores from land-based power plants, in order to have a prime example (regrettably, not the only instance) of a major disaster waiting to happen.

Oil pollution of rivers, groundwater, and oceans seems to be an almost daily occurrence. The recent leakage (July 2000) by Petrobras of oil into the Iguanu

River running south from Brazil to Argentina, the *Exxon Valdez* disaster (Alaska Fish and Game, 1989), the *Amoco Cadiz* tanker shipwreck off the coast of France, and the leakage from submarine oil transport pipelines and from offshore exploration and production platforms, together with the massive sustained (up to around 25%) leakage into the tundra region and groundwater supply of the ancient and decaying Soviet oil pipeline system, all provide examples of the ability to foul the very waters needed to sustain plant, fish, and animal growth, as well as the waters needed directly by humankind to survive.

Foodstuffs. In order to survive, humanity needs to supply itself with foodstuffs. Current estimates indicate that the world as a whole has about 25 to 30 days' reserve supply of food. Yield enhancements of crops have been ongoing for centuries, and the harvesting of fish and animals has occurred as long as civilization has been around.

However, as the world population has risen to its current level of around 6 billion, the pressure to increase food production has risen steadily. Because grains are subject to a wide variety of bacteriologic, fungal, and pestal infections, two major prongs of attack have been made to overcome these problems and so increase yields. First, one can spray growing grains with bactericides, fungicides, and pesticides to limit infestations. The price that is paid is that these various substances are absorbed directly by grains and also contaminate the ground, where they are then absorbed by plants through root feeding. One of the major open concerns at present is whether direct human ingestion of such treated grains has a negative effect on human health and longevity.

Second, one can modify plant materials genetically to make them intrinsically resistant to the various forms of infestation, thereby automatically increasing yield. What is under intense debate at present is the long-term effects on humanity of ingesting such genetically modified foods.

With respect to proteins supplied by animal flesh, perhaps the decades-long major problem of bovine spongiform encephalopathy in English cows (with the attendant transfer to humans in the form of the brain-wasting Jakobs-Creutzfeldt disease) should serve as a very salutary warning that the diseases caused by food supplies to animals can be transmitted to humanity in excruciatingly negative forms, leading to death and to further vector transport of such diseases by infected humans.

In attempts to stimulate yields of crops and fish farms, fertilizers and pesticides are used in abundance throughout the world. Runoff of such residual contaminants by rain or crop irrigation waters pollutes the waters flowing further downstream, including the effect on humanity, fish, plants, and wildlife that depend on such waters for existence. Perhaps DDT, a long-lived chemical used as a pesticide, which causes thinning of eggshells in birds and so leads to a steady loss of bird embryos, is an example of the long-term serious ecologic repercussions that can be caused by application of fertilizers and pesticides. On the one hand, crop yields are indeed increased, and on the other, major ecologic damage is done. An uneasy and very uncomfortable situation exists of attempting to feed the world but at the expense of poisoning some fraction of the biotypes inhabiting the world—including human beings. Without some alternative in terms of adequate food production for the burgeoning world's population, it would seem that humanity will continue with this exceedingly uncomfortable and dangerous situation.

As a further example of contamination of the food supply, consider the example of the release of radioactive strontium from the Windscale (now Sellafields) nuclear reactor in northwest England. This radioactive strontium permeated the ground, was absorbed and concentrated by grasses, which were then eaten by cows, and so the radioactive strontium found its way, in ever more concentrated form, into milk. Because strontium is of the same chemical family as calcium, its properties and affinities are similar. Accordingly, when babies and small children drank the milk, which the human body uses to produce bone tissue, the resulting bone material had highly radioactive strontium as an integral component. Needless to say, the incidences of bone and tissue cancers in such children were out of all proportion relative to other children who did not receive contaminated milk.

Land Use. The ever-increasing world population needs more space for habitation and for agricultural and other uses. In counterpoint, these needs make less land accessible and usable for species survival. For example, the Chernobyl nuclear power plant disaster and subsequent long-term radioactive pollution of the land in both the near and distant vicinities of the massive radioactive release should caution humanity about the need to decommission quickly such unsafe reactors to avoid the risk again of having such a problem. And yet, in countries without the wealth to install safer replacement reactors, continued use of Chernobyl-style reactors persists, with the attendant high risk to human beings, land, air, and foodstuffs being exacerbated as such reactors age.

Again, one needs only to look at the example of the denudation of the Amazon Basin rain forest to recognize a serious threat to many factors on which humanity relies. Because trees absorb carbon dioxide and produce oxygen, the destruction of the Amazon rain forest will eliminate one of the largest single regions of oxygen production. In addition, about 95 to 98% of the nutrients are in the trees so that once the trees are gone, there is nothing to hold the top soil in place, leading to massive soil erosion and transport down the Amazon River to the delta. Thus, while the Amazon burnoff will produce soil with rich nutrients for a few years, over a decade or less the soil will cease to be productive, and a wasteland is created.

Further problems are the loss of potential new drugs and chemicals from plants in the forest, the total erasure from the face of the Earth of animals that have specialized ecologic niches adapted to a rain forest environment, and the continued denudation by water cannons and open-pit mining as procedures for extracting gold and other minerals in a cost-effective but environmentally irresponsible manner.

The loss of societal human cultures that also call the Amazon forest home is also a major factor to be reckoned with, and their loss diminishes the diversity and richness of humankind's disparate evolution as a species.

Severe threats to the land on which humankind lives are provided by the acts of waste production from human and industrial activities, something that humanity is amazingly adept at accomplishing in a remarkably short time. On the industrial side, there are the problems of where to deposit chemical, biologic, nuclear, and general waste products from human activity. Classic examples of such indiscriminate depositories are the Love Canal in the United States, a chemical depository on which, 20 years later, houses were constructed with the concomitant massive increase in all sorts of human illnesses to the point that the Love Canal housing estate had to be abandoned and massive compensation paid to the survivors of the affected families.

On the nuclear waste depository side is the U.S. national low-level waste site at Barnwell, South Carolina, where the influence of subsurface fracturing and faulting, plus the potential for destabilizing earthquakes on the integrity of the waste depository, is raising major concerns. The problem of the safe disposal of high-level radioactive waste is an area of pressing concern around the world in terms of the very definition of *safe,* the length of time such long-lived highly toxic radioactive waste must be kept under control in any depository, and even the question of long-term (1000 years or more) safe record keeping for future generations. The billions of dollars the U.S government has already invested in exploring the feasibility of the Yucca Mountain site as one such possible safe place is a measure of the urgent need to resolve such problems safely and quickly—the Kara Sea dumping by the Soviets has already been mentioned as another lethal problem— an accident waiting to happen.

General-purpose landfill sites and oil pollution of the subsurface, together with the disposal of animal and human sewage waste, are becoming even more critical problems as the world population doubles every 20 to 30 years. The problem is compounded by the ever-increasing cost of disposal of solid wastes, the difficulty in finding appropriate geologic, geographic, and hydrologic sites, and the reluctance of the population to have landfills located near residential areas. There is a limit to the natural recycling rate of such organic wastes, and many have argued that the limit has been reached if, indeed, it has not already been surpassed.

The problem of disposal of biologic waste (be it from hospitals or as the result of bacteriologic warfare products and by-products) is undoubtedly one of the major problems facing the twenty-first century as more and more biologically toxic agents are generated, stored, or released (on purpose or by accident). There seems to be no easy answer to the disposal of such waste.

Mineral and Thermal Pollution.   The ever-increasing search for minerals and metals of economic worth has always had a major role to play in changing the environment. The marble quarries of the island of Thassos, Greece, and the gold, silver, and copper industries of the past on the island have all led to changes of forested areas to ugly eyesores of waste rock deposits next to the mines and quarries. This scenario is repeated around the world time and again as humankind has quarried and tunneled for rock, metals, coal, uranium, etc. Two or three major environmental pollution problems stand out in this regard, with a host of ancillary problems as corollaries.

Perhaps most exemplifying the cumulative problem of tunnel mining over the centuries is the general region around Bochum in the Ruhr Valley of Germany, where coal and metal mining has been carried out for centuries. Many mines were not recorded or the records lost, a large number of these being illegal. At present, the underground is a maze of known and unknown interconnected and cross-connected tunnels. As a result of long-term groundwater infiltration and, in general, the lack of cohesiveness of mechanical support in the residual rock strata, major sinkholes appeared in 1998–1999 in a heavily populated area. Immediate resolution of the problem called for cement to infill the huge sinkholes. However, the general problem is not solved so simply. There remain miles of such tunnels in one of the most heavily populated and industrial sectors of Germany. There is absolutely no idea of the likely occurrence of another such event, when it may occur, or where. The potential environmental impact is truly enormous.

In addition to physical collapse damage, both open-pit (quarry) mining and tunnel mining lead to exposure of waste rock to the elements. The problem of leachate production of acid mine drainage waters is particularly severe in many areas of the world where coal has been the main economic component being sought. Acid mine drainage is a significant water pollution problem in parts of Colorado, Illinois, Indiana, Kansas, Kentucky, Maryland, Missouri, Ohio, Oklahoma, Pennsylvania, Tennessee, West Virginia, and Wyoming (Keller, 2000). The production of heavy-metal residues in river, lake, and ocean waters as a consequence of mine waste piles being indurated is not a problem to be dismissed lightly either.

There are also soil erosion problems that arise both from mining activities and from irresponsible forest clearcut and slash-and-burn techniques. As remarked already, the Amazon Basin rain forest provides a prime example of such uncontrolled exploitation.

As an added element to the environmental problems created by mining activities, there are the major contamination problems caused by oil spillage already referred to. Also, the seemingly inexhaustible appetite for more and more power (usually in the form of electricity generation) leads to thermal pollution problems of rivers, lakes, and oceans from the cooling systems, as well as to generalized thermal pollution from energy-hungry devices. The current estimate of 5% of the U.S. electricity energy output for computers, with a rise to around 12% predicted over the next decade, serves to indicate the compelling need for energy frugality if we are not to thermally pollute water reserves to the point where nothing can survive in such high-temperature waters. How this environmental problem will develop as the world demands ever more power is not at all clear.

Environmental Impacts of Societal Activities.   Humanity is its own worst enemy with respect to environmental pollution and contamination. As already mentioned, natural events contaminate the environment humanity deems desirable to continued species survival, but the cumulative effect of such events is small compared with the scale of anthropogenically generated environmental problems.

The biggest problem is caused by the sheer number of human beings, currently around 6 billion, with an increase estimated to double that number in a decade or two. This large number of people requires air, water, food, shelter, clothing, and land. In return, humankind produces human waste; pollution of the air, water, and land it uses; and also, because of the close proximity of the mass of humanity (almost 30% live in major cities worldwide), contagious diseases. Such diseases can be spread by rodents (plague), by mosquitoes (yellow fever, Nile fever, malaria, etc.), by human-to-human contact or nearness (Ebola, typhus), by human sexual activity (HIV, AIDS, syphilis), and by other transport vectors.

Humankind has always warred with itself. Ever since recorded history began, there are records of war. The reasons for wars are as many and as varied as humankind has been able to invent, ranging from the need for forced labor, minerals, oil, or land through the desire of one religious, political, business, or national group to foster its own brand of system on another people or, indeed, just the massive aggrandizement of raw, naked power. As a consequence of this inherent historical conflict, over the centuries, weapons of greater and greater power have been generated. Indeed, a good fraction of humanity's endeavors has always been in the area of military weapons development.

The end result is that as a race today we possess weapons that can trivially annihilate and obliterate humankind from the face of the planet. Many of the weapons, such as hydrogen bombs, are so dangerous that their very testing in the atmosphere is internationally banned because of the global radioactive fallout problems—a major environmental problem. Many others, such as enhanced biologic anthrax, cannot be tested at all outside the confines of a secure laboratory because of their uncontrollable nature.

Perhaps one of the best recorded events from ancient times is the sack of Carthage by the Roman Empire, where not only was the city put to the torch, but the agricultural fields around the town were sown with salt, making them incapable of producing enough crop yields to support the Carthaginians and leading to a diaspora of the survivors who were not enslaved. Indeed, this event is a clear example of chemical warfare and corresponding environmental destruction.

In the modern era, after the use of mustard gas in World War I, such chemical weapons were banned internationally. The use of Zyklon-B in the concentration camps of Germany during World War II, however, and the more recent gassing of the Kurdish population in Iraq by its own government both show that treaties on chemical weapons are fragile and that careful monitoring may, in some instances, need to be undertaken by the international community.

One further consequence of war is the legacy left afterwards of land mines strewn everywhere. Because land mines are cheap to produce and contain minimal metal content, they are used as a deterrent that is difficult to detect by an opposing force. After a war, however, there are often no records available of where land mines were deployed, their deployment patterns, or the number deployed. Finding such fields of sudden death is often by accident—and normally by human beings being maimed or killed. The land is then not usable until all such land mines are cleared. And in poor countries this clearing is often long delayed or never under-taken before yet another war breaks out with a repeated sowing of death and destruction.

16.3.2   Resources and Environmental Issues

Natural Resources: Depletion and Waste.   Resources that humanity brings to its own domestic and industrial uses are of two basic types: intrinsically renewable on a time frame much shorter than humanity's existence (such as fishing, forestry, plants, animal husbandry) and intrinsically not renewable on such a time scale (such as oil, diamonds, metals, minerals, marble). However, the fact that some resources can be intrinsically renewed does not mean that they will be. For instance, humanity's needs have led to massive overfishing near Newfoundland and in the North Sea, to name but two instances. Thus a potentially renewable resource can be so overutilized by humanity as to destroy the resource!

For intrinsically nonrenewable resources, such as oil, for example, there are two major problems to consider: Is the currently available supply large enough to meet the demand? Are the known reserves and the rate of reserve replenishment by exploration large enough to keep the current mode of civilization in operation, and for how long?

In the cases of both renewable and nonrenewable resources one is faced with the problems of responsible management of the resource. The problems of deple-tion by overutilization or by limitations on the amount of the resource needed to be incorporated in the management so that one is well prepared, well ahead of time, for the eventual loss of the resource. Exacerbation of the resource supply and demand occurs when one or more nations control a supply that other nations need—as in the case of OPEC and the Western world's demand for ever more oil. Indeed, wars have been fought to keep the supply intact and to keep the Western world's economy in operation, with the Gulf War of 1991 serving as a recent example of such practices.

At the same time, resources are often wasted. For example, the main Russian oil pipelines from the West Siberian fields (and earlier from the Rumanian Ploetsi fields and the Azerbaijan oil fields) often had sustained leakage rates of more than 10 to 20%—a serious waste of a potentially valuable commodity. The quest for more resources, both renewable and nonrenewable, in order to maintain and improve the quality of life has a long way to go to become rational and efficient for the good of the entire human population.

Entrepreneurial Developments.   A broad classification of collective human endeavor relevant to environmental considerations can be drawn along the business, political, and national arenas. Within these broad groupings, environmental problems and potential solutions have to be considered. Not all groupings make for ease in containing or eliminating anthropogenically produced environmental problems; indeed, the groupings can themselves be the main source of such problems and often have no intrinsic desire to provide solutions. Consider some illustrative points.

The fundamental purposes of business are to make as much profit as possible at whatever is undertaken and also to grow. Within these two basic paradigms, one also must remember that each major business endeavor produces waste—be it in the form of carbon dioxide emissions, cyanide from gold mining, or whatever—and the waste must be disposed of. One also must keep in mind that businesses have accidents (e.g., the Bhopal India toxic gas release from a Union Carbide plant, oil refinery explosions on a recurring basis, etc.).

From the point of view of a business without any external controls on its operations, it is most cost-effective just to dump the waste material. However, with the modern realization that we all live on a very small planet indeed, increasingly stringent controls are being enacted by governments on local, regional, national, and supranational levels to force businesses to police their waste and to clean up previous waste production where possible. Businesses, in turn, end up passing these additional costs directly onto product prices for consumers where possible, as well as taking a tax writeoff of the costs. The ultimate motive is still the cost/profit drive. And the ultimate environmental problem still remains: Where is the waste to be disposed of? External environmental control does not solve the basic problem.

Methods of political governance of populations have been as many and as varied as humanity has been able to devise, and corruption of and by political officials also has been as equally varied over the course of history. The activities of governments, some secret and some open, also have led to major environmental problems, which often have come to light only years later, if at all. For instance, in Western societies, most of the funds dedicated to remediation projects have been used for partial remediation of military bases with chemical spills into the ground, for remediation of nuclear plants that often were not reporting correctly their accidents, for biologic weapons' testing facilities remediation, etc. When environmental problems of such magnitude are caused by democratic governments, then there is little chance in societies under a variety of less than open national governments of even finding out what the problems are, never mind attempting remediation of such problems.

The common needs of a nation of people and their mutual agreement on societal forms of controls and rights are perhaps the quintessential backbone underpinning how a nation evolves. The interests of the nation as a whole become the paramount concern, with the needs of other nations, subscribing to alternative common good agreements and to alternative forms of societal controls and rights, being of secondary consideration. The conflicts of such different perceptions of worth of humanity have led to major wars over the centuries, to the detriment of the different perceptions, and producing ravages to humanity, countries, and cultures as the bitter fruit.

As with any war, national effort goes into diverting natural resource raw materials to the instruments of war, so not only is there a loss to society of the benefits such resources could bring, but the national effort to improve the environment for the benefit of the nation is thwarted. And the depletion of scarce nonrenewable resources, together with their waste on the pursuit of harming one's neighbor, is hardly a responsible measure in managing such resources.

Recognition that an environmental problem exists is the first stage in attempting to address the issue, irrespective of who instigated the problem. However, dealing with the problem in a remediation sense requires money. This money can come from two major sources: private industry or government.

In the case of private industry, money to remediate self-made environmental problems or to minimize production of such problems generally will be spent under one of four broad conditions: (1) when the government penalties for not so doing are more burdensome a charge to the company than are the remediation costs, (2) when a fine can be levied for not taking sufficient action, (3) when the public perception of an environmentally responsible company helps boost sales to such an extent that they exceed in profit the environmental costs, and (4) when the company needs to recycle waste material to recover components of the waste for reuse because the cost of so doing is cheaper than buying fresh components. In this case, recycling is in the corporation's best interests.

In the case of governments, several factors come to the fore in addressing remediation of environmental problems. First, the only source of money a government has is the taxes it collects from corporations and individuals. Hence any money a government commits (other than from penalties or fines imposed) to remediate environmental problems must come either from the general revenue surplus fund, from the current operational budget, or from a surcharge to taxpayers. Second, identifying the culprit in a particular environmental problem may not be easy; a corporation may long since have become defunct, but its environmental pollution legacy survives; the government may (and often is) the main pollution culprit over the years; or the environmental standards imposed and met in an earlier era may no longer be relevant when viewed in the light of later scientific and technological information. Third, the government rarely performs remedial action itself. Instead, it often prefers to let a contract to private industry to perform remediation, occasionally to the same corporation that produced the problem in the first place. Then some form of independent monitoring of the terms of the contract is required to ensure that proper remediation is carried out.

Whichever way the problem is addressed, the point remains that capital is required. Too little capital invested will not solve an environmental problem but often will exacerbate it. The severity of some environmental problems we are facing today requires, in many cases, funds of such magnitude that it is questionable that societies will agree to the required level of investment.

16.4 REMEDIATION OF ENVIRONMENTAL PROBLEMS

In this overview of natural and anthropogenic environmental problems, it is not appropriate to consider specific technical methods for remediation of every specific type of environmental concern. Such an undertaking is best left to those who are specialists in given areas. However, several factors to consider in general stand out in sharp focus; these factors are discussed here.

Because societies have reached a stage where the environmental problems they create diminish the quality (and quantity) of the very environment required to survive, it would appear that three dominant characteristics need to be addressed. First, and foremost, societies must be both educated and encouraged (with a reward/penalty system most likely imposed) to minimize the anthropogenic environmental problems. Failure to do so means that we will increasingly pollute the planet in all aspects, a less than healthy long-term species survival trait for humankind.

Second, changes to the infrastructures of societies and to the living standards of humanity as a whole should be encouraged at a global level so that some of the vast untapped human potential can be channeled to further environmental improvements. Failure to do so, without altruistic aid from the better-developed nations, will increase the gap between developing and developed countries and just lead to more environmental problems.

Third, the major future problem is surely the explosive growth of human population, which creates even more strain on the world's natural resources and which seriously erodes the global quality of air, water, food, shelter, clothing, and energy that humankind needs to survive. If we do not learn to control our own human population, then it is questionable whether we will be able to address the ever-increasing stresses put on the environment.

Addressing major environmental problems can be broken into two broad categories: avoidance, control, and/or remediation of those problems which have a deleterious impact on our current and near-future use of the environment (such as oil spills, heavy-metal water pollution, carbon dioxide in the atmosphere, etc.) and treatment and disposal of environmentally damaging materials that have already been created and which would have profound long-term consequences for the global environment if not safely contained (such as nuclear waste, biologic organisms, long-lived chemical toxic pollutants, etc.).

Each requires an intermix of national and supranational government protocols, accords, and laws for enforcement; each requires responsible action on the part of individuals, businesses, and governments; and each requires coordinated remediation on a global scale involving all nations. One cannot, for instance, declare a river pollution problem "solved" in a given nation just because the water-borne pollution is transported to a neighboring nation by the river. Such an approach may serve well as political rhetoric, but it does nothing at all to solve global environmental problems.

Perhaps we are making a start on the awareness sensitivity level of particular issues (such as the greenhouse problem or the nuclear waste and nuclear accident problems), but we have a long way to go to address all such major environmental problems to the benefit of humankind.

One of the classic English sayings is "An ounce of prevention is worth a pound of cure." This saying also pertains to the environment. It is better not to create or at least try to control the magnitude of an environmental problem than it is to have to remediate one. However, until we clean up the environmental problems we have already created, we will have to spend enormous amounts of money to remediate. At the same time, prevention of future potential environmental problems by global agreement, national laws, independent agencies, and corporate and individual responsibility surely must become the sine qua non if humankind is to flourish in a healthy environment.

16.5   RULES FOR ADDRESSING ENVIRONMENTAL PROBLEMS

In the ancient Greek days, advice given to doctors was, "First and foremost, do no harm." This adage can be applied equally well to remediation of environmental problems today.

1. First and foremost, study the particular problem thoroughly before proposing a course of remedial action.

2. Be ready with different potential courses of action to be used in parallel or sequentially as the need is seen.

3. Before commencing a major remediation effort, do a pilot study to determine that there are no untoward nasty environmental surprises caused by the suggested remediation procedure itself.

4. Do not do an incomplete or poor remediation treatment. It will just have to be done all over again properly at even more expense.

Humankind as a whole must be trained to stop or minimize its generated environmental problems.

CHAPTER 17
EPILOGUE

While there are many professional papers and books devoted to the scientific analysis of the environmental problems of transport and storage of toxic materials, and while there are equally many contributions to estimates of scientific risk and uncertainty of long-term contamination, there appears not to be available an equally well-developed literature on the economic aspects of transport and storage risk factors. Yet the economic aspects control, to a very large extent, the potential involvement of corporations in such environmental projects. The dominant theme of this book has been to show how one can intertwine the scientific risk and economic risk aspects from a corporate perspective. The main aim of a corporation is to prosper profitably. Thus the question of whether a corporation should become involved in a particular environmental project eventually reduces to whether the corporation estimates it can make a profit. On the way to making a decision, a corporation must grapple with concerns of regulatory agencies and their requirements, with potential leakage and spillage and liability, with insurance cover costs, with fractional working interest estimates, and with a host of ancillary problems. This book has tried to address the majority of these concerns, examine some of the basic problems that arise from a corporate perspective, and provide tools for the analysis of more complex situations.

Nevertheless, there are just too many individual facets in each and every environmental problem for this book to examine all variations; such would make for a very long tome indeed. Instead, we have opted throughout to provide methods that illustrate how one can bring the powerful arsenal of economic probability methods to bear on organizing quantitative procedures for evaluating potential involvement. Often we sacrificed complexity of real situations for simplified rules and regulations in order for the pattern of development to become more transparent than it otherwise would be.

However, the quintessential theme has been to show that one can allow for uncertainty and risk in assessing the economic worth to a corporation of different aspects of environmental problems. It is this theme that has dominated development of the methods and simple illustrations reported here. We will have more than fulfilled our aim if others, more gifted than ourselves, can use the methods and procedures presented here and extend them to encompass a broader spectrum of economic environmental issues.

We would, however, be remiss if we did not also mention that political, social, and other noneconomic issues often have an impact on an integrated scientific-economic evaluation in unforeseen ways. For instance, a recent success of a political party in Europe in taking the reins of power was contingent on coalition with an environmental party. As an outcome of this coalition, one of the first political decisions involved the eventual shutdown of all nuclear power plants in that country. This decision clearly has consequences for disposal of the radioactive material, for contracts already signed, for nuclear power generation and station maintenance, and for replacement sources of power generation.

The point is that the ability to provide, from a corporate perspective, an objective scientific-economic evaluation of a particular environmental project can be easily thrown into disarray by overriding political or social decisions and by the attendant associated effects such decisions bring to projects—often in very unsuspected manners and from unsuspected directions.

We see no way, currently, to incorporate all such possible decisions into assessments of environmental projects. Indeed, given the rapidity with which governments change relative to the much longer lifetimes of typical environmental problems, it seems to us to be difficult to guard against every political or social eventuality that could occur. Perhaps this aspect, too, is one that needs more ability than we are capable of providing.

Recognition that such problems exist and that they color the best scientific-economic assessments made can be allowed for to some extent by adding probabilistic models of rate change—as we have shown with a simple illustration earlier. It would indeed be fascinating to develop this aspect further than we have taken the process. In fact, it would be fascinating to take all the aspects to greater levels of perception and understanding.

In view of the ever-increasing demands that humanity is placing on its environment and the consequent problems that arise, it is our opinion that the next decades will see a remarkable advance in the accuracy and precision of such economic methods coupled with both scientific and political/social risks and uncertainties.

APPENDIX 1
PHASE I, II, AND III SITE ASSESSMENTS

This appendix contains excerpts from the *Civilian Federal Agency Task Force (CFATF) Guide on Evaluating Environmental Liability for Property Transfers* of August 1998 that relates to Phase I, II, and III reporting. The *CFATF Guide* summarizes the requirements for evaluating potential liability from environmental contamination associated with real property transfers. The intent of this appendix is to familiarize the reader with some of the requirements for environmental site assessments that have been referred to in the text. For application to real sites, the full text of the *CFATF Guide* and its appendices should be consulted, together with additional agency policies and guidelines, as well as federal, state, and local regulations. The full *CFATF Guide* and its appendices can be downloaded from the EPA site: *www.epa.gov.*

A1.1 THE ENVIRONMENTAL DUE DILIGENCE AUDIT PROCESS

The process of evaluating proposed transfer properties for potential environmental contamination and liability is referred to in the *CFATF Guide* as the *environmental due diligence audit (EDDA) process.* The focus of the EDDA process is to identify and document proposed transfer properties for potential environmental contamination. The agency's objectives in executing the EDDA process include

- Ensuring that all environmental due diligence requirements are addressed and potential environmental contamination is identified
- Establishing a consistent and defensible approach for addressing necessary environmental actions
- Providing the environmental baseline and assessment of properties to assist in property transaction decision making

- Avoiding costly litigation and environmental remediation liability under the Comprehensive Environmental Response, Compensation and Liability Act (CERCLA), Resource Conservation and Recovery Act (RCRA), or other relevant regulatory statutes

A1.2 PHASES OF THE EDDA

The EDDA process contains three distinct and cumulative phases that are designed to support key decision points. Progress from one EDDA phase to the next is based on the need to further assess property contamination. Thus the EDDA process ends at any point where the agency deems it has sufficient property contamination and liability information to make a decision regarding the property. If all three phases are necessary, they are as follows:

Phase I. Liability assessment

Phase II. Confirmation sampling

Phase III. Site characterization

The intent of the Phase I is to evaluate the *potential* for environmental liability at the site. This is done through interviews, by means of a site visit, and by gathering and analyzing information on current and past site uses and activities. If the Phase I report indicates a potential for environmental liability from contamination, the Phase II assessment is performed using focused field sampling to confirm or deny the suspected contaminants. Once contamination is confirmed, a Phase III EDDA may be initiated to fully characterize the nature and extent of the contamination and develop cleanup options and recommendations. Throughout the process, the EDDA has two overriding objectives: (1) to identify liability from past site uses and (2) to provide technical information to assist in agency decision making.

A1.3 PHASE I: LIABILITY ASSESSMENT

Phase I identifies potential areas of contamination and environmental concern that may result in environmental liability. It consists of

- Preliminary activities
- Site visit
- Records review
- Regulatory review
- Geologic and hydrogeologic review
- Report development

All data gathered during this phase are documented in a Phase I report. The activities and process for the Phase I liability assessment can be found in Chap. 4 of the *CFATF Guide* and are discussed in more detail below.

A1.3.1 Phase I: Description

The purpose of the Phase I Environmental Due Diligence Audit (EDDA) is to identify potential areas of hazardous waste contamination or environmental liability associated with a property to be transferred. This section describes in greater detail the EDDA Phase I Liability Assessment, which consists of the following elements:

- *Preliminary activities.* Coordinating site visit logistics, gathering basic information regarding the subject property, and building a rapport with the site owner and site contact.

- *Site visit.* Observing visual signs of contamination and uncovering evidence of potential liabilities and contamination from past and current operations or from off-site activities.

- *Records review.* Examining applicable documents, records, and aerial photography to supplement site visit findings and to gain additional information regarding prior uses of the property that may indicate that a release of hazardous substances has occurred.

- *Regulatory review.* Examining applicable enforcement records to supplement site-visit findings and to gain additional information regarding past environmental compliance violations, fines, or outstanding liens.

- *Geologic and hydrogeologic review.* Evaluating potential contaminant migration pathways and exposure routes.

- *Report.* Documenting the results of the EDDA Phase I Liability Assessment, documenting that due diligence has been exercised, and as necessary, documenting information to initiate phase II confirmation sampling.

Sampling is not performed during the liability assessment; all the information included in the Phase I report is gleaned from existing documents or inferred from observation made during the site visit. The environmental areas that are examined during this process include

- Hazardous substance release on the subject or adjacent property
- Hazardous material and waste handling practices
- Underground and aboveground storage tanks
- Polychlorinated biphenyls (PCBs)
- Pesticides and herbicides

- Sensitive environments (including wetlands) on the subject or adjacent property
- Historic or cultural significance of the subject or adjacent property
- Asbestos
- Lead
- Radon and indoor air
- Ionizing and nonionizing radiation
- Topographic and natural resource factors

These elements and areas are covered in further detail in the following discussion of the EDDA Phase I Liability Assessment process.

A1.3.2   Phase I: Preliminary Activities

Prior to conducting the site visit, some preliminary activities are necessary, including logistics for the site visit, obtaining basic property information, and contacting the site owner or operator to brief him or her on the purpose, scope, and process of the EDDA Phase I Liability Assessment.

Logistics.   Identifying a primary point of contact for the subject property will facilitate the entire EDDA process. Negotiate an exact date and time for the site visit with this person, and then inform the property owner and operator; the environment, health, and safety manager; other site representatives; and relevant agency officials. As appropriate, invitations to attend the briefing and walk-through should be extended. During this preliminary step, it is also necessary to discuss and resolve escort issues, including access, site security, safety briefings, and the need for specialized equipment, such as personal protective equipment (PPE).

Contact.   Preliminary telephone interviews may include the property owner or operator, adjacent property owners, and state and local authorities.

Questionnaire.   A questionnaire can be used as a tool to gather fundamental information from the site owner and operator or lead point of contact. It may be administered by the assessor during the initial phone contact, or it may even be mailed electronically to the property manager when the property is held by an agency. In either case, the name, phone number, position, and responsibility of the person answering the questions must be documented to allow for later verification as necessary. In addition to gathering basic information about the property, a questionnaire may help focus the site visit and document search to issues of relevance to the particular property. A sample questionnaire is provide in Appendix M of the *CFATF Guide.*

Gather and Review.   The assessors should prepare for the site visit by reviewing available site maps and documentation relevant to site activities and environmental issues. Information gathered during this step will give the assessors a general understanding of the property and site activities, in particular

- The exact location and size of the property
- Identity of current property owners
- A site contact, to provide access to the property during the site visit
- The number of buildings and structures located on the property
- Presence of aboveground storage tanks (ASTs) or underground storage tanks (USTs)
- Current site activities or operational issues that could have an impact on the site visit

All the preliminary activities provide a foundation for conducting the site visit.

A1.3.3   Phase I: Site Visit

The site visit is an essential element of the EDDA Phase I Liability Assessment that allows the assessors to make first-hand observations. In general, it consists of the following activities:

- Visual survey of the subject property and neighboring properties
- Interviews with property owners, on-site employees, and neighboring property owners
- Review of on-site documentation

The visual survey portion is intended to identify visible signs of environmental contamination or evidence of suspected contamination from current or past operations, both on and off the property.

The interviews should include discussion of site management and operations with the property owner, manager, or a designated representative. As warranted and reasonably possible, former facility personnel also may be identified and interviewed. They may have information regarding suspected contamination from past activities conducted at the facility. The input and inquiry of as many personnel as possible will help produce valid and defensible information.

The site visit further includes review of on-site records relevant to the environmental management and history of the property. All three aspects of the site visit are discussed in greater detail in the following sections. Appendix N of the *CFATF Guide* provides samples questions to consider when performing a site visit.

Focus.   Basic environmental considerations, including the items listed below, should be reviewed as part of the site visit. In addition, an assessor should walk the *entire* perimeter of the property to look for potential site contamination issues and to note the presence and condition of any sensitive environments. *Any potential or actual hazardous conditions encountered during the site visit should be reported to the owner and operator or facility manager.*

The following is a summary of issues to be addressed during the walk-through:

- Former and current uses of the subject and adjacent properties
- Adjacent property characteristics, such as zoning, future land use, USTs, and past uses
- Sensitive environmental areas
- Surveys or inspections, past and present, including radiologic, asbestos, radon, and USTs
- National Priorities List (NPL) status of the subject property and properties in the vicinity
- Permits, past and present, including air, Nuclear Regulatory Commission (NRC), National Pollution Discharge Elimination Systems (NPDES), Publicly Owned Treatment Works (POTWs), USTs, and hazardous waste Treatment, Storage, or Disposal Facilities (TSDFs)
- Hazardous releases, including disposal, injection, and discharging
- Hazardous waste handling and storage practices
- Other waste handling practices—solid, sewage, septic, drains, sumps, lagoons, and pits
- USTs or ASTs—operating, closed, leaking, or inactive
- Fuel leaks or releases on both subject and adjacent properties
- Radon
- Potentially hazardous dusts and indoor air quality
- Asbestos-containing materials—use, storage, and research
- Lead-based paints and other lead sources—use, storage, and research
- Ionizing and nonionizing sources, such as radiologic materials and equipment—use, storage, and research
- PCB-containing materials—use, storage, and research
- Pesticides—use, storage, and research

Observations.   Observations made during the site visit will include obvious signs of current or potential contamination. Many hazardous substances will stain soils or other surfaces and may destroy vegetation such as grass or plants. The presence of drums may be an indication of hazardous waste contamination. The site owner and operator or representative should be consulted to identify the contents of unlabeled drums. Material Safety Data Sheets (MSDS) on file at the site also may be helpful in determining hazardous materials present. Additionally, to determine the potential for contamination, inquiries should be made about past practices, such as the disposal of chemicals in sinks.

In conducting a walk-through, EDDA team members should not engage in any activities that could put themselves or others in jeopardy. Certain activities may require specialized training, procedures, or permits in order to conduct them safely and in compliance with regulatory requirements. Such activities include opening drums of known or suspected hazardous materials and entering hazardous areas, such as confined spaces, trenches, or pits 5 ft deep or deeper.

Hazardous Material and Waste Handling Practices. The terms *hazardous material, hazardous waste,* and *hazardous substance* refer to a wide range of chemical, radioactive, and biologic substances or materials.

- *Hazardous material.* Any substance or material that has been determined to be capable of posing an unreasonable risk to health, safety, and property when transported in commerce (49 CFR Part 172, Table 172.101). This includes hazardous substances and hazardous wastes.

- *Hazardous waste.* Under the Resource Conservation and Recovery Act (RCRA), a waste is considered hazardous if it is listed in or meets the characteristics described in 40 CFR Part 261, including ignitability, corrosivity, reactivity, or extraction procedure toxicity.

- *Hazardous substance.* Any element, compound, mixture, solution, or substance defined as a hazardous substance by the Comprehensive Environmental Response, Compensation and Liability Act (CERCLA) and listed in 40 CFR Part 302. If released into the environment, hazardous substances may pose substantial harm to human health or the environment.

Hazardous wastes and the potential for past release or mismanagement present the greatest single area for environmental concern and potential for liability. Details about correct waste handling, storage, and disposal practices should be available from either the owner and operator or the facility environment, health, and safety manager. If the subject property is owned or operated by the agency and is under review for disposal or lease termination, the review team should be extravigilant to account for all hazardous materials and wastes and when they will be transferred. Any facility that disposes of regulated quantities of hazardous materials will have an Environmental Protection Agency (EPA) waste generation identification number on record.

Underground Storage Tanks and Aboveground Storage Tanks. Petroleum products or hazardous substances may be present on site in USTs or ASTs, as well as associated underground pipelines. A leaking UST or AST system presents a potential risk of contaminating surface soils, surface waters, or groundwater. There also may be a potential fire or explosion hazard from a poorly maintained or leaking UST or AST system containing ignitable or reactive materials. The entire system, including sumps and pits, should be inspected visually where possible to identify potential sources of contamination. The questionnaire found in Appendix O of the *CFATF Guide* may be used as a protocol or guide for obtaining additional information on USTs and ASTs. Federal regulatory requirements for managing USTs are found in 40 CFR Part 280.

Polychlorinated Biphenyls. PCBs are organic chemicals that have been determined to be a public health concern. In the United States, PCBs have not been manufactured since 1979; however, they remain prevalent in many types of electronic equipment and hydraulic fluids. Examples of equipment that may contain PCBs include transformers, capacitors, and light ballasts. In addition, fluids associated with heat-transfer systems, hydraulics, and waste oils may also contain PCBs. The regulations under the Toxic Substance Control Act (TSCA) cover PCBs and require prominent labeling and management activities. Once introduced into the environment, PCBs are extremely persistent and do not breakdown. The assessor should establish whether PCBs are associated with the property and pose a potential liability. Assessors also should document all equipment that may contain PCBs. Any suspect equipment that is not marked as "non-PCB equipment" should be considered a potential source. TSCA regulations for PCB management are codified in 40 CFR Part 781.

Pesticides. Pesticides are chemical products developed to control plant or animal life. The term *pesticide* includes insecticides, herbicides, rodenticides, fungicides, disinfectants, and plant growth regulators. The most widely used pesticides share some common traits:

- They tend to be chlorinated hydrocarbons.
- At sufficient levels, they tend to produce a wide range of adverse effects in humans, such as nerve damage, liver damage, and kidney failure.
- They tend to bioaccumulate, meaning that as plants and animals ingest these chemicals and in turn are ingested by other animals, the poisons accumulate up the food chain. Therefore, what starts out as a small, nonharmful release may accumulate into harmful doses for other organisms.

Assessors should document both the use and management of pesticides at the site.

Sensitive Environmental Areas. Sensitive environments encompass a broad spectrum of site characteristics (e.g., wetlands, coastal zone, and parks and recreational areas). Certain ecosystems are considered critical when endangered or threatened species are sustained within that ecosystem. As a result, evaluation of a property requires an awareness of floral and faunal environments, wetlands, and endangered species. Although some issues related to the presence of sensitive environments are outside the scope of EDDA, during the site visit, the assessor should walk the entire perimeter of the site to identify sensitive environments and note any potential for contamination, either on or off site.

In addition to wetlands, all surface water retention ponds, stormwater management units, surface impoundments, or pits should be identified. To the most reasonable extent possible, the use, contents, and characterization reports of these ponds, pits, or impoundments should be analyzed to determine if suspected contamination exists.

Other sensitive environments, such as coastal areas, parks and natural preserve areas, or surface waters (e.g., rivers, streams, or ponds) may also present future limitations to property use or constitute difficult-to-address receptors for contamination issues. These resources should also be noted and specifically observed during the site visit.

Historic and Cultural Significance.   Phase I provides an opportunity to consider historic significance in the property transfer. Districts, sites, buildings, structures, and objects that are significant in American history, architecture, archaeology, engineering, and culture are to be preserved for present and future generations. Assessors should inquire specifically into the historical and cultural significance of the site and adjacent properties. The EDDA Phase I Liability Assessment should note any potential restrictions to property use or development.

Asbestos.   Asbestos is a naturally occurring mineral that is a very effective heat and sound insulator. As a consequence, it was used in many buildings as a fire and noise retardant. However, it has been linked to several diseases, including lung cancer, and since 1987 it has not been used in construction materials. Nonetheless, most structures constructed before 1987 have asbestos-containing materials (ACMs) in insulation, floor tiles, mastic, pipe wrap, roofing, and other materials. Sites that manage friable ACMs should have an asbestos operations and management plan on site that contains a survey of the site ACMs. Assessors should review any site-specific asbestos documentation, assess construction dates, and visually examine building materials to judge whether ACMs may be present at the site. Keep in mind that assessors should *never disturb any objects suspected of containing ACMs*; doing so requires specialized training and certification and, if done improperly, may result in a hazardous situation.

Lead-Based Paint and Other Lead Sources.   Many buildings and structures contain significant amounts of lead-based paints and other lead sources that may pose an environmental hazard at the subject property. Other sources include lead piping and solder that may contribute to high lead content in the drinking water. Lead has been associated with central nervous systems disorders, particularly among children and other sensitive populations. Exposure to lead is usually through *inhalation* during renovations and demolition activities or through *ingestion* of paint chips or lead-contaminated drinking water. Assessors should evaluate the potential for site structures to have lead-based paints and inspect building features and documentation to determine whether lead piping has been used.

Indoor Air Quality, Radon, and Potentially Hazardous Dusts. Radon, potentially hazardous dusts, and other indoor air issues are not readily observable. Radon is a naturally occurring, invisible, odorless, and tasteless radioactive gas. Inside enclosed spaces, radon and other potentially hazardous dusts can accumulate to levels that may pose risks to human health. Assessors should inquire about tests conducted at the subject property and should review area documentation for the presence of radon in and around the subject property.

Facility Documentation. Clues to past and present hazardous material and waste management practices also can be ascertained from facility records, and their review is an important aspect of the site visit. Facility records provide an excellent document trail of the environmental history and current management practices at the site. Records on hazardous waste accumulation, storage, treatment, or disposal (e.g., satellite waste accumulation records and manifests) should be reviewed. Other environmental management plans and reports also may provide information on the use and management of hazardous materials and wastes. These include, but are not limited to, spill prevention, control, and countermeasures (SPCC) plans; pollution prevention (P2) plans; and Emergency Planning and Community Right-to-Know Act (EPCRA) plans and reports. Quantities of hazardous materials used and stored, as well as reportable hazardous substance releases, must be accurately identified for compliance with federal and state requirements (e.g., 40 CFR Part 373). Finally, assessors should review any previous environmental inspection or audit reports, management plans, National Environmental Policy Act (NEPA) documentation, and any other relevant information to gain a comprehensive understanding of the environmental history of the site.

A1.3.4 Records Search and Review

A significant element of a Phase I Liability Assessment is the record search and document analysis. EDDA assessors must analyze documents obtained during the site visit, as well as records from federal, state, and local regulatory entities. In conjunction with the site visit, an assessor should visit local regulatory and county offices to obtain and review additional records that may shed light on the environmental history of the property and, to the extent practicable, on contiguous and adjacent properties. For example, chain-of-title documents, aerial photographs, incident reports, and other key documents that might provide information on past site uses and hazardous materials management and disposal activities. Many records also can be obtained without traveling to the site. Table A1.1 summarizes target records and their potential sources.

Adjacent Property. The purpose of the adjacent property records search and review is to identify the issues that may have adversely affected the environmental

TABLE A1.1 Target Information/Records and Potential Sources

Target information/records	Potential sources
Site ownership history	Title search—local courthouse Agency real estate office Agency historian National Archives and Records Administration (NARA), related regional archives, and state archives
Site use history	Current owner and operator Facility records Previous owner and operator Sanborne Fire Insurance maps Agency historian Agency facility, architecture, or engineering office Agency reports (budget, A-106, FedPlan, annual, etc.) NARA, regional archives, state archives Title search
Aerial photographs	Facility records Current owner and operator Previous owner and operator Agency historian Agency regional or area office Agency facility, architecture, or engineering office Agency reports (budget, A-106, FedPlan, annual reports) Local collections, universities, or museums Local highway or transportation department United States Geological Survey (USGS)
Environmental permits	Current owner and operator Facility records Agency regional or area office Previous owner and operator State and local regulatory authorities
Environmental surveys	Current owner and operator Facility records Agency regional or area office Previous owner and operator
Harardous Materials and Waste	Facility hazardous materials management plans EPCRA reports State environmental agency Environment, health, and safety manager Facility engineering manager Facility records Material safety data sheets Facility environmental compliance audit reports Facility personnel or service contractors Agency regional or area office Agency facility, architecture, or engineering office Agency reports Product manufacturers
Site contamination per National Priorities List, Federal Facilities Docket, or State Contaminated Site List	LandView (mapped environmental and census data tool) IDEA database (1-888-EPA-IDEA or *http://es.inel.gov/oeca/idea*) CERCLA Information System (CERCLIS) Hotline (202-260-0056)

TABLE A1.1   Target Information/Records and Potential Sources (*Continued*)

Target information/records	Potential sources
Site contamination per National Priorities List, Federal Facilities Docket, or State Contaminated Site List	RCRA/Superfund Industry Assistance Hotline (1-800-424-9346) EPA regional offices (Web site: *http://www.epa.gov/*) State environmental agency
Fuel leaks	State environmental agency Local fire and health department Facility records
Aboveground and underground storage tanks	State environmental agency Local fire department Facility records Facility environmental compliance audit reports Facility SPCC plans Facility hazardous materials management plans EPCRA reports Material safety data sheets
PCB equipment, use, and incidents	Facility PCB log or records Current owner and operator Facility personnel or service contractors Local fire department Agency regional or area office Agency facility, architecture, or engineering office Equipment manufacturer Utility company (large transformers or utility-owned)
Wetlands and environmentally sensitive areas	Facility studies Town/county planning and zoning office County soil survey reports Local soil conservation district National/State Wetland Inventory Maps (available from EPA regional offices) United States Army Corps of Engineers Wetlands Protection Hotline (1-800-832-7828)
Asbestos	Building age (pre-1987) Asbestos survey reports Facility records Facility as-built drawings and specifications Current owner and operator or manager Environment, health, and safety manager Facility engineering manager Facility personnel or service contractors Agency regional or area office Agency facility, architecture, or engineering office Product manufacturers
Lead paint and other lead or heavy metal sources	Building age Lead survey reports Construction blueprints and specifications Facility maintenance records or procedures Facility personnel or service operators Water utility service

TABLE A1.1   Target Information/Records and Potential Sources (*Continued*)

Target information/records	Potential sources
Indoor air quality, radon, and potentially hazardous dusts	Facility records and survey reports Environment, health, and safety manager Agency regional or area office County/local health department State Occupational Safety and Health Administration
Hydrogeology and geology	Facility soil studies and groundwater test results USGS United States Department of Agriculture (USDA) State water resources control County planning office Local soil conservation district Local county planning office Aerial photographs

condition of the subject property. Generally, the adjacent property review effort will be limited and not as extensive as the subject property review. The search radius may be left to the discretion of the environmental professional (i.e., EDDA Phase I Liability Assessment review team). Factors that may be considered when evaluating adjacent properties and determining the search radius include

- Density of the setting where the facility is located (e.g., rural, urban, or suburban)
- Distance that hazardous substances or petroleum products are likely to migrate based on local geologic or hydrogeologic conditions
- Adjacent NPL or contaminated sites

Site History: Ownership and Use.   Prior site ownership and use typically are documented in local property ownership and tax records. A chain-of-title review should be conducted to list continuous ownership and use of the property to the present time. A title search, Sanborne Fire Insurance (SFI) map, and special hazard area map review will reveal the past owners, uses of the property, and properties that are subject to flood hazards. The general site history and property owners for the last 50 years should be identified to determine the past property uses and activities.

Title Search.   Chain-of-title records are maintained at the local courthouse and may be researched with the assistance of agency real estate specialists or court clerks. The purpose of the title search and review is to fully identify past owners and research any information that might affect the current environmental condition of the subject property.

Sanborne Fire Insurance Maps.   SFI maps identify past property owners and property uses. Analysis of this information may reveal the types of activities and associated materials that could have been managed at the facility.

Aerial Photographs. Aerial photographs are used to reveal past site uses that raise environmental concerns or may help in documenting the timetable for site improvements and associated activities. Aerial photographs of the subject and surrounding properties should be reviewed for the last 50 years to verify site activities and the activities at neighboring sites. An individual qualified and trained to interpret aerial photographs should perform this review.

Environmental Surveys and Audit Reports. All available environmental survey and audit reports (from current or past owners and operators) should be reviewed to determine if contaminants were or are currently present at the site or adjacent properties. This includes surveys and reports for USTs, lead-based paint, air quality, radiologic, mercury, PCB, and asbestos contaminants. In addition, beneficial information may be obtained from reviewing reports on multimedia environmental compliance status, management practices, NEPA, and P2.

Utility Transformer Records: PCBs. Under 40 CFR Part 761.180, facilities that use or store a total capacity in excess of 45 kg of PCBs, one or more PCB transformers, or 50 or more PCB large capacitors are required to maintain an annual PCB log on site. Records also may be sought from the local utility companies. Note that occasionally more than one utility company will have jurisdiction over a given property.

Special Hazard Area Maps. Special hazard areas denote properties that lie within floodplains and have flood, mudslide, or flood-related erosion hazards. The maps identify properties in terms of 10-, 50-, 100-, and 500-year flood discharges. Such designations may limit the type of activity permitted on a property. In addition, understanding the property's location with respect to floodplain areas will assist in interpreting the potential for on- and off-site contamination impacts.

Nuclear Regulatory Commission (NRC). For properties where there have been permitted radiologic activities (e.g., laboratories, medical facilities, and some commercial research and development applications), the NRC or facility should have information on the facility's radioactive materials license. The NRC license, license conditions, and notices of violation should be obtained and reviewed to determine the nature and types of materials handled, stored, and disposed of. The operating procedures applicable to licensed activities also should be reviewed to determine the potential areas of contamination; equipment and laboratory surface exposure; potential air emissions and heating, ventilation, and air condition (HVAC) duct contamination; and potential contaminated environmental media (e.g., groundwater, surface water, soil). NRC licenses require monitoring and surveys to be maintained by the facility. These surveys should be reviewed to determine the potential levels and locations of radioactive contamination. A list of the four NRC regional offices and their phone numbers is provided in Appendix P of the *CFATF Guide.*

A1.3.5 Regulatory Review

Regulatory review is another essential step in the investigative due diligence process. This activity involves reviewing the permit and compliance history for the subject property, as well as neighboring sites that may have an impact on the property (typically a 1-mile radius). Otherwise unknown environmental concerns can be revealed, such as a history of fines for spills on the adjacent property. Of course, the regulatory search can only reveal the known compliance history; unreported spills and other activities that could contribute to contamination are not part of an official regulatory record.

Records and files should be obtained for applicable environmental enforcement agencies such as the EPA, state environmental protection department, water control board, local fire department, and the health inspector. Each entity can be contacted independently for a search of the necessary records, or commercial vendors can be used to provide regulatory database search services. Examples of federal, state, and local regulatory data sources are provided in the following subsections.

Federal Lists.   Federal regulatory data sources include

- *RCRIS.* The RCRA Information System is an EPA list of permitted hazardous waste facilities and generators.

- *CERCLIS.* The CERCLA Information System is an EPA database with information on Superfund sites on the NPL. CERCLIS is a component of IDEA (see below).

- *SETS.* The Site Enforcement Tracking System (SETS) is an EPA database listing responsible parties at NPL sites.

- *ERNS.* The Emergency Response Notification System (ERNS) for spill and response activity information, which is maintained by the U.S. Coast Guard.

- *IDEA.* The Integrated Data for Enforcement Analysis (IDEA) database contains data from 15 EPA and EPA-related databases, including RCRIS, CERCLIS, SETS, and ERNS. Information on IDEA can be obtained from the hotline (1-888-EPA-IDEA) or the Internet (*http://es.inel.gov/oeca/idea*).

State Agency Lists.   The appropriate state environmental agencies should be contacted for information on fuel or other regulated releases that may have occurred on the subject or adjacent property. Many states maintain lists similar to CERCLIS and RCRIS on environmental site contamination, response actions, and small fuel releases. State enforcement inspection reports should be reviewed for information on potential sources of contamination. In addition, state environmental permits should be obtained and reviewed for specific closure requirements. Other permit areas to be considered include USTs, ASTs, air quality, hazardous waste, industrial and domestic wastewaters, radioactive materials, and hazardous materials.

Local Authorities.   Regulatory records from local authorities should not be overlooked. Local fire and health departments typically conduct enforcement inspections, which could reveal environmental conditions relative to local codes and standards. Fire departments may have information regarding facility hazardous substance use and USTs, as well as past releases or environmental incidents. Health departments may have information on radon levels in the area of the site, as well as site activities that may have an impact on human health and the environment.

All local environmental permits and inspection reports also should be obtained and reviewed, including those for POTWs, sanitary sewers, and stormwater discharges. Permits and inspection reports will assist in determining the potential composition of the hazardous materials used and whether there is cause for concern based on the permit parameters and report findings.

Town and County Planning or Zoning Offices.   Typically, the planning or zoning office is located with the main city or county offices. The applicable entity should be contacted to determine whether the property is zoned for a particular use (e.g., industrial, agricultural, wetland, or sanctuary) and whether the property has any historical or recreational value. The county planning office or the local soil conservation district also may be able to provide a copy of county soil survey reports for the area. This information will be helpful in accurately characterizing the property's features (including wetlands). It also will be useful in determining the limitations of future land use or property transfer. Information obtained regarding the existence and classification of wetlands should be verified with other hotline or national wetlands inventory map data.

A1.3.6 Geology and Hydrogeology Review

The geology and hydrogeology of a property are investigated to provide an understanding of how potential contamination could affect the soil and groundwater of subject or adjacent properties. Both the land and water features of the site will have an impact on the speed and ability for potential contaminants to migrate. Topics to consider in this review are

- Direction of groundwater flow
- Depth to groundwater
- Floodplain
- Water quality
- Soil characteristics
- Site topography

The property owner and operator should be contacted for a copy of any previous soil or groundwater studies, which should be reviewed for general geologic and hydrogeologic information as well as data on suspected contamination. The following subsections provide a list of other available information sources.

United States Geological Survey.   The USGS maintains information on the soil characteristics and hydrogeology for the United States. Reports for the applicable area should be analyzed to determine the groundwater depth and flow and surface water flow.

State Water Resources Control.   State water resources control boards conduct well surveys of groundwater and drinking water. Information on the aquifer type, depth to groundwater, classification, and use is often found in regional reports. Data from these surveys also should be reviewed to characterize and identify existing or formerly operated wells on the site.

United States Department of Agriculture and Local Authorities.   The USDA and regional Soil Conservation Service (SCS) districts generate soil survey reports on regional geology and soil types. The county planning office or the local soil conservation district also may be able to provide a copy of county soil survey reports for the area. This information is helpful in accurately characterizing the property's features (including wetlands). It also will be useful in determining the limitations of future land use or property transfer. Information obtained regarding the existence and classification of wetlands should be verified with other hotline or national wetlands inventory map data.

A1.3.7   Phase I Report

The EDDA Phase I Liability Assessment report is prepared after all the information-gathering activities have been completed. The intent of the report is to document the results of the liability assessment, including the findings, conclusions, and recommendations. By its nature, it also documents that due diligence was exercised.

Report Development.   The Phase I report must document all aspects of the site visit, as well as the record, regulatory, geologic, and hydrogeologic reviews. The report also must include statements of conclusion on the possibility and nature of environmental contamination associated with the site and the potential for liability. Further, the report should recommend appropriate next steps based on intended use of the property and the liability conclusion stated. Any limitations should be stated directly to ensure that the reader and decision maker are aware of what information was not available for assessment of potential liability. Backup documentation also should be provided with the report, including but not limited to inspection notes, property-related reports, completed questionnaires, correspondence with state agencies, and site maps. A suggested outline of the report is provided in Appendix R of the *CFATF Guide* and is included at the end of this appendix.

Phase I Report Review. It is essential for the Phase I report to be reviewed for correctness and completeness. In this role, the technical reviewer ensures that the report is complete and properly worded, but more important, he or she evaluates the assessor's methodology to ensure that the report reflects that due diligence and all appropriate inquiries were applied during the investigation. The reviewer also must ensure that statements of conclusion regarding suspected contamination and liability are correctly derived from, and supported by, the data collected. To do this, the technical reviewer must be qualified, with the relevant technical environmental background, training, and experience (see Appendix K of the *CFATF Guide* for a list of qualifications). Agency legal counsel may want to review the draft reports to ensure that the content is consistent with agency policies.

Use of the Phase I Report. Following approval and acceptance of the Phase I report by the technical reviewer, it is forwarded to the executive decision maker. This individual or group of individuals evaluates the findings, conclusions, and recommendations contained in the report and decides how to proceed with a proposed property transfer. For all transactions, if the Phase I report indicates no evidence of contamination or liability, then the EDDA process is complete; environmental due diligence has been met, and results may be used to satisfy any property disposal obligations under CERCLA Section 120(h)(4).

If the agency is considering an acquisition or lease initiation and the findings of the Phase I report indicate there is the potential for contamination or liability, then decision makers must weigh other property options against the importance or strategic value of the subject property. When the agency has continuing interest in the property, Phase II must be conducted to confirm contamination and liability. If the property transaction is a disposal or lease termination and the findings indicate potential contamination or liability, then Phase II also must be conducted.

A1.4 PHASE II: CONFIRMATION SAMPLING

If the Phase I Liability Assessment indicates possible contamination and the agency owns, occupies, or has sufficient interest in the property, Phase II (or the regulatory-mandated equivalent) will be conducted. [*Note:* For agency property designated for disposal, confirmation sampling is also required by the Community Environmental Response Facilitation Act (CERFA).]

The phase II EDDA involves targeted sampling to confirm or deny the presence of suspected contamination identified during liability assessment. The information contained in the Phase I Liability Assessment report is used to develop a strategy for carrying out Phase II. Depending on the findings and recommendations described in the Phase I report, several activities may be performed under Phase II. Typically, these activities consist of

- Reviewing and evaluating the findings in the Phase I report
- Developing a confirmation Sampling and Analysis Plan (SAP)
- Performing sample collection and analysis
- Evaluating the sampling results against environmental or hazardous waste standards
- Developing the Phase II report

The activities and process for the Phase II EDDA are included in Chap. 5 of the *CFATF Guide* and are discussed in more detail below.

A1.4.1 Phase II: Description

The purpose of the Phase II Environmental Due Diligence Audit (EDDA) process is to confirm the presence or absence of contamination and liability identified in the Phase I Liability Assessment. The Phase II EDDA is accomplished through confirmation sampling, where the suspected areas of concern noted in the Phase I Liability Assessment are physically sampled to determine if actual contamination exists. Phase II procedures are designed specifically to confirm the presence or absence of contamination. This is achieved through targeted field sampling of suspected areas and appropriate laboratory analysis to quantify suspected contaminant compounds. These activities may range from intrusive sampling, such as advancing groundwater monitoring wells, to simple surface soil samples readily taken by hand augers. In some cases, the sampling may consist only of taking asbestos sampling or setting radon canisters. The range of required sampling will influence the scope of the Phase II activities and associated resources to complete the investigation. If Phase II activities show that contamination exists, then Phase III activities may be undertaken to fully characterize site contaminants. However, if Phase II shows that contamination does not exist, the EDDA process is concluded.

Following an indication of possible contamination from a Phase I Liability Assessment, the motivation for proceeding to Phase II differs by property transaction type. For agency property targeted for disposal, any suspected contamination must be further investigated and, as necessary, remediated in accordance with applicable regulations (see Fig. A1.1). For acquisitions, the agency interest in a given property must out weigh the expense of further investigation and other property alternatives (see Fig. A1.2). For other transactions, the decision to proceed with confirmation sampling will depend on numerous factors, including

- The type of transaction
- The level and severity of suspected contamination
- The price and availability of alternate sites
- The cooperation of the subject property owner for the investigation to continue

PHASE I — Conduct Liability Assessment to evaluate property condition and potential environmental concerns

Contamination suspected? — NO → EDDA Complete Proceed with transfer → Address CERFA compliance

YES

Property owned by agency? — NO* → Likely that agency contributed to contamination? — NO* → Inform owner of findings and implications

YES   YES

Likely that cleanup requirements apply? — YES → Continue under Federal or State cleanup requirements

NO

PHASE II — Conduct Confirmation Sampling to verify suspected contamination areas

Contamination exists? — NO → EDDA Complete Proceed with transfer → Address CERFA compliance

YES

Qualifies for regulatory process?** — YES → Continue under Federal or State cleanup requirements

NO

Plan for site characterization and budget requests

PHASE III — Conduct Site Characterization to determine extent of contamination and select remedial alternative

Select remedial technology and secure budget

EDDA Process Complete

Clean up property → Address CERFA compliance → Transfer property***

* Lease termination only
** Consult experts for guidance on regulatory contamination and cleanup requirements
*** Properties may—with regulatory approval—be transferred before cleanup is complete

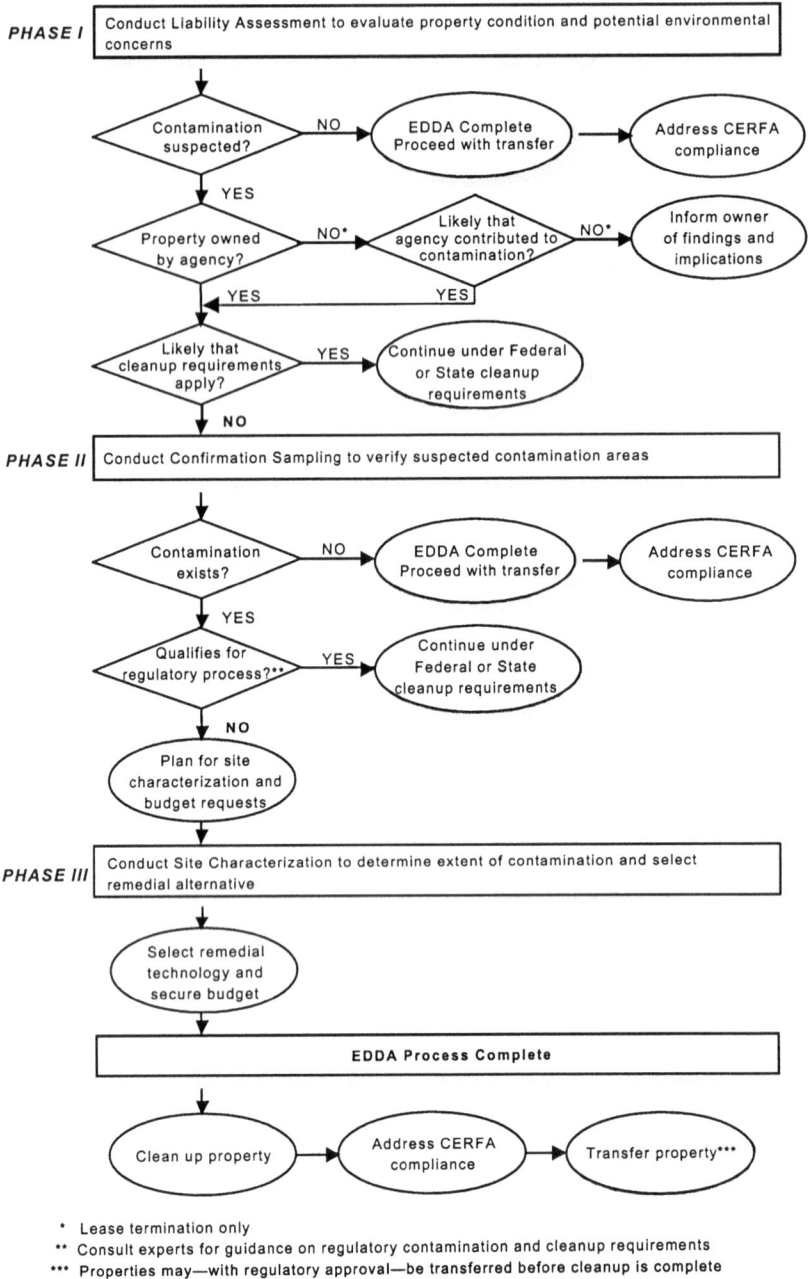

FIGURE A1.1 Decision process for disposal and lease termination actions.

PHASE I | Conduct Liability Assessment to evaluate property condition and potential environmental concerns

Contamination suspected? — NO → EDDA Complete Proceed with transfer

YES

Agency as a continuing interest? — * NO → Inform owner of findings and implications → Pursue other property opportunities

YES

Inform Owner / Negotiate further investigation issues

PHASE II | Conduct Confirmation Sampling to verify suspected contamination areas

Contamination exists? — NO → EDDA Complete Proceed with transfer

YES

Agency as a continuing interest? — * NO → Inform owner of findings and implications → Pursue other property opportunities

YES

Inform Owner / Negotiate further remediation

PHASE III | Conduct Site Characterization to determine extent of contamination and select remedial alternative

Agency as a continuing interest? — * NO → Pursue other property opportunities

YES

Inform Owner / Negotiate further investigation issues — ** NO → Negotiations are satisfactory? — NO ↑

YES

EDDA Process Complete

Clean up property → Acquire/Lease*** Property

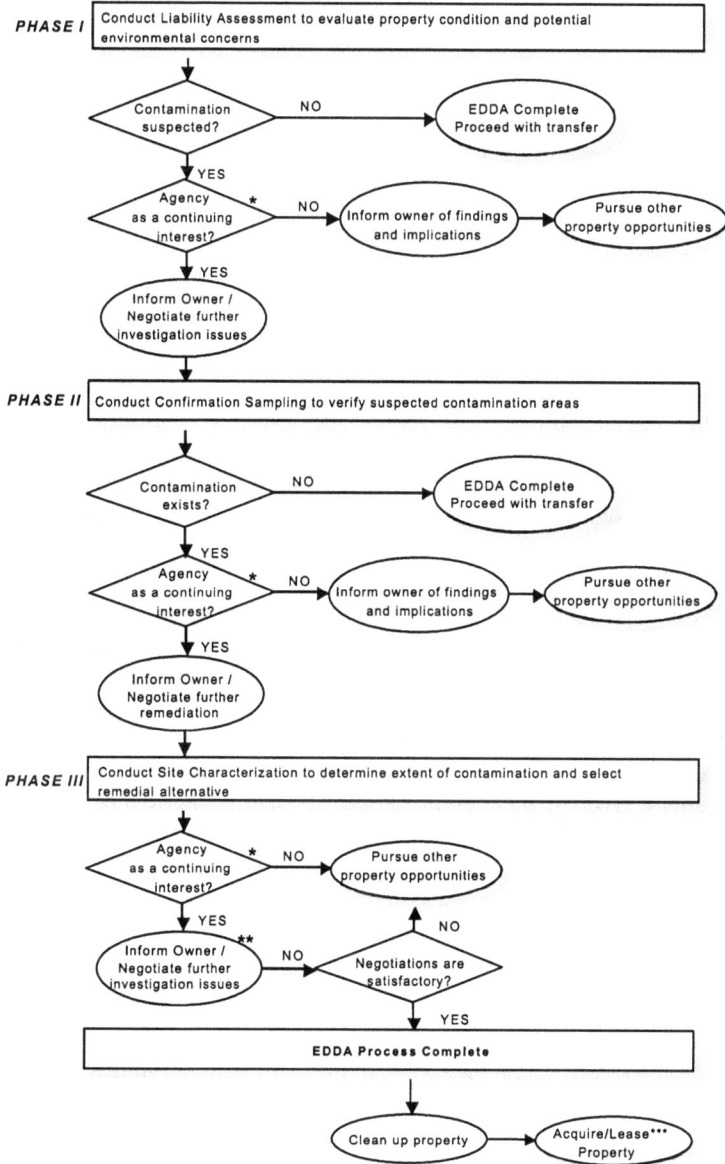

* Value of property to the agency outweighs the scope of potential liability or is of strategic interest.

** Property owners or agency may complete EDDA at their cost, provide for cost offsets against property value, or agree on other cost allocation position as part of negotiated settlement.

*** Acquisition / lease may proceed before cleanup activities are completed.

FIGURE A1.2 Decision process for acquisition and lease execution actions.

These issues are discussed in Chap. 3 of the *CFATF Guide* and should be fully considered as part of the agency's decision-making process.

Due to the technical requirements and potential liability issues raised by Phase II confirmation-sampling activities, the use of certified contractors is strongly recommended. The agency is responsible for selecting a *qualified* contractor with a licensed professional engineer (PE) or a licensed professional geologist/hydrologist (PG/PH) on staff to supervise and approve the work. Sample specifications for Phase II and Phase III EDDA contractors are provided in Appendix S of the *CFATF Guide.* When a contractor is engaged to design and perform Phase II confirmation-sampling field activities, the agency role in the field will be to provide oversight and logistical support (e.g., site access). Additionally, the agency will oversee the progress of the investigation to ensure that it is completed within budget and on time. Depending on the scope and planned activities, Phase II confirmation sampling can address one or more issues and require different levels of field activities and analytic procedures. Consequently, this can become a very expensive and time-consuming process unless it is properly planned, managed, and controlled. Monitoring and oversight of these activities are paramount and present the agency with an opportunity to ensure that the Phase II confirmation sampling is fully executed.

Agency personnel are responsible for reviewing and accepting the contractor's plans and reports. Accordingly, the personnel involved with the review process should be familiar with the sampling strategy and understand the implications of the sampling results and recommendations of the report. The Phase II EDDA is a critical element in developing specific knowledge about the presence or absence of site contaminants and, if confirmed, generating an initial understanding of potential future site implications. The duration of the Phase II confirmation-sampling process depends on the specific activities planned and the scope of the confirmation-sampling program. Phase II confirmation-sampling activities consist of the following four steps (a description of each of these activities is provided in the subsequent paragraphs):

- Reviewing and evaluating the findings in the Phase I report
- Developing and implementing a confirmation-sampling and analysis plan (SAP)
- Identifying site risk based on the results of the confirmation sampling
- Developing the Phase II report

A1.4.2 Review and Evaluation of Phase I Report

Prior to initiating any Phase II confirmation-sampling activities, the Phase I report should be reviewed thoroughly to gain a complete understanding of what is currently known about the site and the suspected contamination. The Phase I report provides valuable information on the environmental condition of the property. Specifically, the conclusions and recommendations section documents the potential areas of concern and provides recommendations for performing Phase II confirmation-sampling activities. These areas can include the site structures, site grounds, or information on suspected sources on neighboring properties. Several types of surfaces and environmental media may need to be sampled. Additionally, the Phase I report will contain valuable background information that will be pertinent to designing and conducting Phase II confirmation sampling. This information should be reviewed to ensure that the Phase II confirmation-sampling contractor has a full understanding of what is known about the site and the specific areas of suspected contamination. Relevant background information may include

- Recommended locations of investigation and issues supporting the suspected types of contamination and sources
- Potential sources of contamination based on prior site use
- Past site operations and practices
- Physical characteristics of the site, such as soil types, depths to groundwater, geologic and hydrogeologic features
- Noted background (or ambient) levels of contaminants of potential concern
- Previous hydrologic, testing, or assessment report identified and reviewed in Phase I that supports the recommendations or provides additional site detail and characteristics

It is important for the Phase II confirmation-sampling contractor and applicable federal agency staff to become familiar with the contents of the Phase I report. This information forms the basic building blocks for designing, planning, and performing Phase II confirmation-sampling activities. These activities must address all the issues raised in the phase I report. Therefore, the success of the Phase II confirmation-sampling activities in part rests with having a thorough knowledge of the site conditions and areas of concern noted in the Phase I report. This information is used to develop the background and understanding as well as to specifically set forth the objectives of a site SAP.

A1.4.3 Development and Implementation of the SAP

The purpose of the SAP is to establish an agreed-upon sampling strategy that will fully address each potential liability area through confirmation sampling and analysis. The SAP contains two distinct elements. The first is the *field sampling plan* (FSP) that specifically discusses the sampling activities, scope, analysis, health and safety activities, and rationale for each. The second is the *quality assurance project plan* (QAPP) that identifies the quality assurance/quality control (QA/QC) procedures used in the field sample collections and analyses to ensure accuracy and precision of the sampling results. The use of an independent contractor is always encouraged to demonstrate that an objective and defensible Phase II confirmation-sampling approach is executed and accurate results are obtained. The SAP must be developed by the Phase II contractor and approved by the agency before Phase II confirmation sampling commences.

Element 1, the FSP, should consist of field sampling and analysis procedures, a safety and health plan, and a project management plan. The FSP must describe the following activities:

- Sampling objectives
- Site background
- Site characteristics
- Potential contaminants of concern
- Types of media being sampled
- Sample type and the location, number, and frequency of samples being taken
- Sample collection, handling, designation, numbering, and preservation techniques
- Field quality assurance and quality control procedures

A description of each of these activities is provided in Appendix T of the *CFATF Guide.* A safety and health plan is also developed to ensure adequate precautions and planning for on-site activities. This portion of the plan must adhere to the Occupational Safety and Health Administration (OSHA) regulations in 29 CFR Parts 1910 (General Industry Standards) and 1926 (Construction Safety).

The safety and health plan delineates the roles and responsibilities of site personnel, site-specific hazards, safety precautions, and regional medical response facilities. Contact the facility's environment, health, and safety manager for additional information. The overall objective of the plan is to ensure the safety and health of workers performing confirmation-sampling activities. The agency must require contractors to have their own OSHA-compliant safety program to comply with OSHA multiemployer work-site regulations.

Element 2, the QAPP, establishes the quality management system for all environmental programs performed by or for the agency. Specific policies and program requirements involving QA/QC activities will depend on internal agency policies. A program should be in place to define in detail how specific QA/QC activities will be implemented during a specific project. The four general quality assurance elements are

- Project management
- Measurement and data acquisition, including sampling analysis, data handling, and quality control
- Assessment and oversight
- Data validation and usability

These elements correspond to planning, implementation, and assessment. QA/QC applied to a project will be commensurate with the following:

- The purpose driving environmental data collection (e.g., enforcement, research and development)

- The type of work to be done (e.g., site characterization, baseline of site conditions)
- The intended use of the results

The best means of achieving the appropriate content and level of detail in the quality management program may be through having the agency's QA/QC requirements reviewed and confirmed by the agency's project manager and documented through a QAPP.

The QAPP is usually submitted with the FSP; it describes the steps and procedures that will be used to ensure quality information for field sampling and laboratory analysis. The plan usually demonstrates that

- The project technical and quality objectives are identified, and there is concurrence.
- The intended measurements or data-acquisition methods are appropriate for achieving project objectives.
- Assessment procedures are sufficient for confirming that the type and quality of data needed are obtained.
- Any limitations on the use of the data can be identified and documented.

Both the field (e.g., FSP) and quality (e.g., QAPP) components of the SAP are used as a management tool to monitor the field and analytic laboratory performance of the Phase II confirmation-sampling activities. Typically, a project manager will develop the work schedule, milestones, and associated costs based on the requirements identified in these documents. Site sampling may begin once the SAP has been developed and accepted.

The Phase II confirmation-sampling contractor will be responsible for completely executing the field sampling program specified in the SAP and meeting the field, laboratory, and analytic objectives described in the QAPP. Federal agencies will be responsible for providing oversight during Phase II confirmation-sampling activities and coordinating with the contractor and the landowner to provide site access as appropriate. Federal managers should not provide field direction to on-site contractors, since this type of activity may compromise the integrity of the approved SAP. Where unexpected field or technical issues arise during the course of the sampling activities, federal oversight managers should work with the contractor to amend and document changes to the SAP and, where necessary, add change orders to the contract.

A1.4.4 Phase II Report

The Phase II findings, results, and recommendations must be formally documented in a report. Typically, the report includes

- A summary of the Phase I findings

- The results of the confirmation sampling and analysis
- Discussion of potential risk to human health and the environment
- Discussion of potential remedial alternatives
- Recommendations for performing follow-on Phase III site-characterization activities or concluding the EDDA

Appendix U of the *CFATF Guide* is a sample outline/table of contents for a Phase II report and is included at the end of this appendix.

The report should clearly document the findings and conclusions of the Phase II confirmation sampling. It is essential that the results of the confirmation sampling and analysis be reviewed against the specifications of the QAPP to ensure that the data are accurate and will support drawing meaningful conclusions. Data also should be evaluated specifically against the QA/QC parameters, and the report should show an accounting for all deviations from designated sample quality standards. Additionally, the sample results must be evaluated against established Applicable or Relevant and Appropriate Requirements (ARARs) to compare the contaminants against established permissible levels. ARARs include federal, state, and local standards that apply to the contamination compounds and issues at the site. Additionally, contamination areas may be compared against appropriate background samples or information to help determine the source and impacts of the contamination areas.

When contamination is confirmed, the report should document the locations and types of contamination found and provide the specific contamination levels. Information on the steps and types of analysis necessary to further investigate the contamination area is often appropriate at this point and provided in the Phase II report. In the event that the confirmation sampling determines that no contamination is present, the report should fully document the sampling activities, analytic results, and justification for determining that contamination is absent or below levels of concern.

Preliminary identification of remedial alternatives may be included in the Phase II report based on the types and locations of noted contamination. Any estimates necessarily will be precursory and intended only to assist in decision making based on best judgment and potential extent of contamination confirmed in the Phase II. The full range of contamination will not be known until a comprehensive site investigation (Phase III site-characterization EDDA) has been completed. The preliminary remedial alternatives are used to make property management decisions in situations such as acquisition or to form a basis for refinement if site characterization is required in situations such as disposal. Examples of some of the more common remediation technologies are listed in Appendix V of the *CFATF Guide,* which is included at the end of this appendix.

In addition to the fundamental components of a Phase II report, any deviations from the SAP, the rationale for deviations, and a strong justification and supporting information for the conclusions and recommendations are essential. The Phase II report must be reviewed and approved for content and accuracy by oversight personnel. The Phase II confirmation-sampling report is the decision-making tool to assist agency managers in understanding the actual presence of site contaminants and need to conduct further study through Phase III. Agency personnel responsible for property transfer, such as the program manager, property transfer manager, safety, health, and environmental manager, facility engineering, and legal and real estate representatives, should review the report. Their review must

- Evaluate the accuracy of the conclusions and recommendations relative to the data gathered
- Determine whether the investigation actually was carried out in accordance with the SAP
- Ensure consistency between field samples and the QA/QC samples
- Evaluate the field data against the appropriate and relevant criteria
- Approve or concur with the conclusions and recommendations in the Phase II report

A1.5   PHASE III: SITE CHARACTERIZATION

A Phase III EDDA may be necessary when contamination has been confirmed by the Phase II EDDA. The purpose of Phase III is to fully characterize and assess the nature (i.e., types) and extent (i.e., magnitude or distribution) of site contamination. In addition, site characterization involves identifying appropriate cleanup technologies based on the nature and extent of contamination, potential cleanup goals, technology applications, and cost. Typically, phase III activities include

- Evaluating prior EDDA reports to develop a site-characterization–sampling strategy
- Performing more extensive sampling to assess the full extent of contamination
- Evaluating the contamination risk in relation to future land use
- Evaluating the technological viability and cost of cleanup alternatives
- Developing the Phase III report

The activities and process for Phase III are presented in Chap. 6 of the *CFATF Guide* and are discussed in more detail below.

A1.5.1   Phase III: Description

The Phase III site-characterization process provides information to agency decision makers regarding the extent and magnitude of contamination liability. This phase is initiated when a subject property is of continuing interest to the agency and the Phase II Environmental Due Diligence Audit (EDDA) results confirmed contamination at concentration levels equal to or above regulatory limits or risk levels. During Phase III, site contamination is fully characterized, and cleanup alternatives are developed. This is the final step in the EDDA process; thus any subsequent remediation activities follow solely Resource Conservation and Recovery Act (RCRA), Comprehensive Environmental Response, Compensation and Liability Act (CERCLA), or other statutory and regulatory processes.

Continuing a property investigation through Phase III site characterization is extremely rare in proposed acquisition or lease initiation transactions. Such actions would be motivated only by steadfast agency interest, probable funding reimbursement, and estimated cost of cleanup to overall property value. Conversely, when the agency owns the property or is otherwise responsible for the contamination, prior to disposal or lease termination, the agency inevitably will be required to conduct a phase III site characterization or a regulatory equivalent. (Refer to Figs. A1.1 and A1.2 for Phase III decision making for acquisition and disposal scenarios.)

Much like the Phase II confirmation-sampling process, Phase III site characterization consists of numerous activities. The Phase III process builds on previously generated information to develop a comprehensive assessment of all site contamination areas. Appropriate cleanup standards also must be identified, based on the site-specific human health risk, ecologic risk, or regulatory requirements. Information on the nature (i.e., types of contaminants found) and extent (i.e., magnitude across media) of site contamination also is used in the Phase III process to develop and recommend cleanup technology alternatives.

Phase III activities include the following (discussion of each of these activities is provided in subsequent paragraphs):

- Review and evaluation of the Phase II report
- Development and implementation of a sampling and analysis plan (SAP) to fully characterize contamination at the site
- Assessment of risk and future land use options
- Evaluation and selection of remedial alternatives
- Development of the Phase III report

Phase III site-characterization activities are conducted by independent contractors experienced in site characterization and remediation. The selection of Phase III contractors is based on the contractors' qualifications, experience, and ability to conform to the contractor procurement specifications (see Appendix S of the *CFATF Guide* for a list of qualifications).

A1.5.2 Review and Evaluation

Phase II report information provides the necessary background and building blocks for designing, planning, and performing Phase III site-characterization activities. As such, it is important for the agency technical reviewers and the Phase III contractors to review and evaluate the contents of the Phase II confirmation-sampling report.

A1.5.3   Development and Implementation of a Full-Characterization SAP

The Phase III full-characterization SAP is similar to the SAP process described for Phase II. Both contain a project management plan, a safety and health plan, sampling and analysis procedures, and quality assurance and quality control (QA/QC) requirements for the quality assurance project plan (QAPP).

The major difference between Phase II confirmation sampling and the Phase III SAP is the objective. The objective of the Phase II SAP is to confirm the presence of contamination. The objective of the Phase III SAP is to determine the extent and severity of the contamination and to provide the technical basis for establishing a site cleanup strategy. Due to the expanded objective of the Phase III SAP, the scope and number of samples collected likely will increase during this phase of the EDDA process and result in higher costs. Additionally, the Phase III SAP typically will call for higher-resolution sampling and analytic procedures to evaluate performance limits of potential cleanup technologies. Extreme care and professional judgment must be exercised to ensure that excessive sampling is not performed and excessive costs are not incurred. As in Phase II, the QAPP elements in the SAP need to be implemented in proportion to the project.

Contractor-developed SAPs are submitted to the agency for review and approval. Once approved, the contractor initiates site activities, and samples are collected and sent to a laboratory for analysis. Rigorous and documented sampling procedures (refer to Appendix T of the *CFATF Guide* for a description of the procedures) should be followed to ensure that the results of the sampling are accurate and representative of site conditions.

When the analytic results from the sampling are obtained, they are compiled and analyzed. This information is used to determine the nature and extent of site contamination and to assess the risk posed to human health and the environment from the contamination.

A1.5.4   Risk Assessment and Future Land Use Options

The analytic results provide the data needed to assess the risk posed to potential human and ecologic receptors. These data, in turn, can be used to develop appropriate risk-based cleanup levels in the absence of specific media criteria. Acceptable risk levels are typically in the 10^{-4} to 10^{-6} (1 in 100,000 to 1 in 1 million) range for potentially impacted populations. The risk assessment depends on source characterization, exposure assessment, dose-response evaluation, and risk characterization. A description of each of these components is provided in the following list:

• *Source characterization.* Identifies the contaminants of concern and their rates of release.

• *Exposure assessment.* Identifies the potentially exposed populations, pathways of exposure, and extent of exposure.

- *Dose-response evaluation.* Assesses the type of effects that could occur and the magnitude of the effects.

- *Risk characterization.* Determines the amount of exposure involved, its associated risks, and the relative significance of the risk.

Each of these components must be evaluated to determine the overall risk posed to human and ecologic receptors.

Often the CERCLA (or Superfund) process is the applicable regulation for federal site contamination. As such, when assessing risk during Phase III activities, it is appropriate to reference the following EPA guidance documents [which can be obtained by calling the Superfund Hotline at 1-800-424-9346 or by contacting the U.S. National Technical Information Service (NTIS)]:

- *Risk Assessment Guidance for Superfund,* Vol. I: *Human Health Evaluation Manual* (Parts A, B, and C) provides guidance for developing human health risk information at Superfund sites.

- *Risk Assessment Guidance for Superfund,* Vol. II: *Environmental Evaluation Manual* provides guidance for developing environmental assessments at Superfund sites.

In addition, the EPA recently published the *Revised Guidelines for Ecological Risk Assessment* (NTIS Publication PB98-117849), which provides a framework for evaluating past and future impacts to ecologic resources. Effective April 30, 1998, this guideline may be useful in determining ecologic-based cleanup goals or assessing the potential impact of selected remedial actions at a site.

Depending on the types of contaminants involved and the information available, risk assessments can be qualitative or quantitative. Although quantitative risk assessments are normally performed, qualitative risk assessments may be required if (1) regulators consider it appropriate, (2) cost and timeliness are an issue, (3) toxicity data on chemicals are not available, or (4) some other phenomena are not quantifiable. When conducting qualitative risk assessments, risk-management decisions must be based on prudence and best professional judgment.

Future land use options are also considered when determining risk. There are four commonly recognized future land use options: industrial-commercial, agricultural, recreational, and residential. When considering the impacts of future land use on the overall risk of the property, the industrial-commercial option is usually the least conservative, whereas the residential option is the most conservative. Future land use options must be evaluated in conjunction with risk to determine the appropriate level of risk reduction and cost-effectiveness during the cleanup process. Keep in mind that the EPA and state and local environmental regulatory agencies often select the most restrictive land use option—the residential scenario—in setting and approving risk-based cleanup goals.

A1.5.5   Remedial Alternatives: Evaluation and Selection

Remedial alternatives are screened against evaluation criteria to reduce the number of remedial alternatives available for selection and implementation. [See Appendix V of the *CFATF Guide* for a sample listing of remedial technologies (also included here).] Only the remedial alternatives most representative of the evaluation criteria should be placed on the short list of alternatives. In general, the evaluation criteria consist of

- Overall protection of human health and the environment
- Compliance with Applicable or Relevant and Appropriate Requirements (ARARs)
- Long-term effectiveness and permanence
- Reduction of toxicity, mobility, or volume through treatment
- Short-term effectiveness
- Ability to implement
- Cost
- State acceptance
- Community acceptance

Developing and assessing a (simple to complex) range of technically appropriate alternatives are important to evaluate their relative feasibility. The short list of remedial alternatives may include the following: *no action* (i.e., natural attenuation), *institutional controls* (such as deed restrictions or perpetual federal ownership), *technological solutions* (involving remedial, demolition, or decontamination activities), or some *combination* of these.

Using the preceding criteria, the contractor-proposed short list of applicable alternatives is evaluated by the agency. The ultimate solution for the site, however, must be selected in coordination with regulators and with consideration to public concerns.

A1.5.6   Phase III Report

The Phase III report documents all pertinent site information in one place for agency decision makers, including the nature and extent of contamination, activities performed, risk assessment results, cleanup goals, remedial alternatives, and recommendations. Specifically, the phase III site-characterization report should be a comprehensive statement delineating

- Prior site activities
- Efforts leading up to site characterization
- Sampling rational and activities

- Final sampling results, clearly displayed with a complete vertical and horizontal distribution of site contaminants and concentrations
- A comparison of these results with site ARARs, background levels, or risk-based action levels
- Appropriate cleanup goals
- Results from the analysis of applicable cleanup alternatives
- Recommended alternatives and their rational, technical implementation issues, and costs

The Phase III report will constitute the guideline and technical basis for any further (and possibly costly) remediation activity planned at the site and, as such, must be closely reviewed and understood by agency technical staff and decision makers. Appendix W of the *CFATF Guide,* which is included at the end of this appendix, contains a suggested outline of the Phase III report.

A1.6 EDDA SCHEDULE

Depending on the type of property transfer, organizations involved, location importance, environmental condition of the property, and other agency-specific issues, the EDDA process may include one or more phases. Likewise, the schedule for the EDDA may vary. For example, an assessor budgets 60 to 80 working hours to complete Phase I; however, issues regarding availability and accessibility of information result in delays. Such delays in the schedule are likely to defer the completion date, although they should not unduly increase the overall level of effort.

Figure A1.3 provides a conceptual breakout of time allocations and activities for the Phase I Liability-Assessment process. Managers should anticipate the complexity of the real property transfer process and the unique nature of the property in determining the time needed to collect information and address site logistics. For all EDDA phases, a flexible schedule is often appropriate.

A1.7 RISK AND FLEXIBILITY IN THE EDDA PROCESS

Identifying potential risk and liability involved in property transactions is the heart of the EDDA process. As discussed previously, these risks will vary greatly depending on the type of property transaction and the type of property to be transferred. For instance, industrial properties typically have a higher probability for potential contamination than property that is used solely for office space.

Figure A1.4 provides a general risk framework for different transaction and property types. When approaching an EDDA project, it is useful to consider this model—and to plan the EDDA with consideration for the appropriate risk level.

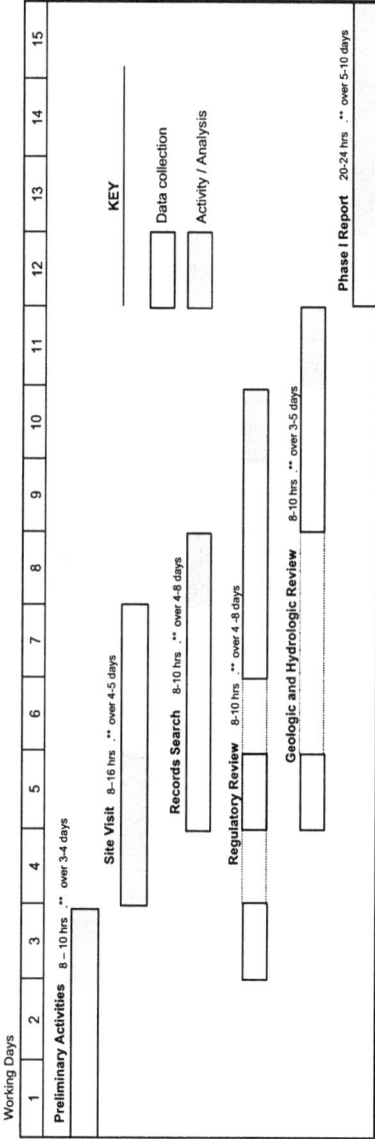

Timeline (Gantt chart):

	1	2	3	4	5	6	7	8	9	10	11	12	13	14	15

Preliminary Activities 8 – 10 hrs .** over 3-4 days

Site Visit 8–16 hrs .** over 4-5 days

Records Search 8-10 hrs .** over 4-8 days

Regulatory Review 8-10 hrs .** over 4 –8 days

Geologic and Hydrologic Review 8-10 hrs .** over 3-5 days

Phase I Report 20-24 hrs .** over 5-10 days

KEY
- Data collection
- Activity / Analysis

* Minimum times will vary based on property and transaction types.
** Displayed time estimate based on requirement of approximately 60 to 80 hours for a basic Phase I EDDA

	Preliminary Activities	Site Visit	Records Search	Regulatory Review	Geologic/Hydrologic Review	Phase I Report
Principal Activities	• Obtain basic property use/activities Information • Property map • Contact Owner and establish logistics for site visit • Develop and distribute questionnaire	• Visual survey of the site and neighboring properties • Interview property owners, on-site employees and neighboring owners • Review on-site documentation	• Site ownership and use history (title search, facility records) • Aerial photos, site photos, permits, previous environmental surveys • Fuel storage tanks, hazardous material and waste management, sensitive environments	• Review permit and compliance history for subject and neighboring properties • Access the regulatory databases (IDEA, RCRIS, and CERCLIS)	Obtain basic information on: • Direction of groundwater flow • Depth to groundwater • Water Quality • Soil characteristics • Site topography • Site hydrology	•Draft report •Technical and quality revisions •Publication and delivery
Potential Factors Affecting Schedule	• Receipt of pre-audit questionnaire and materials	• Site access and security issues • Availability of site personnel • Site size and complexity • Travel time to site	• Availability of records • Proximity to source of records (e.g. county seat)	• Access and responsiveness to local regulators • Document requests (e.g. FOIA)	• Availability of documents	• Revisions and response to comments

FIGURE A1.3 Time line for Phase I Liability Assessment activities.

Even though all EDDAs cover a standard set of investigative areas, they are not all equal, and inherently, the EDDA process needs to be flexible to reflect the varying degree of risk associated with the different types of property transactions and property types. For example, the assessment of a former industrial site will likely require deeper investigation and evaluation than an EDDA for previously undeveloped woodlands in a remote location.

Flexibility is built into the EDDA process to allow the assessors to target high-risk areas and allocate resources (and time) appropriately. Particularly in the Phase I Liability Assessment, the EDDA process requires that the assessor investigate a broad range of areas and exercise professional judgment in order to focus on those issues which present the highest likelihood for risk. The assessor's understanding of risk is continuously refined as additional information is gathered and assessed during the EDDA process. Based on this emerging understanding of potential risk, the assessor continues to make adjustments with each step in the process.

Inevitably, there will be exceptions to the general risk framework provided in Fig. A1.4, and it is the role of the assessor to recognize these exceptions and modify the EDDA process accordingly. For instance, the model shows a low risk associated with the disposal of an office building property; however, if an industrial operation located on the site prior to environmental regulation was forgotten, soil contamination may remain. An experienced assessor should see clues when reviewing the title search results and other Phase I material. This example also emphasizes how prior site history and use provide a baseline for evaluating the nature of risk at the site, whether or not the agency owns the property.

Overall, the strength of the EDDA process is in its flexible approach to balancing the agency's need to uncover and define risk areas, to manage liabilities

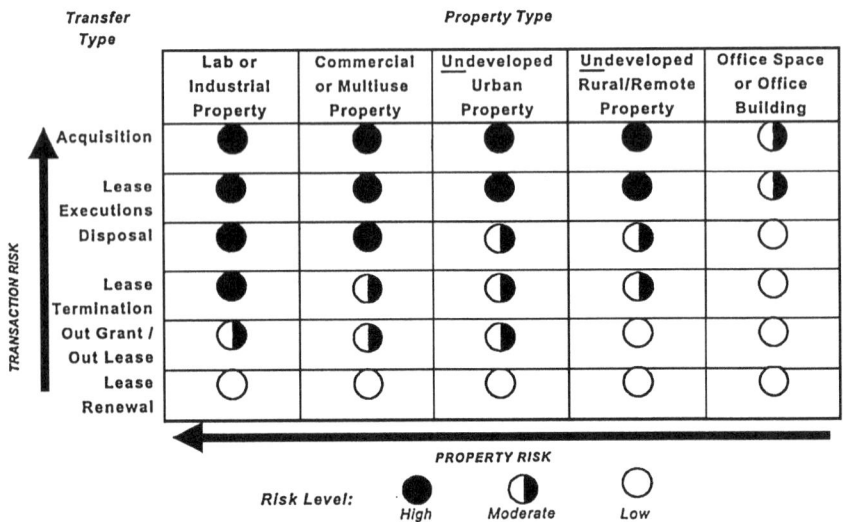

FIGURE A1.4   General risk framework.

associated with property transactions, and to accomplish due diligence. Readers should keep in mind that the materials, techniques, and information sources presented are to be used as a "guide" in how the process should be applied. The background and experience the assessor and EDDA team bring to the process play a significant roll in successfully managing the flexibility provided by the EDDA process.

A1.8 REGULATION OF CONTAMINATED SITES

CERCLA, RCRA, the Toxic Substance Control Act (TSCA), the Clean Water Act (CWA), the Oil Pollution Act (OPA), and state regulations prescribe for responsible parties the procedures to investigate and remediate environmental contamination. Alternately, evaluating for potential environmental contamination and liability was a process developed originally by lenders in reaction to CERCLA. Thus the EDDA "phases" were developed to provide a practical acquisition focus to property transfer rather than to provide a framework for environmental cleanup. Addressing the CERCLA considerations of "due diligence" and "all appropriate inquiry" for environmental risks is the basis for Phase I Liability Assessment activities. When activities beyond Phase I are indicated on agency-held property, the agency may be compelled to follow a regulatory-based process for addressing contamination. In such instances, *compliance with applicable CERCLA, RCRA, TSCA, CWA, OPA, or state requirements takes priority over EDDA in dictating the process to address specific types of liabilities or contamination.* When the property is not held by the agency and the agency is not a potentially responsible party, the Phase II and Phase III EDDA processes exist as guidance to confirm and characterize potential contamination and liability. The statutes and regulations governing the requirement to investigate and remediate environmental contamination will vary depending on the nature of the contamination and when the release occurred. In such situations, the Phase II and III chapters of the *CFATF Guide* provide only a broad reference to the parallel regulatory procedures. For additional information on the CERCLA and RCRA processes, refer to Appendix H, "Regulatory Overview," of the *CFATF Guide* (also included here in Sec. A1.11).

In addition to the regulations on investigation and remediation of environmental contamination, the National Environmental Policy Act (NEPA) also may apply. A facility closure or property transfer disposal action often will qualify as a "major federal action" under NEPA. As such, the agency's NEPA compliance efforts for this action (environmental impact assessment and public participation process) also should consider the EDDA activities. Depending on the timing, the agency's NEPA documentation may reference the completed EDDA Phase I and II reports or may mention these as planned activities.

Along this line, NEPA public outreach activities may be sequenced with the evaluation and selection of site cleanup options. Public outreach may involve agency meetings with the community to address their concerns, agency grants to the public for their evaluation of the remedial alternatives, or the creation of fact sheets for the public on the contamination and remediation at the site.

To determine the applicability of NEPA, CERCLA, RCRA, or state regulations to a given property transaction, refer to Appendix H of the *CFATF Guide,* the program offices, and legal counsel before proceeding.

A1.9 ROLES AND RESPONSIBILITIES

The EDDA process typically is undertaken by a team of technical and management staff responsible for overseeing and managing the process, conducting the EDDA, performing the technical reviews, and developing a decision based on report findings and input from agency real estate and legal staff. Roles and responsibilities for conducting the EDDA process and decision making are determined on an agency-by-agency basis. However, in all cases, an agency representative should be identified to manage and oversee execution of the EDDA.

Regardless of the nature of the property transaction, the EDDA should be performed by qualified individuals who have the relevant technical environmental background, training, and experience. Contractors selected to assist or conduct Phase I activities also should satisfy the contractor specification guidelines (see Appendices K and L of the *CFATF Guide*).

The following description of typical EDDA participants and roles provides a generic interpretation of how agencies may structure an EDDA team. Depending on internal agency policies, at least three primary team roles will be carried out during Phase I: assessors, technical reviewers, and agency decision makers. In addition, there are collaborating roles for real estate and legal staff. The specific roles of the EDDA team include

- *Assessors.* Technical environmental staff who conduct the EDDA, develop recommendations, and draft reports. Assessors may be either agency or contractor staff; often Phase II and III EDDAs are performed exclusively by contractor resources.

- *Technical reviewers.* Technically qualified agency personnel who review the EDDA report for technical accuracy in methodology, scope, depth, and findings. A technical reviewer, who concurrently is the project leader, also may work with the assessors up front to determine the scope and work plan—in addition to reviewing and accepting the EDDA report. Technical reviewers always should include, but may not be limited to, agency personnel.

  - *Agency decision makers.* Agency management involved with overseeing the full scope of the property transfer activities and vested with authority to determine the agency's ongoing interest and responsibility for a given property. Agency executive decision makers typically are briefed throughout the EDDA process to maintain an understanding of site issues, parameters, and implications of the EDDA process. This central role on the EDDA team should be fulfilled by one or more individuals who are collectively vested with the authority for determining the agency's position on the property transaction and committing necessary resources.

- *Real estate personnel.* Professionals responsible for executing the property transaction on behalf of the agency. In acquisition lease execution situations, real estate personnel will screen and identify possible candidate sites. During the EDDA process, real estate staff remain involved, facilitating the EDDA process by providing basic site information, executing the title search, and coordinating with the property owner and operator.

- *Legal representatives.* Agency staff who participate in property transactions to ensure that the EDDA process is conducted appropriately and the report findings demonstrate due diligence. This offers legal protection to both the agency and the transacting entity. Lawyers typically provide advice on the EDDA process and are involved in the document reviews to ensure that the final report meets legal objectives. In some cases, the EDDA document may be conducted as attorney-directed work to ensure future protection of the documents for future landowners.

- *Site owner and operator.* Whether initiating occupancy (acquisition, lease executions) or vacating a site (disposal, lease termination), it is important to involve the current landowner or operator early in the EDDA process planning. Site owners and operators need to be fully apprised of the scope, intent, and specific activities of the EDDA process and understand the implications of assessing and determining environmental liabilities. Owners and operators typically are included in a review capacity for preliminary EDDA report findings and are often a recipient of the final EDDA documents. When the subject property is owned or operated by the agency, there is also a critical role for the facility manager and staff. In such cases, the facility manager needs to identify the relevant personnel to facilitate the interview process, and all must provide accurate information to the EDDA team. Likewise, out-of-agency landowners and operators are excellent sources of basic site information and are an important source of both interview information and current site documentation.

A1.10 USE OF EDDA IN DECISION MAKING

The EDDA process is used to document the results of the investigation, document that due diligence has been exercised, and provide a basis to evaluate potential and actual environmental liabilities to aid in property transaction decisions. Professional judgment decisions are an integral part of the EDDA process, from deciding whether a full or partial Phase I Liability Assessment should be conducted to deciding whether or not to proceed with the property transaction based on EDDA investigation results.

The EDDA process differs significantly from CERCLA in this decision-making aspect. While CERCLA directs a structured process from identifying contamination through site cleanup, the focus of the EDDA process is to manage liability. This is particularly important for property acquisition and lease execution. Figures A1.1 and A1.2 are decision-making flowcharts for acquisition and disposal transactions.

To further illustrate the decision-making process, consider that the EDDA Phase I Liability Assessment is conducted for most property transactions. The need for additional EDDA phases for acquisition or lease transactions, however, will depend on the importance or strategic nature of the property. If the location of the property to be acquired or leased is relatively unimportant, then the agency decision maker choosing to minimize agency liability would pursue a property without suspected contamination. Conversely, if the location of the property to be acquired or leased is important, then the agency may decide to gather additional information from a Phase II EDDA to further assess the likelihood of liabilities before making a decision about acquiring or leasing the property. In some situations, if an EDDA report documents contamination at a property of strategic interest, the agency may elect to lease rather than acquire the property. In such a case, the EDDA has allowed the agency to avoid "acquiring the liability" and to structure a "managed approach" to insulate intended operations from known or suspected contamination.

During a Phase II EDDA, if contamination is not confirmed, the property transaction can proceed without adding undue risk for environmental liability. If, however, the presence of contamination is confirmed during the phase II EDDA, decision makers must determine whether the importance of the site outweighs the potential liability that would accompany acquisition of the property. At this point, the agency may enter into negotiations with the owner to address the contamination, or the agency may choose to pursue further assessment of the extent of the contamination on the property with a Phase III EDDA. The Phase III EDDA information will allow the agency to make a decision by weighing the potential liability costs against the value of the property. The agency could decide to take on these costs—possibly even using the information to lower the purchase price of the property. If the findings of the Phase II or Phase III EDDA appear significantly adverse, then other acquisition or lease opportunities may become more acceptable. There is no requirement to continue to a Phase II or Phase III EDDA for acquisition or lease transactions.

For disposal of federally owned or operated properties, the decision-making process is fairly direct. Those properties that the agency releases to organizations outside the federal government are required to be free of contamination or have an authorized remediation in process before the transfer process can be completed.

Where the Phase I Liability Assessment indicates a potential for contamination, federal agencies must confirm and, as necessary, characterize and clean up contaminated property.

A1.11 APPENDIX H OF THE CFATF GUIDE: REGULATORY OVERVIEW OF MAJOR ENVIRONMENTAL STATUTES AND DIRECTIVES

Introduction

To ensure compliance, the regulatory framework, status, and permits for a given property should be identified and clearly understood as part of an environmental due diligence audit (EDDA). One or a combination of the following regulatory requirements may dictate the scope of property transfer activities for the facility or a portion of the facility:

- Comprehensive Environmental Response, Compensation and Liability Act (CERCLA)
- Resource Conservation and Recovery Act (RCRA)
- Toxic Substance Control Act (TSCA)
- Clean Air Act (CAA)
- Clean Water Act (CWA) and Safe Drinking Water Act (SDWA)
- Atomic Energy Act (AEA)
- National Environmental Policy Act (NEPA)

These laws have prescriptive requirements and protocols for facilities regulated, permitted, or licensed by these authorities. An overview of each regulatory requirement is provided in the following sections.

Comprehensive Environmental Response, Compensation and Liability Act (CERCLA)

This section discusses how to address CERCLA requirements if the Phase I EDDA finds suspected areas of contamination regulated under Superfund. Federal facilities can clean up hazardous substance contamination pursuant to CERCLA, but the cleanup may not be financed through the Superfund. Superfund was created to pay for response actions where the responsible parties cannot be found or are unable to pay. Superfund financing is reserved for nonfederal facilities on the National Priorities List (NPL). The NPL is a group of sites with hazardous substance contamination substantial enough to warrant federal attention and money for cleanup. Congress required EPA to create the NPL to identify the most serious sites, ensuring that Superfund monies are spent on the most serious problems. The purpose of the NPL is to notify the public of sites that need remedial action and may present a long-term threat to public health or the environment. Federal facilities may be placed on the NPL even though they cannot receive Superfund money.

Removal actions and enforcement actions are not limited to NPL sites. CERCLA authority may be used for responding to releases of hazardous substances into the environment. Understanding these terms is essential to understanding the scope of CERCLA. The definitions of these terms are provided in Section 101 of CERCLA.

In some cases the PTM may need to determine whether a CERCLA response is warranted. The removal site evaluation process provides flexibility to determine whether a CERCLA response is warranted or another appropriate federal or state response is available. For example, a CERCLA response may not be necessary for a facility licensed by the NRC and being closed in conformance with an NRC-approved decommissioning plan, for a facility being closed in compliance with a RCRA permit or order, or if a release or a substantial threat of a release is not present at the facility or the amount of hazardous substances present does not warrant federal response.

Under CERCLA Section 120, each federal agency is responsible for carrying out most response actions at facilities under its own jurisdiction, custody, or control. Section 120(a) states that federal departments, agencies, and instrumentalities are subject to CERCLA just like nongovernment entities, including CERCLA's liability provisions. Pertinent guidelines, rules, regulations, and criteria apply in the same manner and to the same extent, with the exception of requirements pertaining to bonding, insurance, and financial responsibility.

Special requirements and timetables are established under Section 120. For example, Section 120(c) requires establishment by EPA of a Federal Agency Hazardous Waste Compliance Docket that lists federal facilities that have reported managing hazardous substances or releases of hazardous substances. Section 120(h) is presented in Appendix D of the *CFATF Guide* and addresses in detail guidelines for the property transfer by federal agencies.

Release Reporting, Removal, and Remedial Authority.   CERCLA gave EPA the authority to require reporting of certain releases of hazardous substances and the authority to require cleanup of those releases through a short-term removal action and/or a long-term remedial action. CERCLA response actions should be determined on a site-by-site basis and in consultation with EPA enforcement officials and the state environmental agency as appropriate. CERCLA response actions include removal (emergency, time-critical, or non-time-critical) and remedial actions.

Discovery and Notification of a Release.   In order for a site to be considered eligible for Superfund response, a release of a hazardous substance must be discovered and reported to the government. If a release of hazardous substances is discovered during phase II EDDA activities, the owner or operator may need to make a notification about the release and can be held liable for any contamination. Personnel in charge of the facility should carefully examine any records on site for information about what types of chemicals have been used at the facility and which may have been released. If a hazardous substance has been released into the environment in a quantity equal to or greater than its reportable quantity (RQ) (specified in the list of hazardous substances found in 40 CFR Part 302.4) within a 24-hour period, upon discovery the owner or operator of the facility must immediately notify the National Response Center. This involves a notification to the National Response Center by telephone at (800) 424-8802 providing detailed information about the facility and the nature of the release (40 CFR Part 302.6). If the owners/operators are unsure of whether an RQ of the hazardous substance was released and whether it occurred within a 24-hour period, they should still report the release to the National Response Center as a precautionary measure (55 FR 8676, March 8, 1990). Even if a release does not warrant notification, this does not mean that the owner or operator of the facility will not be held liable for the release and any cleanup costs pursuant to the release.

After discovery or notification of a hazardous substance release, a preliminary assessment (PA) is conducted to decide if the release is a threat to human health and the environment. If further investigation is warranted, EPA will conduct a site inspection (SI). The information gathered during the PA and the SI is used to develop a Hazard Ranking System (HRS) score. The HRS evaluates relative risks to human health and the environment posed by uncontrolled hazardous waste sites by assessing four pathways of potential human exposure to contamination (i.e., groundwater, surface water, soil, and air). EPA uses a site's HRS score to determine if the site should be placed on the NPL. Information gathered on these sites is maintained by EPA in the CERCLA Information System (CERCLIS), a comprehensive national database that inventories and tracks releases that may need to be addressed by the Superfund program.

Removal Actions. If a release presents a serious immediate threat, the federal agency may take a removal action to stabilize or clean up the release. Typical removal actions include removing leaking tanks or drums of hazardous substances, installing security measures such as a fence at a site, or providing a temporary alternate source of drinking water to local residents. A removal action may be taken at any time necessary during the response process.

There are three types of removal actions (emergency, time-critical, and non-time-critical), each with different regulatory requirements, depending on the urgency of the need for a response to the release. All removal actions have a time and spending restriction of 12 months and $2 million, respectively. The time and spending limits may be exceeded when continuing the removal action is necessary to prevent, limit, or mitigate an immediate risk to public health or the environment that will not be acted upon by another party or when continuing the removal action is consistent with a remedial action that will be taken at the site.

An emergency removal action requires on-site activities to commence within hours of the lead agency's determination that a removal action is appropriate.

A time-critical removal action occurs when, based on the site evaluation, the lead agency determines that a removal action is appropriate and that there is less than 6 months available before on-site activities must be initiated. For time-critical removal actions, the community must be involved and an administrative record of the removal action must be created [40 CFR Parts 300.415(n)(2)/300.820(b)].

Non-time-critical removal actions are those where EPA determines a removal action is appropriate and a planning period of more than 6 months is available before on-site activities must commence. In accordance with Part 300.415(b)(4), the lead agency must conduct an engineering evaluation/cost analysis (EE/CA) for a non-time-critical removal action. The EE/CA is an analysis of removal alternatives for a site. More information on the procedures and activities involved in conducting an EE/CA can be found in EPA's document entitled *Guidance on Conducting Non-Time-Critical Removal Actions Under CERCLA* [Office of Solid Waste and Emergency Response (OSWER) Directive 9360.0-32]. Specific administrative record requirements for non-time-critical actions are specified in 40 CFR Part 300.820(a).

Remedial Actions.   If a hazardous substance release does not pose an immediate threat to human health and the environment, the federal agency may take a remedial action after further evaluation of the site. Remedial actions are long term and aimed at achieving a permanent remedy. Examples of typical remedial actions include removing buried drums from a site, thermally treating wastes, pumping and treating groundwater, and applying innovative technologies such as bioremediation to contaminated soil.

A remedial action has two main phases: the remedial investigation/feasibility study (RI/FS) phase and the remedial design/remedial action (RD/RA) phase. The purpose of the RI/FS is to study conditions at the site, identify contaminants, and evaluate cleanup alternatives. The RI entails collecting and analyzing information to determine the nature and extent of contamination at the site. Specific alternatives are then evaluated during the FS. After the RI/FS, the federal agency focuses on designing the selected cleanup alternative in the remedial design stage. The remedial action stage follows, with a varying time frame according to the complexity of the remedy.

A site is considered "completed" or cleaned up once the chosen remedy is operational and functional and meets its designated environmental, technical, legal, and institutional requirements. At this stage, operation and maintenance activities are implemented to monitor the effectiveness of the remedy and to ensure that no new threat to human health and the environment arises.

Liability.   CERCLA imposes liability when there is a release or threatened release of hazardous substances. CERCLA Section 107(a) casts an extremely broad net in defining the scope of persons who can be liable for paying the costs of responding to a release of hazardous substances. The types of parties that can be held liable are (1) the current facility or vessel owners or operators, (2) former facility or vessel owners or operators, (3) those who arrange for treatment or disposal of hazardous substances at a facility, and (4) those who accept hazardous substances for transport to treatment or disposal sites.

There are three defenses to liability outlined in CERCLA Section 107(b):

- An act of God

- An act of war

- An act or omission of a third party who is not an employee or agent of the defendant and does not have a contractual relationship with the defendant

The third-party defense, often called the *innocent landowner provision,* rebuts the presumption of liability associated with ownership of the land by claiming that the landowner made a good-faith effort to discover any contamination. In addition, the third-party defense may come into play when a person is the victim of a so-called midnight dumper. To the third-party defense, the court scrutinizes the defendant's relationship to the property, specifically whether the defendants knew or had reason to know of the disposal of hazardous substances at the facility. The elements of the defense are found in CERCLA Sections 107(b)(3) and 101(35). The defendant raising the third-party defense must be free of both actual or inferred knowledge and any contractual relationship concerning the property, except as allowed under Section 101(35)(A).

If, during closure of a federal facility, hazardous substance contamination is discovered but the federal agency is clearly not responsible for the release, it is possible that the agency may not be held liable for cleanup of the contamination. Determinations will be made on a site-specific basis. Conditions where hazardous substances have come to be located on or in a property solely as the result of subsurface migration in an aquifer from a source or sources outside the property, EPA will not take enforcement actions against the owner of such property to require the performance of response actions or the payment of response costs (60 FR 34790, July 3, 1995).

Enforcement. CERCLA is a strict liability statute, which means that responsible parties are liable without regard to negligence or fault. The concept of joint and several liability applies in situations where more than one potentially responsible party (PRP) is involved and it is difficult to determine each PRP's contribution to the release. In these situations, the courts have held that an owner, operator, or waste generator or transporter may be held liable for the entire cost of site cleanup, unless each party's contribution can be identified. Federal facilities often have their own mandates for responding to hazardous substance releases.

Community Involvement. Community involvement opportunities are tailored to each Superfund site and are an integral part of every Superfund response. The

National Contingency Plan (NCP) provides the public with the opportunity to comment on, and provide input to, decisions about response actions. Interested persons are provided with accurate and timely information about response plans and progress, and their concerns about planned actions are heard by the lead agency.

CERCLA/RCRA Interface.   If a facility that is closing has chemical contamination, a response action may be taken pursuant to CERCLA or RCRA regulations. The authority chosen will depend on factors such as the timeliness of a response and the substances involved. If CERCLA authority is used for the cleanup, the facility will need to follow procedures under the CERCLA regulations to ensure proper cleanup of the site. If the hazardous substance released is also an RCRA hazardous waste, the federal agency may use RCRA authority rather than CER-CLA authority when cleaning up the site. Generally, sites that may be cleaned up under RCRA or certain other laws will not be placed on the NPL.

Resources.   The following resources may assist in complying with closure requirements under CERCLA:

- *CERCLA/Superfund Orientation Manual,* EPA/542/R-92/005, October 1992
- *Questions and Answers on Release Notification Requirements and Reportable Quantity Adjustments,* EPA/540-R-94-005, January 1995

Resource Conservation and Recovery Act (RCRA)

This section focuses on RCRA Subtitle C and the applicable closure requirements associated with hazardous waste management activities under EDDA. In addition to these guidelines, the state and local environmental authority also should be consulted to determine if additional or more stringent requirements exist. To ease the complexity and confusion of complying with RCRA, as amended by the Hazardous and Solid Waste Amendments (HSWA) of 1984, all applicable hazardous waste requirements will be referred to as RCRA requirements from this point forward.

Subtitle C.   The RCRA Subtitle C regulations embody a "cradle to grave" philosophy in that hazardous waste is managed from the time it is generated through its ultimate disposal. Hazardous waste always must be managed by a responsible party, be it the generator, transporter, or treatment, storage, or disposal facility (TSDF). Operations at facilities or laboratories generally are limited to generator functions; however, there may be instances where facilities or laboratories are designated as TSDFs.

Generators.   There are three types of hazardous waste generators:

- Conditionally exempt small-quantity generators (CESQG), which generate less than 100 kg of nonacute hazardous waste or less than 1 kg of acute hazardous waste in a calendar month

- Small-quantity generators (SQGs), which generate between 100 and 1000 kg of nonacute hazardous waste or less than 1 kg of acute hazardous waste in a calendar month

- Large-quantity generators (LQGs), which generate greater than 1000 kg of nonacute hazardous waste or 1 kg or more of acute hazardous waste in a calendar month

Generators must conduct closure activities based on their size (i.e., CESQG, SQG, or LQG) and the type of unit (e.g., containers, tanks, or containment buildings) the waste is stored in on site. These activities involve removing and managing residues and waste, rinsing and decontaminating temporary storage units, and decontaminating equipment. These closure activities are codified in 40 CFR Parts 261, 262, and 265.

TSDFs. Federal facilities or laboratories also can be designated as TSDFs, depending on their activities and the amount of time hazardous waste is stored on site. There are various types of TSDFs used to manage hazardous waste, including containers, tanks, surface impoundments, waste piles, land treatment units, landfills, incinerators, injection wells, corrective action management units (CAMUs), drip pads, miscellaneous units, and containment buildings. The definitions of each of these units is codified in 40 CFR Part 260.10.

There are two sets of closure requirements for TSDFs under RCRA, including the general requirements and the unit-specific requirements. The general closure requirements include performance standards, the closure plan, time allowed for closure, disposal or decontamination of equipment, structures, soil, and closure certification. The unit-specific closure requirements include removal and management of all wastes and residues, decontamination of containment systems and the unit, and decontamination of structures and equipment. If a facility is designated as an interim-status *TSDF* (defined in 40 CFR Part 265), the general closure requirements in 40 CFR Part 265, Subpart G, and the unit-specific closure requirements in 40 CFR Part 265, Subparts I through R, W, or DD, apply. If the facility is designated as a permitted *TSDF* (defined in 40 CFR Parts 264 and 270), the requirements in the facility permit apply. These requirements should be reviewed, referenced, and complied with during the EDDA, if applicable.

Additionally, if protection of human health or the environment is still in question after closure activities have been completed, the facility cannot "clean close" the unit and must conduct postclosure care activities. Clean close is not defined under RCRA; however, the March 19, 1987, *Federal Register* (52 FR 8706) provides language which, in essence, states that clean closure is achieved through removing all remaining wastes and residues from the TSDF or ensuring the contaminants do not pose a threat to human health or the environment. Clean closure is typically left to the discretion of the federal or state enforcement authority. If clean closure is not possible, the facility must perform postclosure care. These activities include the sampling and monitoring of environmental media such as groundwater, surface water, soil, or sediments. Once again, there are general requirements and unit-specific requirements. These requirements should be referenced if clean closure is not feasible. These postclosure care requirements are codified in 40 CFR Part 265, Subparts G, I through R, S, or WW, for interim status facilities and the permit for permitted facilities.

Underground Storage Tanks (USTs). Approximately 5 million USTs across the United States contain petroleum and other chemicals potentially hazardous to human health and the environment. Many of these tanks and associated piping systems are not protected from corrosion or overfill and therefore are leaking product or have lost product while in service. HSWA, Subtitle I of RCRA, established a comprehensive program for new and existing USTs of certain size and use and which hold regulated substances. A regulated substance is a CERCLA hazardous substance excluding hazardous wastes and petroleum, as defined in 40 CFR Part 280.12.

USTs currently in use must meet technical standards to ensure that regulated substances will not leak or spill out of the tank and cause contamination. Specifically, all USTs must have spill, overfill, and corrosion protection by December 22, 1998 or be closed and/or replaced. To prevent spills during delivery of regulated substances, by December 1998, USTs must have catchment basins to contain spills. A tank often can be overfilled, causing a large-volume spill. To prevent this from occurring, USTs must have overfill protection devices, such as automatic shutoff devices, and overfill alarms by the December 1998 deadline. Federal regulations also require corrosion protection for USTs because unprotected steel USTs can corrode and release substances through corrosion holes.

The requirements of 40 CFR Part 280 are currently applicable to about 1.2 million USTs. However, there are some statutory exclusions to the definition of UST under the federal regulatory program, including

- Farm or residential tanks of 1000 gallons or less capacity used for storing motor fuel for noncommercial purposes
- Tanks used for storing heating oil for consumptive use on the premises where stored
- Septic tanks
- Pipeline facilities regulated under federal or state pipeline safety acts, surface impoundments, pits, ponds, or lagoons
- Stormwater or wastewater collection systems
- Flow-through process tanks
- Liquid trap or associated gathering lines directly related to oil or gas production and gathering operations
- Storage tanks situated in an underground area (such as a basement, cellar, or tunnel) if the storage tank is situated on or above the surface of the floor

If a facility with a regulated UST is closing, a determination must be made on future use of the UST, as well as whether the UST is the source of any contamination. If the decision is to close the UST, the appropriate closure requirements must be conducted in accordance with the regulations promulgated at 40 CFR Part 280, Subpart G.

A UST may be considered suspect or an area of potential contamination in the agency Phase I EDDA report for a variety of reasons, such as lack of information on the tank contents and status or physical evidence of stressed or stained vegetation. In addition, a former UST location also may be considered suspect unless documentation, such as sampling results, a letter documenting the state's approval, or a state-certified closure report, is made available during the phase I EDDA confirming the "clean closure" of a former UST. In cases when the UST or former UST location is considered suspect, the Phase II sampling and analysis plan should incorporate a strategy for characterizing the suspected area. Representative sampling should be conducted accordingly to characterize the soil and other surrounding media of the tanks and pipelines, to determine whether the tanks and pipelines contain product, to characterize the contents of the tanks and pipelines, and to verify the integrity and operability of the tank and pipelines. When a UST is suspect, this general approach to characterizing the UST location should be applied to all USTs, regulated and nonregulated.

UST Closure.   There are two types of closure for USTs per 40 CFR Part 280, including temporary and permanent closure. When deciding which closure option to follow, facility owners and operators should consider whether there is a potential for future use of the UST, the use planned for the facility, and the condition of the surrounding land in general.

During temporary closure, tanks may either continue to store regulated substances or be emptied. Temporary closure of a UST may last up to 12 months before it has to be closed permanently. When a UST system is closed temporarily, owners and operators must continue to comply with normal operating requirements, such as the corrosion protection and release detection requirements of 40 CFR Part 280, Subpart D. Release detection is not required if the UST system is empty, meaning that all materials have been removed using commonly employed practices so that no more than 2.5 cm (1 in) of residue, or 0.3 percent by weight of the total capacity of the UST system, remains in the system [40 CFR Part 280.70(a)].

When a UST system is closed temporarily for 3 months or more, owners and operators must comply with additional requirements. Vent lines must be left open and functioning, and all other lines, pumps, manways, and ancillary equipment must be capped and secured [40 CFR Part 280.70(b)].

If a UST system is closed temporarily for more than 12 months, it must be closed permanently. Permanent closure includes emptying and cleaning the UST by removing all liquids and accumulated sludges. All permanently closed tanks also must be either removed from the ground or filled with an inert solid material. Owners and operators must test for the presence of a release from the UST before completion of closure by conducting a site assessment. Records of the site assessment must be maintained for 3 years after the tank is closed. If the owner or operator is vacating the facility, the site-assessment records should be forwarded to the federal agency's environment, health, and safety office to be maintained in an official document management system.

Sampling and measurement methods must be appropriate for the characteristics of the site and the regulated substance. If contaminated soils, contaminated groundwater, or free product liquids or vapors are discovered, owners and operators must begin corrective action in accordance with Subpart F of 40 CFR Part 280.

Release Reporting.   Owners and operators of UST systems must report any suspected or known releases from a UST within 24 hours or another appropriate time period specified by the implementing agency. The implementing agency may direct the owner or operator to determine whether the release has caused any off-site contamination. A suspected release must be investigated within 7 days through either a system test or a site check. If a release is confirmed, the owner or operator must begin corrective action.

Corrective Action.   Releases from USTs pose a serious environmental and human health threat in the United States. Because USTs, by definition, are largely hidden from view, areas surrounding USTs must be inspected carefully for any signs of contamination. The federal corrective action regulations for USTs found at 40 CFR Part 280, Subpart F, provide a flexible framework for owners/operators and implementing agencies to work within and achieve cleanup levels protective of human health and the environment. Corrective action consists of a series of steps that vary depending on the severity of the release.

Short-Term Corrective Actions.   Once a release is detected, immediate response activities such as release reporting, immediate containment, and monitoring of explosive hazards should be taken (40 CFR Part 280.61). Following the immediate response activities, the facility begins abatement measures [40 CFR Part 280.62(a)].

Examples of such measures are

- Performing a site check to evaluate the extent of the release
- Containment of the regulated substance to prevent continued release
- Continued monitoring and mitigation of explosive hazards
- Mitigating hazards posed by soils excavated during response activities
- Determining the presence of free product in groundwater

The owner or operator must submit a report to the implementing agency within 20 days of confirmation of the release describing the extent of initial abatement activities [40 CFR Part 280.62(b)]. The owner or operator must submit a more detailed site-characterization report to the implementing agency within 45 days of confirmation of release [40 CFR Part 280.63(b)]. After reviewing the results, the implementing agency may decide that the release warrants further response activities. If further corrective action is required, the owner or operator must submit detailed corrective action plans, including provisions to remediate contaminated soils, groundwater, and surface water to the implementing agency (40 CFR Part 280.66).

Long-Term Corrective Actions. Regulations for the UST corrective action program cleanup levels or administrative procedures are left to the discretion of the implementing agency (generally, the state). The federal regulations require that state or local cleanup programs be protective of human health and the environment. Although the corrective action technologies are not specified in the federal regulations, there are several commonly employed remediation options for soil and groundwater contamination. Available options for soil remediation include in situ soil vapor extraction, in situ bioremediation, excavation and off-site treatment, and natural attenuation. Technologies typically selected for groundwater remediation include in situ air sparging with soil vapor extraction, pump and treat, and biosparging.

Resources. The following resources may assist in complying with closure requirements under RCRA:

- Underground Storage Tanks; Technical Requirements: Final Rule. September 23, 1988, *Federal Register* (53 FR 37082)
- *What Do We Have Here? An Inspector's Guide to Site Assessment at Tank Closure—Video and Companion Booklet,* video, 510-K-92-006, booklet, 510-K-92-006
- *Tank Closure Without Tears—Video and Companion Booklet,* video, 510-V-92-817, booklet, 510-K-92-817
- *RCRA Orientation Manual,* 1990 edition, EPA/530-SW-90-036

Toxic Substance Control Act (TSCA)

By enacting TSCA on October 11, 1976, Congress established a number of requirements and authorities for identifying and controlling toxic chemical hazards to human health and the environment. This section will focus on TSCA polychlorinated biphenyl (PCB) regulations and the Asbestos Hazard Emergency Response Act (AHERA) as they pertain to agency property transfer and the Phase II EDDA.

Polychlorinated Biphenyls. PCBs are a group of industrial chemicals that were used widely as coolants, insulating materials, and lubricants in electrical equipment such as transformers and capacitors. PCBs were used from their introduction in the mid-1920s until 1979, when the manufacture and distribution of PCBs in the United States was severely restricted due to adverse health effects from exposure.

  PCBs are oily liquids or solids, clear to light yellow in color, with no smell or taste. Because of PCBs' widespread distribution and persistence in the environment, they not only may be found in use but also in general storage items and products or as contamination from prior spills or leaks. It is critical that all PCBs and PCB-containing materials and equipment that may be in service or in storage at the facility are properly identified, managed, and in some cases mitigated. PCBs were used in paints, inks, lubricants, sealants, plasticizers, and carbonless copy paper, as well as the following commonly encountered materials:

- Transformers
- Fluorescent lighting fixtures (ballasts)
- Hydraulic fluids
- Small capacitors (with less than 3 lb of dielectric fluid)
- Switches
- Large capacitors (with 3 lb or more of dielectric fluid)
- Vacuum pumps
- Liquid-cooled electric motors
- Lab samples
- Microscopy mounting media and immersion oil
- Voltage regulators

  If there is suspected PCB contamination identified from the Phase I investigation, the proper procedures for characterizing the extent of contamination should be made part of the sampling and analysis plan for Phase II activities. EPA documents *Verification of PCB Spill Cleanup by Sampling and Analysis* (EPA 560/5-85-026) and *Field Manual for Grid Sampling of PCB Spill Sites to Verify Cleanup* (EPA 560/5-86-017) provide insight on sampling methods used to characterize PCB contamination. There are also field screening techniques to test for the presence of PCBs, such as Clor-N-Soil and Chlor-N-Oil, but laboratory analysis should be instituted to confirm PCB concentration before any remediation or disposal actions are taken. The field screening techniques are a good indicator used for confirming the presence of PCBs rather than performing an all-encompassing PCB sampling event, which can be costly. For additional sources of information on PCB sampling, refer to "Resources" at the end of this section.

PCB Spills.   Disposal of PCBs is defined in 40 CFR Part 761.3 as intentionally or accidentally discarding, throwing away, or otherwise completing or terminating the useful life of PCBs and PCB-containing materials. Disposal includes spills, leaks, and other uncontrolled discharges of PCBs as well as actions related to containing, transporting, destroying, degrading, decontaminating, or confining PCBs and PCB-containing materials. Any release of PCBs to the environment greater than 50 parts per million (ppm) is considered a prohibited act of disposal as defined in the regulations. Facilities are required to report spills of more than 10 lb (4.56 kg) of PCBs of concentrations of 50 ppm to the EPA regional office. Spills of greater than 1 lb (0.45 kg) must be cleaned up.

The federal regulations stipulate a Spill Cleanup Policy at 40 CFR Part 761, Subpart G, that is applicable to spills that occurred after May 4, 1987. For old spills that were discovered after the effective date of this policy (e.g., discovered as a result of a Phase I investigation) but could have occurred before the effective date of the policy, cleanup requirements are established at the discretion of EPA, usually through its regional offices.

There are two types of PCB spills: a low-concentration spill from a source concentration of PCBs from 50 to 500 ppm and a high-concentration spill from a source of 500 ppm or greater. In the event that the source of a PCB spill is unknown, the spill is cleaned up based on the concentration of the contaminated material.

- For low-concentration spills, refer to the cleanup requirements at 40 CFR Part 761.125(b).

- For high-concentration spills, refer to the cleanup requirements at 40 CFR Part 761.125(c).

It is important to contact the local or state environmental authority. In many cases, the local and state authorities have cleanup requirements that are more stringent than federal regulations. Most likely, this is reflected in the definition of "PCB contaminated." TSCA's range is 50 to 500 ppm, whereas some states define this as 5 to 500 ppm.

Compliance with the Spill Cleanup Policy may prevent enforcement action and any need for additional cleanup under TSCA. However, if the cleanup is required under RCRA, CERCLA, or other statutes, then different standards, other than those imposed by TSCA, may be applicable.

Disposal Requirements. A uniform hazardous waste manifest, EPA Form 8700-22, must be prepared if PCB waste is being transported off site. For each shipment of manifested PCB waste a disposal facility accepts, the owner or operator of the facility must prepare a Certificate of Disposal. Refer to 40 CFR Parts 761.60 and 761.218 for specific disposal requirements.

Resources. The following resources may assist in complying with TSCA regulations:

- CFR Part 761, "Polychlorinated Biphenyls Manufacturing, Processing, Distribution in Commerce, and Use Prohibitions"

- *PCB Q&A Manual,* EPA, Office of Pollution Prevention and Toxics, 1994
- CFR Part 761, Subpart G, "PCB Spill Cleanup Policy"
- *Lighting Fixture Management Options, Quick Reference Fact Sheet,* EPA/200-f-94-008, September 1994
- TSCA Assistance Informational Hotline: (202) 554-1404

Asbestos. Under TSCA, EPA regulates the use of asbestos in commerce and has issued standards for both controlling its handling and restricting its use. Congress amended TSCA in 1986 by adding a new Title III, AHERA, which required EPA to conduct a study to determine the extent of human health risks posed by asbestos in public and commercial buildings. The EPA responded in February 1988 by sending Congress a study on asbestos-containing materials (ACMs) in public buildings. On November 28, 1990, the Asbestos School Hazard Abatement Reauthorization Act (ASHARA) was enacted. Section 15 of ASHARA amended AHERA to require accreditation for any person who inspects for ACMs in a public or commercial building or who designs or conducts a response action with respect to friable ACMs in such a building.

AHERA defines *public and commercial buildings* as the interior space of any building that is not a school building, except that the term does not include any residential apartment building of fewer than 10 units or detached single-family homes. Interior space includes exterior hallways connecting buildings, porticos, and mechanical systems used to condition interior space. Examples of public and commercial buildings are government-owned buildings, colleges, museums, airports, hospitals, churches, preschools, stores, warehouses, and factories.

Federal regulations define an *inspection* to mean those activities undertaken to specifically determine the presence and/or location or to assess the condition of friable or nonfriable asbestos-containing building materials (ACBMs) or suspected ACBMs by either visual or physical examination or by collecting samples of such material. Therefore, if asbestos surveys or sampling is conducted at a federal facility, the individual(s) performing the survey/sampling should be accredited. Training requirements include the following:

- Individuals performing asbestos related work must take a 4-day, 32-hour EPA-approved training course consisting of topics such as potential health effects of asbestos exposure, the use of personal protective equipment, and state-of-the-art work practices.
- A contractor/supervisor must take a 5-day, 40-hour EPA-approved course.
- Inspectors take a 3-day course.
- Management planners take a 2-day course.
- Project designers take a 3-day course.

ASHARA does not require building owners to conduct inspections for asbestos-containing materials in public and commercial buildings. However, should the owner decide to conduct an inspection, then he or she must use an inspector who is accredited.

Clean Air Act (CAA)

In addition to TSCA, regulations under other laws apply to asbestos. The Clean Air Act requires the EPA to develop and enforce regulations to protect the general public from exposure to airborne contaminants known to be hazardous to human health. The EPA established National Emission Standards for Hazardous Air Pollutants (NESHAP) and promulgated the asbestos NESHAP in 40 CFR Part 61, Subpart M. The subpart addresses demolition and renovation of facilities and asbestos waste transport and disposal. The regulations require owners/operators to notify the applicable state and local agencies and/or EPA regional offices before demolition or renovation of a building occurs that contains a certain threshold amount of asbestos.

Although the NESHAP has not been revised to alter its applicability to friable and nonfriable ACMs, nonfriable asbestos materials are now classified as either Category I or Category II materials:

- *Category I* material is defined as asbestos-containing resilient floor covering, asphalt roofing products, packings, and gaskets. Asbestos-containing mastic is also considered a Category I material

- *Category II* material is defined as all remaining types of nonfriable ACMs not included in Category I that, when dry, cannot be crumbled, pulverized, or reduced to powder by hand pressure. Nonfriable asbestos-cement products such as transite are an example of Category II material.

The asbestos NESHAP specifies that Category I materials that are not in poor condition and not friable prior to demolition do not have to be removed, except where demolition will be by intentional burning. However, regulated asbestos-containing materials (RACM) and Category II materials that have a high probability of being crumbled, pulverized, or reduced to powder as part of demolition must be removed before demolition begins.

Surveys and Sampling. If ACMs are identified as a suspected area of concern in the Phase I EDDA report, the ACMs in the building must be disclosed to the future occupant. Prior to real property transfer, all available information on the existence, extent, and condition of ACMs should be incorporated into the EDDA reports or other appropriate documents to be provided to the landlord or future occupant.

The EDDA reports should include

- Reasonably available information on the type, location, and condition of asbestos in any building or improvement on the property

- Any results of testing for asbestos

- A description of any asbestos control measures taken for the property
- Any available information on costs or time necessary to remove all or any portion of the remaining ACMs (however, special studies or tests to obtain this material are not required)
- Results of a site-specific update of the asbestos inventory performed to revalidate the condition of ACM

If the presence of asbestos is suspected and an asbestos survey has not been performed or is not available, an asbestos survey should be completed. The occupancy agreement will determine who is responsible for performing the survey. The resources provided to perform an asbestos survey at a federal agency owned or managed facility will depend on the scope of the survey.

ACMs should be remedied prior to real property transfer only if they are of a type and condition that is not in compliance with applicable laws, regulations, and standards or if they pose a threat to human health at the time of transfer of the property. An agreement may be reached with the landlord or future occupant on the appropriate asbestos abatement measures to be taken.

If asbestos is suspected, it is important that the potential health risk is addressed appropriately in the sampling and analysis plan and in, particular, in the health and safety plan. Therefore, individuals performing any demolition or renovation activities as part of Phase II or III activities must be aware of the potential hazards during abatement. As mentioned previously in this section, all individuals performing asbestos surveys or sampling must be certified in accordance with AHERA. There are a number of resources listed at the end of this section to help plan for asbestos sampling and preabatement activities. The manual, *Demolition Practices Under the Asbestos NESHAP* (EPA 340/1-92-013), can assist in planning for demolition activities.

Resources.   The following resources may assist in complying with closure requirements concerning asbestos:

- CFR Part 1910.1001, which applies to all occupational exposures to asbestos in all industries covered by the Occupational Safety and Health Act, except for construction work as defined in 29 CFR Part 1910.12(b). Exposure to asbestos in construction work is covered by 29 CFR Part 1926.1101.
- CFR Part 763, "Asbestos Abatement Projects."
- EPA 340/1-92-013, *Demolition Practices Under the Asbestos NESHAP.* This manual is designed to assist the asbestos NESHAP inspector in identifying practices that normally do or do not make Category I nonfriable ACMs become RACMs.
- CFR Part 1926.58, "Asbestos Standard of the Occupational Safety and Health Administration."
- EPA Region 5 Asbestos Program Overview, which can be accessed via the Internet at *http://www.epa.gov/reg5foia/asbestos/index.html.*

- *Asbestos/NESHAP Regulated Asbestos Containing Materials Guidance,* EPA Publication No. 340/1-90-018, December 1990.

- *Managing Asbestos in Place,* EPA Publication No. 20-T-2003, July 1990.

Clean Water Act (CWA) and Safe Drinking Water Act (SDWA)

This section discusses how to address CWA and SDWA requirements if the Phase I EDDA finds suspected areas of contamination regulated by water management programs. This section specifically addresses requirements under the federal CWA and SDWA programs (see Tables A1.2 and A1.3). Individual state programs should be consulted to determine the applicability of different or more stringent regulatory standards.

In addition to CWA requirements, the facility may be subject to water management activities regulated under the SDWA. Table A1.2 summarizes the major regulatory programs under the CWA that may impact federal agency facilities.

Per Executive Order 12088, "Federal Compliance with Pollution Control Standards," October 13, 1978, federal facilities are required to comply with applicable regulations under CWA and SDWA. For example, the National Vehicle Fuel and Emissions Laboratory in Ann Arbor, Michigan, operates oil-water separators, and the Annapolis Central Regional Laboratory used septic systems prior to connecting to the municipal sewage treatment works. These facilities would be subject to CWA and SDWA requirements. The Phase I report will have identified situations such as these examples that may prompt further review under the Phase II EDDA.

National Pollutant Discharge Elimination System Permits (NPDES).   When transferring a federal agency's real property, any permits acquired to operate the facility must be terminated or transferred. Permit termination (40 CFR Part 122.64) generally will follow prescribed administrative procedures. For a federal NPDES permit, a notice of termination must be filed with the CWA Permitting Division. Contact the Office of Water, Permits Division, at 202-260-9545, for more information if a federal permit needs to be terminated. If a state NPDES permit requires termination, contact the Department of Environmental Quality or equivalent agency where the permit was obtained.

In addition to administrative procedures, any equipment used for wastewater treatment to fulfill NPDES permit conditions must be decontaminated, such as cleaning out and sampling any sludge (as with elementary neutralization tanks) and disposing of equipment or waste products in accordance with pertinent federal and state solid and hazardous waste management rules. Additional information on equipment decontamination and decommissioning is provided in Sec. A.03 of the *CTATF Guide.*

If stormwater discharges, discharge points, or discharge transport pipelines are noted as a potential concern in Phase I, then sampling activities should be conducted as part of Phase II to determine the cause of the problem, and the remedy

TABLE A1.2 Clean Water Act Regulatory Guide

Topic	Action involved	Regulatory citation
Oil discharges	Reporting is required for discharges of oil into navigable waters that Violate water quality standards Cause a film or sheen on the water or shoreline	40 CFR Part 110
Spill prevention control and countermeasures (SPCC) plan	SPCC plans must be developed when petroleum is being stored in quantities greater than 42,000 gallons underground, 1320 gallons total aboveground, or 660 gallons in any single aboveground container.	40 CFR Part 112
Hazardous substance release reporting	Reporting is required for releases of hazardous substances that exceed CWA reportable quantities (listed in 40 CFR Part 116) within a 24-hour period.	40 CFR Part 117
National Pollutant Discharge Elimination System (NPDES) permits	NPDES permits are required for point source discharges of wastewaters into navigable waters of the US	40 CFR Part 122

Permit/Standard	Description	Reference
Stormwater discharge permits	These requirements apply to stormwater discharges from specific activities into navigable waters (e.g., agency facilities having RCRA permits, new construction involving more than five acres of land).	40 CFR Part 122.26
National general pretreatment standards	Discharges of wastewater and sanitary waste to the sewer system are subject to the national general pretreatment standards, which prohibit discharges of certain wastes to the sewer system.	40 CFR Part 403.5(b)
National categorical pretreatment standards	These standards regulate discharges of wastewater to the sewer system from specific categories of industrial activities.	40 CFR Parts 405–471
Local pretreatment standards permit or local	Discharges of wastewater and sanitary wastes to the sewer system will be regulated by a municipal discharge permit or a local sewer use ordinance issued by the local publicly owned treatment works (POTWs).	Municipal discharge permit or local sewer use ordinance
Part 404 dredging permits	Potential discharges, water quality impairment, hydrologic, and navigational impacts associated with dredge and fill activities require review, approval, and permiting by the U.S. Army Corps of Engineers.	40 CFR Part 144.31

TABLE A1.3 Safe Drinking Water Act Regulatory Guide

Topic	Action involved	Regulatory citation
General applicability of SDWA	This subpart establishes key definitions under the national primary drinking water regulations (NPDWR) program, scope of coverage, variances and exemptions, and regulatory effective dates.	40 CFR Part 141, Subpart A
Maximum contaminant levels (MCLs) for organic, inorganic, turbidity, and certain radioactive material	Public drinking water systems providing water for widespread consumption must meet specific maximum contaminant levels to ensure drinking water quality and protect public health.	40 CFR Part 141, Subpart B and Subpart G
Monitoring and analytical requirements for public water systems	Periodic testing and monitoring for coliform bacteria, turbidity, and certain organic and inorganic contaminants is a key aspect of EPA's NPDWR program. The effective dates for these monitoring requirements has been phased in over a period of time.	40 CFR Part 141, Subpart C

Reporting, public notification, and record keeping	Reporting and public notification must be conducted for noncompliance with SDWA requirements for public water systems.	40 CFR Part 141, Subpart F
Filtration and disinfection	Specific filtration and disinfection requirements are established for public water systems and supplied by a surface water source or groundwater influenced by surface water sources.	40 CFR Part 141, Subpart H
Control of lead and copper in drinking water	New action levels of 0.015 mg/liter for lead and 1.3 mg/liter for copper were established in 1991. If these values are exceeded at the tap in 10 percent of the public water system subject to monitoring programs, corrective actions must be initiated.	40 CFR Part 141, Subpart I
Underground injection control	Discharges or introduction of wastewaters, industrial wastes, and hazardous wastes into injection wells require specific approvals and permits.	40 CFR Parts 146–149

should be addressed in Phase III. Sections A.06 and A.07 of the *CTATF Guide* should be referenced for additional information on planning, implementing, and reporting the sampling and analysis activities associated with stormwater discharges, discharge points, or discharge transport pipelines.

Potable Water.   Although most facilities are served by public water systems and typically are not subject to SDWA regulations for delivery of treated drinking water, facilities that have on-site wells supplying water for consumption by an average of at least 25 people daily for at least 60 days of the year are subject to requirements applicable to nontransient, noncommunity water systems. If the facility is considered a drinking water supplier, it is required to be within the federal and state maximum contaminant levels (MCLs) for any pollutants in its potable water. If the Phase I determines that the facility is a drinking water supplier and that the water supply is likely polluted, the Phase II activities will consist of drinking water sampling and documentation.

Septic Systems.   Administrative closure procedures for septic systems need to be addressed with state authorities and the local municipality. Although not directly part of CWA or SDWA requirements, some municipalities have septic tank abandonment procedures that are administered through the local health department. For example, when a sanitary system is attached to a municipal system, the existing septic tank is required to be "caved in" and filled with an inert material. This will ensure that the abandoned tank does not present a safety issue when transferring the property. In general, there may not be any Phase II sampling activities associated with the septic system unless the facility Phase I investigation indicated discharges of industrial wastewaters or other nonsanitary wastes.

Wells and Groundwater.   Federal and state requirements may be most applicable to underground injection control wells. Accordingly, Phase II should document any sampling activities to verify adherence of past practices to pertinent standards. The facility should ensure that the potential for wells to become a route of transport for contaminants is eliminated.

  If groundwater sampling is needed because of suspected contamination, Phase II analysts may use existing wells, install monitoring wells, or use hydropunch or other sampling methods. Any activities performed and their results should be documented.

Wetlands.   Requirements for dredging or filling wetlands are issued and enforced by the Army Corps of Engineers under Section 404 of the CWA. State Section 404 programs can be enforced on waters not susceptible to interstate commerce, including tidal waters and wetlands. If the facility obtained a Section 404 permit, it must be terminated with the issuing authority, and any dredging and filling activities must cease. Cases of contamination resulting from these activities and associated sampling operations would likely fall under the federal or state

Superfund statutes.

Spill Containment, Control, and Countermeasures (SPCC) Plan. Some areas with suspected contamination may stem from oil stored on site. If the facility stored more than 42,000 gallons underground, 1320 gallons aboveground total, or 660 gallons in any single container aboveground, then the facility should have an SPCC plan in place in accordance with 40 CFR Part 112. The Phase II investigation should include a review of the SPCC plan to determine what engineering controls and systems are in place, an analysis of this equipment to identify whether it is functioning effectively to eliminate the potential for future spills if oil is still stored on site, and an inspection to make sure that any incidental spills that may have occurred were cleaned up properly.

Resources. The following resources may be helpful in obtaining further information about clean water management:

- SDWA Hotline: *hotline-sdwa@epamail.epa.gov*
- Office of Water Resources Center: 202-260-7786, or *waterpubs@epamail.epa.gov*
- Oil Pollution Act information exchange under the EPA RCRA/Superfund /Emergency Planning and Community Right-to-Know Act (EPCRA) Industry Assistance Hotline: 1-800-424-9346, or 703-412-9810 for the Washington, DC, metropolitan area
- Wastewater Sampling Computer-Based Training and Manual, a set of six computer-based training modules and complementary manual being developed for region 1 by the EPA SHEMD Multimedia Laboratory in conjunction with the Office of Water. Contact the Multimedia Laboratory at 202-260-2215.

Atomic Energy Act (AEA) and Nuclear Regulatory Commission Licenses

Federal agencies operate laboratories nationwide that may use nuclear material as part of their experiments. Depending on the laboratory's research mission and the extent to which nuclear materials are used, the laboratory obtains a license from the Nuclear Regulatory Commission (NRC) as part of its compliance with NRC regulations under the Atomic Energy Act (AEA). In most cases, the use of nuclear material is in a sealed-source state. When facilities are being prepared for property transfer or consolidation, NRC licensees must terminate their licenses.

Overview of License Termination Process. When a licensed facility terminates its operations and ceases to use or handle radioactive materials, the facility must notify the NRC to terminate its license. The licensee submits a request for termination to the NRC, and in cases where contamination is significant, the facility must develop a decontamination and decommissioning (D&D) plan to reduce radioactivity to acceptable levels.

After NRC approval of the licensee's termination request or D&D plan, the facility must carry out the D&D process, if applicable. This process can be simple, where only sealed sources or short-lived materials are handled, or it can entail extensive efforts for large-scale nucleotide users. A decision tree of required steps in the license termination process is shown in Fig. A1.5. Although not required, an initial radiation survey and documentation review serve as good starting points. Results from these reviews should be included with the termination request letter to the NRC.

The NRC will only terminate the user's license by written notice after the user

- Terminates the use of radioactive materials
- Properly removes and disposes of radioactive wastes
- Remediates the site, if D&D is required
- Submits NRC Form 314, a copy of which is provided in Fig. A1.6
- Conducts and submits the results of a final radiation survey to confirm decontamination

To aid in the decontamination determination, the licensee submits the results of a radiation survey of the facility and common use areas to the NRC. If the results are satisfactory, the NRC provides written confirmation of license termination. The licensee remains under license to the NRC, and thus subject to NRC requirements, during D&D activities. After license termination, the facility is no longer subject to NRC requirements regarding further unrestricted use of the facility.

Recent NRC efforts are being studied to streamline the licensing process and extend it to the D&D process. NRC's Office of Nuclear Material Safety and Safeguards has undertaken a pilot program to determine the feasibility of performing computer-assisted review of license applications. This system presents consolidated licensing guidance on the Internet and significantly reduces the turn-around time on license applications. Presently, the system is capable of handling new portable gauge license applications only. However, in the future, it is hoped that the system capabilities will be expanded to include other licensing classes and decommissioning and decontamination.

License Termination Letter.   The NRC examines license termination on a case-by-case basis. There are no defined criteria provided for significant contamination determination and subsequent D&D plan development.

A license termination request starts by submitting a notification of termination intentions to the NRC. This usually takes the form of an official letter that contains a description of the facility's nuclear material usage, spill records, site plans with pre- and postconstruction modifications, a list of all regulated areas requiring doc-

Decide to Terminate
NRC License

Notify NRC of
Decision to Terminate

Conduct Initial Radiation
Survey and Document Review

Submit Termination
Request Letter and Report

Develop D&D Plan including:

• D&D Activities
• Radiation H&S Plan
• Final Survey Plan
• Records and Data Plan
  Approval
• Update D&D Cost Estimate

No

NRC Deems
Contamination
Insignificant

Yes

Terminate
By-Product Use

Execute D&D Plan
Upon NRC Approval

Remove Radioactive
Wastes

Properly Dispose
of Wastes

Submit
NRC Form-314

Conduct and Submit Results
from Final Radiation Survey

Await NRC Notification
of License Termination

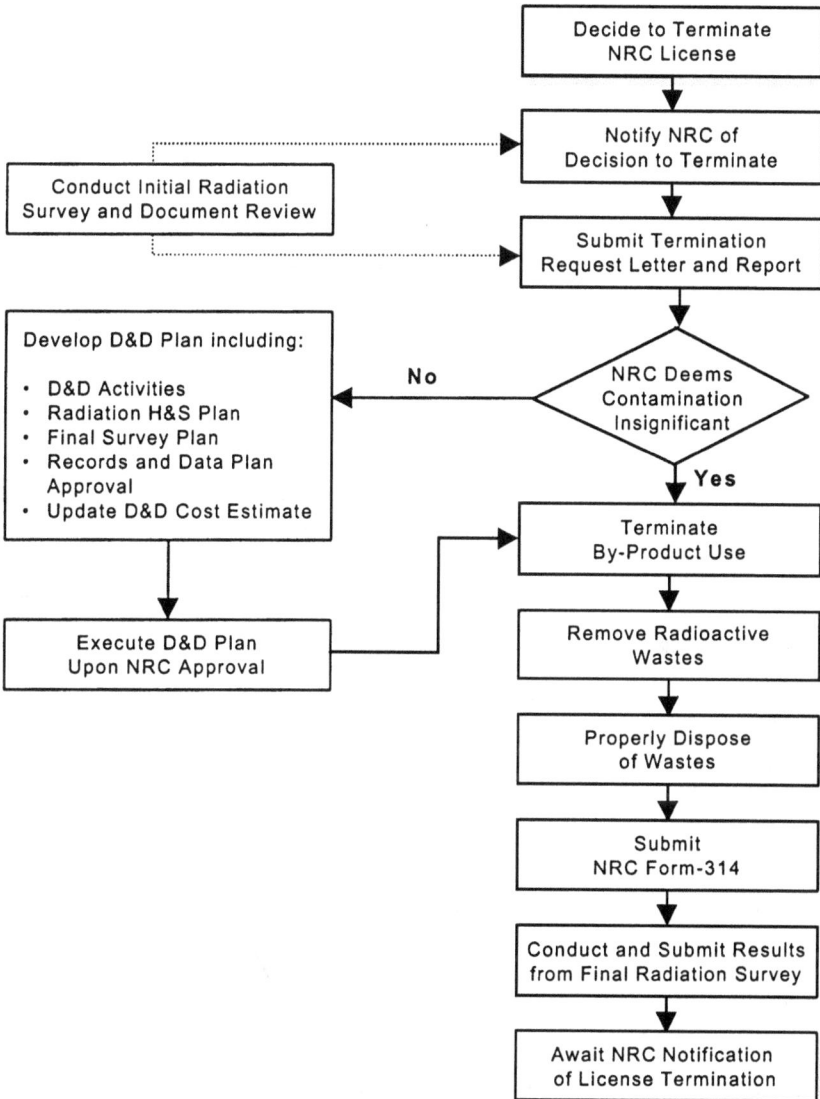

FIGURE A1.5   Decision tree: NRC license termination process.

umentation, and a previously prepared cost estimate for decommissioning. An example of an NRC license termination request letter appears in Fig. A1.6.

The facility may work with the NRC to identify additional information needs in making its determination for a decommissioning plan submittal. Because many facilities use minor quantities of radioactive material, the information request generally will be straightforward and limited. For those larger radioactive material

Nuclear Regulatory Commission (Address to NRC regional office)
Office of Nuclear Material Safety and Safeguards
Division of Low-Level Waste Management and Decommissioning
Street Address
City, State, Zip

Dear Sir or Madam:

In accordance with 10 CFR Part 30, Subpart 36b, this letter shall serve as notice of intent to terminate NRC license (*number*), which will return the property to unrestricted use. This NRC license enabled the facility to use, store, and dispose of unsealed, radioactive by-products which were used for scientific research and testing. Please find enclosed:

- A record of all spills or other unusual occurrences involving the spread of contamination
- As-built drawings and modifications of buildings where the unsealed sources were used and stored and locations of possible inaccessible contamination
- A list of all regulated areas that require documentation
- Cost estimates initially performed to implement a decommissioning plan (if necessary).

This facility has experienced no spills or other unusual occurrences and has detected no residual radioactive readings from an initial site survey, per recommended procedures. The facility expects NRC to conclude that a decommissioning plan is not required because contamination should be deemed insignificant.

Upon NRC's response to this termination request, the facility will proceed with termination of by-product use, properly dispose of remaining materials, and conduct a final radiation exit survey in anticipation of NRC's own exit survey validation.

If you have any questions about this termination request, please contact the Radiation Safety Officer at (*phone number*).

Sincerely,

Radiation Safety Officer

Enclosures

FIGURE A1.6   Sample NRC license termination request letter.

users, all information should be obtained from radiation safety officers' reporting requirements files.

Laboratory Decontamination Guidance.   The NRC regulations, at 10 CFR Part 20, Subpart E, establish criteria for the remediation of contaminated sites or facilities that will allow their release for future use with or without restrictions. The criteria include a total effective dose equivalent limit of 15 mrem/year, meaning that the average individual should not be exposed above this level from residual activity within the decommissioned facility. The criteria also require a licensee to reduce any residual radioactivity to as low as reasonably achievable (ALARA).

The NRC developed the *Regulatory Guide on Release Criteria for Decommissioning* (NUREG-1500) to assist facilities in following acceptable procedures for determining the predicted dose level (PDL) from any residual radioactivity at the site. The criteria describe the basic features of the NRC's models and acceptable parameters to factor into PDL calculations.

The NRC has not developed specific guidance on acceptable procedures for decontaminating laboratory equipment. For most facilities, decontamination will use suitable solvents. Note that all solvent disposal also must conform strictly to requirements under CERCLA and RCRA. In cases where ductwork, drains, or fumehoods are contaminated beyond the decontamination abilities of certain solvents, they may have to be decommissioned, removed, and disposed of as low-level radioactive waste.

Conducting a Final Radiation Exit Survey. The extent of residual contamination will depend on the type and quantity of nuclear material used at the facility. The NRC has developed a guidance document entitled, *Manual for Conducting Radiological Surveys in Support of License Termination* (NUREG/CR-5849), to assist all types of facilities in executing final radiation exit surveys. This document contains procedures for conducting radiologic surveys to demonstrate that residual radioactive material satisfies release criteria. Survey methodologies describe the state-of-the-art instrumentation and procedures for conducting radiologic surveys. The document also incorporates statistical approaches for survey design, evaluation, and quality assurance.

Resources. In preparing for facility transfer or closure, the following resources may assist in complying with NRC regulations and NRC closure protocols:

- CFR Part 30, Subparts 35 and 36

- Manual chapter, *NRC Protocols for Decommissioning a Facility* (NRC Internal Draft),

- NUREG-1500, *Working Draft Regulatory Guide on Release Criteria for Decommissioning,* staff draft

- NUREG-1501, *Background as a Residual Radioactivity Criterion for Decommissioning,* draft

- NUREG-5512, *Residual Radioactive Contamination from Decommissioning*

- NUREG/CR-5849, *Manual for Conducting Radiological Surveys in Support of License Termination*

- Regulatory Guide 1.86, *Termination of Operating License for Nuclear Reactors*

- Task DG-3001, *Records Important for Decommissioning for Licensees under 10 CFR Parts 30, 40, 70, and 72*

- NRC-7590-01, *Action Plan to Ensure Timely Cleanup of Site Decommissioning Management Plan Sites*

Numerous examples of laboratory closure materials are available through the NRC's Public Document Room, 2120 L Street, NW, Washington, DC 20555.

National Environmental Policy Act (NEPA)

The substantive and procedural requirements of NEPA must be followed for all major federal actions, including some activities under the EDDA. In regard to federal agency real property transfers, potential major federal actions applicable for NEPA review include lease termination, building consolidation, mission change, or construction of a new facility or laboratory. In working with the General Services Administration (GSA) and other federal agencies to execute the NEPA process, the responsible official, such as a property transfer manager, should investigate the potential completion of NEPA documentation by other federal agencies. If NEPA documentation does not exist for the proposed real property transfer, then the federal agency should initiate the NEPA process.

Federal agencies follow a three-tiered procedural review process when an action that could affect the environment is proposed. The NEPA process chart (Fig. A1.7) gives an overview of the process. Tier 1 determines whether the project qualifies for a categorical exclusion (CX). Tier 2 determines whether the project qualifies for a finding of no significant impact (FNSI) after performing an environmental assessment (EA). If no significant impacts are discovered in the EA process, the project qualifies for an FNSI. If significant impacts are discovered in the EA process, an environmental impact statement (EIS) must be prepared. Tier 3 entails preparing an EIS and issuing a record of decision (ROD).

Some examples of real property transfer activities that require the preparation of NEPA document include, but are not limited to, activities that

- Significantly affect the pattern and type of land use or growth and distribution of human population
- Conflict with local, regional, or state land use plans or policies
- Significantly affect cultural resource areas, endangered or threatened species, or environmentally important natural resource areas such as wetlands, floodplains, or coastal waters
- Significantly have an adverse effect on local ambient air quality, noise level, surface water or groundwater quality or quantity, water supply, aquatic life, wildlife, and their natural habitats

The results of the Phase I and II EDDA will assist in determining the appropriate level of NEPA review and documentation through the disclosure and characterization of environmental conditions of the property.

The following provides a suggested approach for identifying the NEPA requirements and applicability when closing or addressing environmental contamination at a federal agency's property transfer project. At a minimum, the PTM, laboratory director, or Phase II or III oversight official should consult with NEPA specialist and the facility's engineering services division to determine the appropriate and required NEPA activities. The PTM, laboratory director, or Phase II and

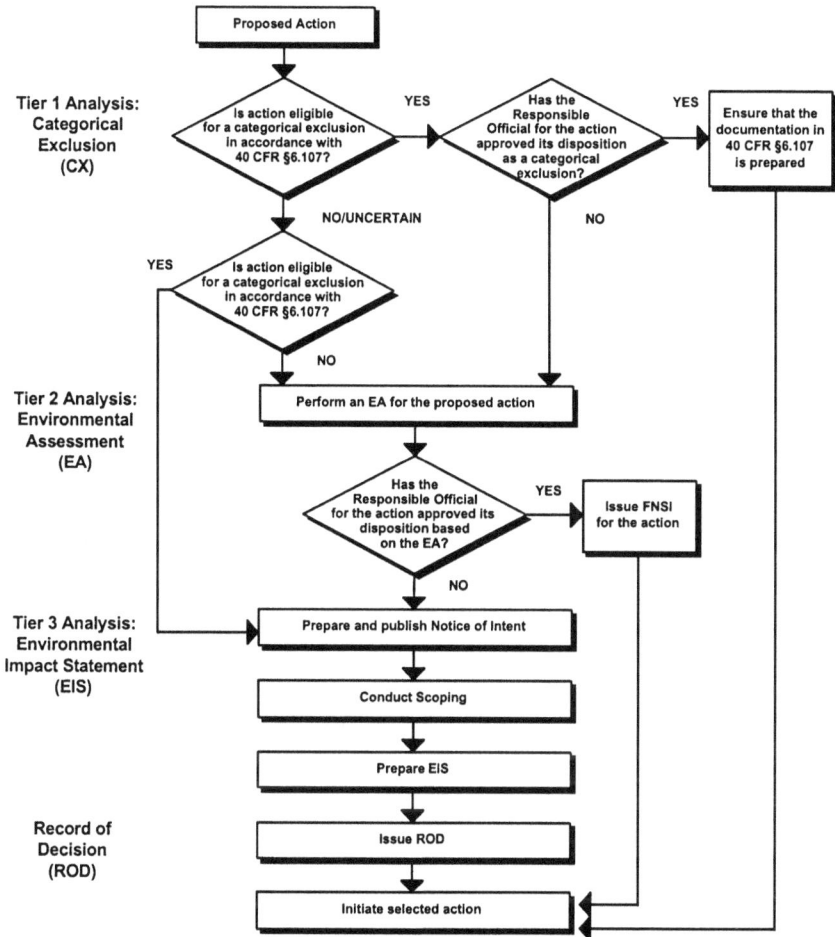

FIGURE A1.7   The NEPA process.

III oversight official will be responsible for ensuring the NEPA review process is executed, including, but not limited to, evaluation, document development, public notification, and mitigation measures. Finally, all supporting materials, reports, and NEPA documentation should be submitted to the federal agency's headquarters to be maintained indefinitely in an official document management system.

A1.12 APPENDIX R OF THE CFATF GUIDE: EXAMPLE PHASE I REPORT OUTLINE

1.0 Introduction

2.0 Site Location and Description

3.0 Site Ownership and Use

    3.1 Site Ownership
    3.2 Site Use—Historical
    3.3 Site Use—Current

4.0 Site Inspection

    4.1 Site Buildings
    4.2 Site Grounds
    4.3 Underground Storage Tanks
    4.4 Aboveground Storage Tanks
    4.5 Transformers
    4.6 Asbestos
    4.7 Indoor Air
    4.8 Radioactive Materials
    4.9 Motor Pools, Shops and
        Laboratories Operations,
        Analyses and Experiments
    4.10 Other Information
    4.11 Facility Records

5.0 Regulatory Review

    5.1 Federal Records
    5.2 State Records
    5.3 Local Records

6.0 Adjacent and Neighboring Properties

7.0 Hazardous Materials and Waste Management

    7.1 Hazardous Waste Generation,
        Storage and Disposal Practices
    7.2 Hazardous and Regulated
        Materials Management
    7.3 Nonhazardous Waste Management

8.0 Sensitive Environmental Areas

    8.1 Wetlands
    8.2 Historic Value
    8.3 Recreational Land Use
    8.4 Future Use and Zoning

9.0 Supplemental Information and Previous Studies

10.0 Conclusions and Recommendations

APPENDICES

- Site Location Map
- Site Plan
- Site Photographs
- Site Ownership Records
- Federal, State, Local, and EPA Facility Records
- Aerial Photographs
- List of Chemicals Used and In Use at the Facility
- Material Safety Data Sheets
- Groundwater Monitoring Results
- Previous Site Work Repairs
- Groundwater Monitoring Results
- Soil Boring Data

A1.13 APPENDIX U OF THE CFATF GUIDE: EXAMPLE PHASE II REPORT OUTLINE

1.0 Introduction

    1.1 Summary of Known Information Findings of Phase I Assessment

    1.2 Scope of Investigation

2.0 Site Map/Building Plans

3.0 Likely Sources of Contamination

    3.1 Likely Sources of Contamination

    3.2 Location of Likely Sources

    3.3 Approximate Date/Type/Quantity of Release

4.0 Soil Sampling and Analysis

    4.1 Sampling Overview

        4.1.1 Types of Samples

        4.1.2 Location of Samples

        4.1.3 Sampling Objective and Justification

        4.1.4 Analytical Parameters Including Justification

    4.2 Sampling Methods

        4.2.1 Sampling Methods and Procedures

        4.2.2 Boring Logs

        4.2.3 Field Screening Data

    4.3 Analytical Methods

        4.3.1 Analytical Parameters Including EPA Method Number and Detection Limit

        4.3.2 Maps and Diagrams Showing the Extent of Contamination

    4.4 Results

        4.4.1 Data Presentation Including Tables

        4.4.2 Notation of Results Above Applicable Standards

        4.4.3 Maps and Diagrams Showing the Extent of Contamination

5.0 Ground Water Sampling and Analysis

    5.1 Sampling (QA/QC) Overview

        5.1.1 Number of Wells

        5.1.2 Surveyed Well Location

        5.1.3 Well Placement

        5.1.4 Well Depth and Screened Interval

        5.1.5 Well System Justification

        5.1.6 Analytical Parameters Including Justification

    5.2 Sampling Methods

        5.2.1 Well Drilling Methods

          a. Screening Data in Cuttings/Soil Samples

          b. Drilling Logs

          c. Well Construction Descriptions and Diagrams

          d. Well Development Methods

          e. Well Stabilization Period

        5.2.2 Sampling Methods and Procedures

          a. Sampling Rationale

          b. Field Testing Results

          c. Sampling Frequency

    5.3 Analytical Methods

        5.3.1 Analytical Parameters Including EPA Method Number and Detection Limit

        5.3.2 Name and Certification of Laboratory

    5.4 Results

        5.4.1 Data Presentation Including Tables

        5.4.2 Notation of Results Above Applicable Standards

        5.4.3 Maps and Diagrams Showing the Potentiometric Surface and Extent of Contamination

A1.14 APPENDIX V OF THE CFATF GUIDE: SAMPLE LIST OF REMEDIATION TECHNOLOGIES

Technology	Description
Asbestos containment/removal	Immobilization of asbestos fibers to ensure that particles cannot become friable. Traditional abatement can also include complete removal of asbestos fibers which can be an extremely expensive procedure.
Incineration/thermal treatment	Heat is used to concentrate or alter the concentration of soil contaminants. The technology is effective in treating organic wastes; however, it is expensive.
Capping/slurry wall	Wastes are isolated to prevent migration. This remedial solution is not recommended because it simply contains wastes and does not eliminate significant hazards.
Excavation/off-site disposal	Material is transported off-site for disposal at an approved facility. Limited landfill capacity and RCRA Land Disposal Restrictions have made this option increasingly expensive.
Bioremediation	Microorganisms are used to consume and render the waste less hazardous. The process is limited by several factors, including salinity of soils, presence of metals, use of designer microorganisms, and toxicity of contaminant substrates. This technology tends to be among the least expensive ones and is gaining public acceptance.
Soil venting	Air is drawn through wells drilled around the treated area and assists in removing volatile chemicals. Soil venting is inexpensive to install and effective for volatile chemicals only.
Soil washing	A washing solution is applied to an excavated area of contaminated soil which removes persistent wastes. The cleaned soil is then returned to the excavation site. This process is effective in treating most contamination, but is expensive and time-consuming for small quantities.
Groundwater pump and treat	Wells are installed around the site of contamination and water is pumped to retrieve contaminants. This technology tends to be expensive and require many years to effectively remove site contamination.

A1.15   APPENDIX W OF THE CFATF GUIDE: EXAMPLE PHASE III REPORT OUTLINE

1.0 Introduction
   1.1 Summary of Findings from Phase I and II Investigations
   1.2 Scope of Phase III

2.0 Site Maps/Building Plans

3.0 Sources of Contamination
   3.1 Description of Contaminant Sources
   3.2 Locations of Contaminant Sources
   3.3 Approximate Date/Type/Quantity of Release

4.0 Sampling and Analysis Results
   4.1 Soil Sampling
      4.1.1 Sample Types
         4.1.1.1 Procedures for Each Type
         4.1.1.2 Boring Logs
         4.1.1.3 Physical and Chemical Field Screening Data Quality Objectives
         4.1.1.4 Field Analytical Procedures
         4.1.1.5 Field Deviations from Sampling and Analysis Plan
      4.1.2 Sample Locations
         4.1.2.1 Objectives and Sampling Rationale
         4.1.2.2 Distribution and Density
         4.1.2.3 Deviations from Sampling and Analysis Plan
      4.1.3 Laboratory Analysis
         4.1.3.1 Laboratory Analysis Data Quality Objectives
         4.1.3.2 Analytical Parameters (include EPA Method Number and Detection Limit)
         4.1.3.3 Quality Assurance Sample Analysis Results
      4.1.4 Sampling and Analysis Results
         4.1.4.1 Data Presentation including Tables
         4.1.4.2 Discussion of Results
         4.1.4.3 Maps, cross-sections, and other diagrams depicting extent of contamination
   4.2 Groundwater Sampling
      4.2.1 Sample Types
         4.2.1.1 Procedures for Each Type
         4.2.1.2 Physical and Chemical Field Screening Data Quality Objectives
         4.2.1.3 Field Analytical Procedures
         4.2.1.4 Deviations from Sampling and Analysis Plan
      4.2.2 Sample Locations
         4.2.2.1 Objectives and Well Placement Rationale
         4.2.2.2 Surveyed Well Locations, Depths, and Screened Intervals
         4.2.2.3 Sampling Depths
         4.2.2.4 Field Deviations from Sampling and Analysis Plan
      4.2.3 Laboratory Analysis
         4.2.3.1 Laboratory Analysis Data Quality Objectives
         4.2.3.2 Analytical Parameters (include EPA Method Number and Detection Limit)
         4.2.3.3 Quality Assurance Sample Analysis Results
      4.2.4 Sampling and Analysis Results
         4.2.4.1 Data Presentation including Tables
         4.2.4.2 Discussion of Results
         4.2.4.3 Maps, cross-sections, and other diagrams depicting extent of contamination
   4.3 Surface Water/Sediment Sampling
      4.3.1 Sample Types
         4.3.1.1 Procedures for Each
         4.3.1.2 Physical and Chemical Field Screening Data Quality Objectives
         4.3.1.3 Field Analytical Procedures
         4.3.1.4 Deviations from Sampling and Analysis Plan

5.1.2 Quantitative Risk Assessment
    5.1.2.1 Rationale for Performing Quantitative Risk Assessment
    5.1.2.2 Assessment of Carcinogens
    5.1.2.3 Assessment of Noncarcinogens
    5.1.2.4 Results of Quantitative Risk Assessment
5.1.3 Assessment on Overall Risk of the Property
5.2 Future Land-Use Options
  5.2.1 Discussion on Future Land-Use Options
  5.2.2 Rationale for Selection of Future Land-Use Option
  5.2.3 Relationship Between Future Land-Use and Overall Risk at the Property

6.0 Final Technology or Alternative
6.1 Final Technology or Alternative Selection
  6.1.1 Description of Screening Process
  6.1.2 Description of Technologies or Alternatives Under Consideration
  6.1.3 Comparative Analysis of Technologies or Alternatives
    6.1.3.1 Overall Protectiveness of Human Health and the Environment
    6.1.3.2 Compliance With ARARs
    6.1.3.3 Long-Term Effectiveness
    6.1.3.4 Reduction of Toxicity, Mobility, or Volume Through Treatment
    6.1.3.5 Short-Term Effectiveness
    6.1.3.6 Implementability
    6.1.3.7 Cost
    6.1.3.8 State Acceptance
    6.1.3.9 Community Acceptance
  6.1.4 Detailed Discussion of Selected Alternatives
  6.1.5 Implementation Plan and Schedule for Selected Alternatives
  6.1.6 Statutory Determination for Selected Alternatives

7.0 Conclusions and Recommendations
7.1 Conclusions
  7.1.1 Summary of Site Characterization
  7.1.2 Discussion on Success in Meeting Remediation Goals and Objectives
7.2 Recommendations
  7.2.1 Discussion on follow-up Actions

APPENDICES (as needed)
- Site location and topography
- Sampling locations and designation (excepts from FSP)
- Site safety and health plan
- Field and boring logs
- QA/QC program and documentation (e.g. chain of custody and laboratory QA/QC program)
- Laboratory analytical results
- Mapping of contamination plumes and zones
- Remedial alternative technical/cost specifications

APPENDIX 2

BRIEF OVERVIEW OF THE FEDERAL STATUTES OF THE UNITED STATES

This appendix highlights some of the statutes governing a variety of the more common environmental prevention measures enacted by Congress. The majority of the statutes can be retrieved directly from government Internet sites. In addition, more detailed information including the actual text of the various acts can be obtained by searching the Environmental Protection Agency (EPA) Internet sites. The reason for including a summary version of many environmental policies and regulations here is so that consulting firms with interest in particular components of environmental cleanup or wishing to partake in environmental projects can have some idea of the basic laws that would influence their assessment of costs and conditions. While these statutes pertain only to the United States, many other countries have enacted similar statutes so that by perusal of individual nations' regulations, a corporation involved in multinational environmental projects can obtain some idea of likely liability, costs, and insurance demands prior to deciding whether to become involved in any particular environmental project.

The summary listing of the EPA-mandated acts that follows relies heavily on information at existing government Internet sites and in particular the EPA's web site (*www.epa.gov*), which should be consulted for more detailed and accurate representation of individual acts, statutes, and regulations. The summary presented here is intended to guide the reader to fuller and more accurate descriptions through U.S. government sources and should not be used as a definite statement of the law.

A2.1 BACKGROUND

Laws and regulations are a major tool in protecting the environment. Congress passes laws that govern the United States. To put those laws into effect, Congress authorizes certain government agencies, including the EPA, to create and enforce regulations.

In general, the steps for the creation of a law in the United States are as follows:

Step 1. A member of Congress proposes a bill. A *bill* is a document that, if approved, will become law. The texts of bills Congress is considering, or has considered, can be retrieved at the Library of Congress's Thomas Web server.

Step 2. If both houses of Congress approve a bill, it goes to the President, who has the option to either approve it or veto it. If approved, the new law is called an *act,* and the text of the act is known as a *public statute.* Some of the better-known laws related to the environment are the Clean Air Act, the Clean Water Act, and the Safe Drinking Water Act.

Step 3. Once an act is passed, the House of Representatives standardizes the text of the law and publishes it in the *United States Code.* The *U.S. Code* is the official record of all federal laws.

The *United States Code* database (1994 edition) is available from the Government Printing Office (GPO). The GPO is the sole agency authorized by the federal government to publish the *U.S. Code.* The *U.S. Code* database contains the texts of laws in effect as of January 16, 1996. Cornell University also offers the 1996 version of the *U.S. Code.*

Laws often do not include all the details. The *U.S. Code* would not specify, for example, what the speed limit is at specific locations. In order to make the laws work on a day-to-day level, Congress authorizes certain government agencies—including the EPA—to create regulations.

Regulations set specific rules about what is legal and what is not. For example, a regulation issued by EPA to implement the Clean Air Act might state what levels of a pollutant—such as sulfur dioxide—are safe. It would tell industries how much sulfur dioxide they can legally emit into the air and what the penalty will be if they emit too much. Once the regulation is in effect, EPA then works to help Americans comply with the law and to enforce it.

A regulation is formulated when an authorized agency (such as EPA) decides that such a measure may be needed. The agency researches it and, if necessary, proposes a regulation. The proposal is listed in the *Federal Register* so that members of the public can consider it and send their comments to the agency. The agency considers all the comments, revises the regulation accordingly, and issues a final rule. At each stage in the process, the agency publishes a notice in the *Federal Register.* These notices include the original proposal, requests for public comment, notices about meetings where the proposal will be discussed (open to the public), and the text of the final regulation. (*Federal Register* notices related to the environment are available on EPA's Web site). Most of these are notices issued by EPA. A complete record of *Federal Register* notices issued by the entire federal government is available from the GPO.

Twice a year, each agency publishes a comprehensive report that describes all the regulations it is working on or has recently finished. These are published in the *Federal Register,* usually in April and October, as the "Unified Agenda of Federal and Regulatory and Deregulatory Actions."

Once a regulation is completed and has been printed in the *Federal Register* as a final rule, it is "codified" by being published in the *Code of Federal Regulations* (CFR). The CFR is the official record of all regulations created by the federal government. It is divided into 50 volumes, called *titles,* each of which focuses on a particular area. Almost all environmental regulations appear in Title 40. The CFR is revised yearly, with one-fourth of the volumes updated every 3 months. Title 40 is revised every July 1.

Among the environmental laws enacted by Congress through which EPA carries out its efforts are

1938 Federal Food, Drug and Cosmetic Act

1947 Federal Insecticide, Fungicide and Rodenticide Act

1948 Federal Water Pollution Control Act (also known as the Clean Water Act)

1955 Clean Air Act

1965 Shoreline Erosion Protection Act

1965 Solid Waste Disposal Act

1970 National Environmental Policy Act

1970 Pollution Prevention Packaging Act

1970 Resource Recovery Act

1971 Lead-Based Paint Poisoning Prevention Act

1972 Coastal Zone Management Act

1972 Marine Protection, Research and Sanctuaries Act

1972 Ocean Dumping Act

1973 Endangered Species Act

1974 Safe Drinking Water Act

1974 Shoreline Erosion Control Demonstration Act

1975 Hazardous Materials Transportation Act

1976 Resource Conservation and Recovery Act

1976 Toxic Substances Control Act

1977 Surface Mining Control and Reclamation Act

1978 Uranium Mill-Tailings Radiation Control Act

1980 Asbestos School Hazard Detection and Control Act

1980 Comprehensive Environmental Response, Compensation and Liability Act

1982 Nuclear Waste Policy Act

1984 Asbestos School Hazard Abatement Act

1986 Asbestos Hazard Emergency Response Act

1986 Emergency Planning and Community Right to Know Act

1988 Indoor Radon Abatement Act

1988 Lead Contamination Control Act

1988 Medical Waste Tracking Act

1988 Ocean Dumping Ban Act

1988 Shore Protection Act

1990 National Environmental Education Act

A2.2   MAJOR ENVIRONMENTAL LAWS

More than a dozen major statutes or laws form the legal basis for the programs of the EPA.

A2.2.1   National Environmental Policy Act of 1969 (NEPA); 42 USC 4321–4347

NEPA is the basic national charter for protection of the environment. It establishes policy, sets goals, and provides a means for carrying out the policy.

The full text of the National Environmental Policy Act of 1969, as amended [P.L. 91-190, 42 USC 4321–4347, January 1, 1970, as amended by P.L. 94-52, July 3, 1975, P.L. 94-83, August 9, 1975, and P.L. 97-258, Section 4(b), September 13, 1982] can be obtained from the EPA Web site (www.epa.gov).

NEPA provides the purposes of the Act as follows: "To declare a national policy which will encourage productive and enjoyable harmony between man and his environment; to promote efforts which will prevent or eliminate damage to the environment and biosphere and stimulate the health and welfare of man; to enrich the understanding of the ecological systems and natural resources important to the Nation; and to establish a Council on Environmental Quality.

(a) The Congress, recognizing the profound impact of man's activity on the interrelations of all components of the natural environment, particularly the profound influences of population growth, high-density urbanization, industrial expansion, resource exploitation, and new and expanding technological advances and recognizing further the critical importance of restoring and maintaining environmental quality to the overall welfare and development of man, declares that it is the continuing policy of the Federal Government, in cooperation with state and local governments, and other concerned public and private organizations, to use all practicable means and measures, including financial and technical assistance, in a manner calculated to foster and promote the general welfare, to create and maintain conditions under which man and nature can exist in productive harmony, and fulfill the social, economic, and other requirements of present and future generations of Americans.

(b) In order to carry out the policy set forth in this Act, it is the continuing responsibility of the Federal Government to use all practicable means, consistent with other essential considerations of national policy, to improve and coordinate Federal plans, functions, programs, and resources to the end that the Nation may

(1) fulfill the responsibilities of each generation as trustee of the environment for succeeding generations;

(2) assure for all Americans safe, healthful, productive, and aesthetically and culturally pleasing surroundings;

(3) attain the widest range of beneficial uses of the environment without degradation, risk to health or safety, or other undesirable and unintended consequences;

(4) preserve important historic, cultural, and natural aspects of our national heritage, and maintain, wherever possible, an environment which supports diversity, and variety of individual choice;

(5) achieve a balance between population and resource use which will permit high standards of living and a wide sharing of life's amenities; and

(6) enhance the quality of renewable resources and approach the maximum attainable recycling of depletable resources.

(c) The Congress recognizes that each person should enjoy a healthful environment and that each person has a responsibility to contribute to the preservation and enhancement of the environment.

Sec. 102 [42 USC § 4332]. The Congress authorizes and directs that, to the fullest extent possible: (1) the policies, regulations, and public laws of the United States shall be interpreted and administered in accordance with the policies set forth in this Act, and (2) all agencies of the Federal Government shall

(A) utilize a systematic, interdisciplinary approach which will insure the integrated use of the natural and social sciences and the environmental design arts in planning and in decision-making which may have an impact on man's environment;

(B) identify and develop methods and procedures, in consultation with the Council on Environmental Quality established by Title II of this Act, which will insure that presently un-quantified environmental amenities and values may be given appropriate consideration in decision-making along with economic and technical considerations;

(C) include in every recommendation or report on proposals for legislation and other major Federal actions significantly affecting the quality of the human environment, a detailed statement by the responsible official on —

(i) the environmental impact of the proposed action,

(ii) any adverse environmental effects which cannot be avoided should the proposal be implemented,

(iii) alternatives to the proposed action,

(iv) the relationship between local short-term uses of man's environment and the maintenance and enhancement of long-term productivity, and

(v) any irreversible and irretrievable commitments of resources which would be involved in the proposed action should it be implemented."

Title II of the Act establishes the Council of Environmental Quality as follows: "Sec. 201 [42 USC § 4341]. The President shall transmit to the Congress annually beginning July 1, 1970, an Environmental Quality Report (hereinafter referred to as the "report") which shall set forth (1) the status and condition of the major natural, manmade, or altered environmental classes of the Nation, including, but not limited to, the air, the aquatic, including marine, estuarine, and fresh water, and the terrestrial environment, including, but not limited to, the forest, dryland, wetland, range, urban, suburban and rural environment; (2) current and foreseeable trends in the quality, management and utilization of such environments and the effects of those trends on the social, economic, and other requirements of the Nation; (3) the adequacy of available natural resources for fulfilling human and economic requirements of the Nation in the light of expected population pressures; (4) a review of the programs and activities (including regulatory activities) of the Federal Government, the State and local governments, and nongovernmental entities or individuals with particular reference to their effect on the environment and on the conservation, development and utilization of natural resources; and (5) a program for remedying the deficiencies of existing programs and activities, together with recommendations for legislation.

Sec. 202 [42 USC § 4342]. There is created in the Executive Office of the President a Council on Environmental Quality (hereinafter referred to as the "Council"). The Council shall be composed of three members who shall be appointed by the President to serve at his pleasure, by and with the advice and consent of the Senate. The President shall designate one of the members of the Council to serve as Chairman. Each member shall be a person who, as a result of his training, experience, and attainments, is exceptionally well qualified to analyze and interpret environmental trends and information of all kinds; to appraise programs and activities of the Federal Government in the light of the policy set forth in Title I of this Act; to be conscious of and responsive to the scientific, economic, social, aesthetic, and cultural needs and interests of the Nation; and to formulate and recommend national policies to promote the improvement of the quality of the environment."

A2.2.2   Chemical Safety Information, Site Security and Fuels Regulatory Relief Act [Public Law 106-40, January 6, 1999; 42 USC 7412(r) Amendment to Section 112(r) of the Clean Air Act]

The Chemical Safety Information, Site Security and Fuels Regulatory Relief Act was established: "To amend the Clean Air Act to remove flammable fuels from the list of substances with respect to which reporting and other activities are required under the risk management plan program, and for other purposes."

Under Sec. 112(r) of the Clean Air Act (CAA), by June 21, 1999, certain facilities were required to have in place a risk-management program and submit a summary of that program—called a *risk-management plan* (RMP)—to the EPA. On August 5, 1999, President Clinton signed legislation that removes from coverage by the RMP program any flammable fuel when used as fuel or held for sale as fuel by a retail facility. The legislation also limits access to off-site consequence analysis (OCA) data that are reported in RMPs by covered facilities. For 1 year beginning August 5, 1999, OCA information will not be available to the public except in certain ways. During that 1-year period, the federal government will conduct an assessment and issue regulations governing future public access to OCA data.

What is new? The recently enacted Chemical Safety Information, Site Security and Fuels Regulatory Relief Act establishes new provisions for reporting and disseminating information under Sec. 112(r) of the Clean Air Act. The law has two distinct parts that pertain to

- Flammable fuels

- Public access to OCA (also known as *worst-case scenario*) data.

Flammable Fuels. Flammable fuels used as fuel or held for sale as fuel at a retail facility are removed from coverage by the RMP program. However, flammable fuels used as a feedstock or held for sale as fuel at a wholesale facility are still covered. A *retail facility* is a facility "at which more than one-half of the income is obtained from direct sales to end users or at which more than one-half of the fuel sold, by volume, is sold through a cylinder exchange program."

Despite removal of flammable fuels from the RMP program, firefighters and other local emergency responders should receive information on the potential off-site effects of accidents involving flammable fuels. EPA and industry are working with the National Fire Protection Association (NFPA), a group that develops fire protection codes and standards, to ensure that local responders receive that information. The new law directs the General Accounting Office (GAO) to assess in 2 years whether this goal has been accomplished.

Public Access to OCA Data. The law exempts OCA data from disclosure under the Freedom of Information Act (FOIA) and limits its public availability for at least 1 year. By August 5, 2000, the federal government is to (1) assess the risks of Internet posting of OCA data and the benefits of public access to those data and (2) based on that assessment, publish regulations governing public access to OCA data. In the meantime, EPA is to make publicly available the OCA data without facility identification information, and covered facilities must conduct public meetings to provide summaries of their OCA data. If the government fails to issue regulations by August 5, 2000, the FOIA exemption expires.

Major Provisions. The law

- Exempts OCA information from public disclosure under FOIA for at least 1 year.
- Makes OCA data available to federal, state, and local officials, including members of local emergency planning committees, for emergency planning and response purposes.
- Provides for a system for making OCA data available to qualified researchers.
- Prohibits federal, state, and local officials and qualified researchers from publicly releasing OCA data except as authorized by the law.
- Calls for an assessment and regulations regarding public access to OCA data within 1 year.
- Preempts state FOIA laws regarding public access to OCA data unless data are collected under state law.
- Requires reports to be submitted to Congress describing the effectiveness of the RMP regulations in reducing the risk of criminally caused releases, the vulnerability of facilities to criminal and terrorist activity, and the security of transportation of substances listed under CAA Sec. 112(r).

Facility Requirements. The new law requires every covered facility to

- Hold a public meeting to share information about the local implications of its RMP, including a summary of the OCA portion of its plan. Small businesses can meet this requirement by publicly posting the OCA summary.
- Notify the Federal Bureau of Investigation (FBI) by June 5, 2000 that it held such a meeting or posted such a notice within 1 year before or 6 months after August 5, 1999.
- Tell EPA if it distributes its OCA data to the public without restrictions. EPA is to maintain a public list of the facilities that have so distributed their OCA data.

Penalties. The law includes criminal penalties of up to $1 million for violating the prohibition on unauthorized disclosure of OCA data.

A2.2.3 The Clean Air Act (CAA); 42 USC s/s 7401 et seq. (1970)

A guide to the CAA is provided in EPA-400-K-93-001(April 1993) and reproduced here.

The Role of the Federal Government and the Role of the States. Although the 1990 Clean Air Act is a federal law covering the entire country, the states do much of the work to carry out the Act. For example, a state air pollution agency holds a hearing on a permit application by a power or chemical plant or fines a company for violating air pollution limits.

Under this law, EPA sets limits on how much of a pollutant can be in the air anywhere in the United States. This ensures that all Americans have the same basic health and environmental protections. The law allows individual states to have stronger pollution controls, but states are not allowed to have weaker pollution controls than those set for the whole country.

The law recognizes that it makes sense for states to take the lead in carrying out the Clean Air Act because pollution-control problems often require special understanding of local industries, geography, housing patterns, etc.

States have to develop state implementation plans (SIPs) that explain how each state will do its job under the Clean Air Act. A state implementation plan is a collection of the regulations a state will use to clean up polluted areas. The states must involve the public, through hearings and opportunities to comment, in the development of each state implementation plan. EPA must approve each SIP, and if a SIP is not acceptable, EPA can take over enforcing the Clean Air Act in that state.

The U.S. government, through EPA, assists the states by providing scientific research, expert studies, engineering designs, and money to support clean air programs.

Interstate Air Pollution. Air pollution often travels from its source in one state to another state. In many metropolitan areas, people live in one state and work or shop in another; air pollution from cars and trucks may spread throughout the interstate area. The 1990 Clean Air Act provides for interstate commissions on air pollution control, which are to develop regional strategies for cleaning up air pollution. The 1990 Clean Air Act includes other provisions to reduce interstate air pollution.

International Air Pollution. Air pollution moves across national borders. The 1990 law covers pollution that originates in Mexico and Canada and drifts into the United States and pollution from the United States that reaches Canada and Mexico.

Permits. One of the major breakthroughs in the 1990 Clean Air Act is a permit program for larger sources that release pollutants into the air. A source can be a power plant, factory, or anything that releases pollutants into the air. Cars, trucks, and other motor vehicles are sources, and consumer products and machines used in industry can be sources too. Sources that stay in one place are referred to as *stationary sources*; sources that move around, like cars or planes, are called *mobile sources.*

Requiring polluters to apply for a permit is not a new idea. Approximately 35 states have had statewide permit programs for air pollution. The Clean Water Act requires permits to release pollutants into lakes, rivers, or other waterways. Now air pollution is also going to be managed by a national permit system. Under the new program, permits are issued by states or, when a state fails to carry out the Clean Air Act satisfactorily, by EPA. The permit includes information on which pollutants are being released, how much may be released, and what kinds of steps the source's owner or operator is taking to reduce pollution, including plans to monitor (measure) the pollution. The permit system is especially useful for businesses covered by more than one part of the law, since information about all of a source's air pollution will now be in one place. The permit system simplifies and clarifies businesses' obligations for cleaning up air pollution and, over time, can reduce paperwork. For instance, an electric power plant may be covered by the acid rain, hazardous air pollutant, and nonattainment (smog) parts of the Clean Air Act; the detailed information required by all these separate sections will be in one place—on the permit.

Permit applications and permits are available to the public and can be obtained at state or regional air pollution control agencies or EPA. Businesses seeking permits have to pay permit fees much like car owners paying for car registrations. The money from the fees will help pay for state air pollution control activities.

Enforcement.   The 1990 Clean Air Act gives important new enforcement powers to EPA. It used to be very difficult for EPA to penalize a company for violating the Clean Air Act. EPA has to go to court for even minor violations. The 1990 law enables EPA to fine violators, much like a police officer giving traffic tickets. Other parts of the 1990 law increase penalties for violating the Act and bring the Clean Air Act's enforcement powers in line with other environmental laws.

Deadlines.   The 1990 Clean Air Act sets deadlines for EPA, states, local governments, and businesses to reduce air pollution. The deadlines in the 1990 Clean Air Act were designed to be more realistic than previous versions of the law, so it is more likely that these deadlines will be met.

Public Participation.   Public participation is a very important part of the 1990 Clean Air Act. Throughout the Act, the public is given opportunities to take part in determining how the law will be carried out. For instance, the public can take part in hearings on the state and local plans for cleaning up air pollution. A citizen can sue the government or a source's owner or operator to get action when EPA or the state has not enforced the Act. The public can request action by the state or EPA against violators.

The reports required by the Act are public documents. A great deal of information will be collected on just how much pollution is being released; these monitoring (measuring) data will be available to the public. The 1990 Clean Air Act ordered EPA to set up clearinghouses to collect and give out technical information. Typically, these clearinghouses will serve the public as well as state and other air pollution control agencies.

Market Approaches for Reducing Air Pollution: Economic Incentives.   The 1990 Clean Air Act has many features designed to clean up air pollution as efficiently and inexpensively as possible, letting businesses make choices on the best way to reach pollution cleanup goals. These new flexible programs are called *market* or *market-based approaches.* For instance, the acid rain cleanup program offers businesses choices as to how they reach their pollution-reduction goals and includes pollution allowances that can be traded, bought, and sold.

The 1990 Clean Air Act provides economic incentives for cleaning up pollution. For instance, gasoline refiners can get credits if they produce cleaner gasoline than required, and they can use those credits when their gasoline does not quite meet cleanup requirements.

Smog and Other Criteria Air Pollutants. A few common air pollutants are found all over the United States. These pollutants can injure health, harm the environment, and cause property damage.

EPA calls these pollutants *criteria air pollutants* because the agency has regulated them by first developing health-based criteria (science-based guidelines) as the basis for setting permissible levels. One set of limits protects health (primary standards); another set of limits is intended to prevent environmental and property damage (secondary standards). A geographic area that meets or does better than the primary standard is called an *attainment area*; areas that do not meet the primary standard are called *nonattainment areas.*

Although EPA has been regulating criteria air pollutants since the 1970 CAA was passed, many urban areas are classified as nonattainment for at least one criteria air pollutant. It has been estimated that about 90 million Americans live in nonattainment areas.

Smog. What is typically called *smog* is primarily made up of ground-level ozone. Ozone can be good or bad depending on where it is located. Ozone in the stratosphere high above the earth protects human health and the environment, but ground-level ozone is the main harmful ingredient in smog.

Ground-level ozone is produced by a combination of pollutants from many sources, including smokestacks, cars, paints, and solvents. When a car burns gasoline, releasing exhaust fumes, or a painter paints a house, smog-forming pollutants rise into the sky.

Often, wind blows smog-forming pollutants away from their sources. The smog-forming reactions take place while the pollutants are being blown through the air by the wind. This explains why smog is often more severe miles away from the source of smog-forming pollutants than it is at the source.

Weather and geography determine where smog goes and how bad it is. When temperature inversions occur (warm air stays near the ground instead of rising) and winds are calm, smog may stay in place for days at a time. As traffic and other sources add more pollutants to the air, the smog gets worse.

Since smog travels across county and state lines, when a metropolitan area covers more than one state (e.g., the New York metropolitan area includes parts of New Jersey and Connecticut), their governments and air pollution control agencies must cooperate to solve their problem. Governments on the East Coast from Maine to Washington, D.C., will have to work together in a multistate effort to reduce the area's smog problem.

Procedures through which the 1990 Clean Air Act reduces pollution from criteria air pollutants, including smog, are as follows. First, EPA and state governors cooperate to identify nonattainment areas for each criteria air pollutant. Then, EPA classifies the nonattainment areas according to how badly polluted the areas are. There are five classes of nonattainment areas for smog, ranging from marginal (relatively easy to clean up quickly) to extreme (that will take a lot of work and a long time to clean up).

The 1990 Clean Air Act uses this new classification system to tailor cleanup requirements to the severity of the pollution and set realistic deadlines for reaching cleanup goals. If deadlines are missed, the law allows more time to clean up, but usually a nonattainment area that has missed a cleanup deadline will have to meet the stricter cleanup requirements set for more polluted areas.

Not only must nonattainment areas meet deadlines, states with nonattainment areas must show EPA that they are moving on cleanup before the deadline, making reasonable further progress. States usually will do most of the planning for cleaning up criteria air pollutants, using the permit system to make sure power plants, factories, and other pollution sources meet their cleanup goals.

The comprehensive approach to reducing criteria air pollutants taken by the 1990 act covers many different sources and a variety of cleanup methods. Many of the smog cleanup requirements involve motor vehicles (cars, trucks, buses). Also, as the pollution gets worse, pollution controls are required for smaller sources.

Other Criteria Pollutants (Carbon Monoxide and Particulates).   The carbon monoxide (CO) and particulate matter (PM-10) cleanup plans are set up like the plan for smog, but only two pollution classes are identified for each (instead of the five for ozone). Getting rid of particulates (soot, dusts, smoke) will require pollution controls on power plants and restrictions on smaller sources such as wood stoves, agricultural burning, and dust from fields and roads. Because so many homes have woodstoves and fireplaces, this summary of the Clean Air Act includes a section on woodstoves and fireplaces, providing information on how the Clean Air Act will affect these home-heating systems.

1997 Changes to the Clean Air Act.   EPA recently reviewed the current air quality standards for ground-level ozone (commonly known as *smog*) and particulate matter (or PM). Based on new scientific evidence, revisions have been made to both standards. At the same time, EPA is developing a new program to control regional haze, which is largely caused by particulate matter.

Offsets.   What if a company wants to expand or change a production process or otherwise increase its output of a criteria air pollutant? If an owner or operator of a major source wants to release more of a criteria air pollutant, an offset (a reduction of the criteria air pollutant by an amount somewhat greater than the planned increase) must be obtained somewhere else so that permit requirements are met and the nonattainment area keeps moving toward attainment. The company also must install tight pollution controls. An increase in a criteria air pollutant can be offset with a reduction of the pollutant from some other stack at the same plant or at another plant owned by the same or some other company in the nonattainment area. Since total pollution will continue to go down, trading offsets among companies is allowed. This is one of the market approaches to cleaning up air pollution in the Clean Air Act.

Hazardous Air Pollutants.   Some air pollutants can cause cancer, problems with having children, and other very serious illnesses as well as environmental damage. Air pollutants have killed people swiftly when large quantities were released; the 1984 release of methyl isocyanate at a pesticide-manufacturing plant in Bhopal, India, killed approximately 4000 people and injured more than 200,000. EPA refers to chemicals that cause serious health and environmental hazards as *hazardous air pollutants* (HAPs) or *air toxics.*

Air toxics are released from sources throughout the country and from motor vehicles. For example, gasoline contains toxic chemicals. Gases escape from liquid gasoline and form a vapor in a process called *vaporization* or *evaporation.* When cars and trucks burn gasoline, air toxics come out of the tailpipes (these air toxics are combustion products—chemicals that are produced when a substance is burned).

Air toxics are released from small stationary sources, such as dry cleaners and auto paint shops. Large stationary sources, such as chemical factories and incinerators, also release hazardous air pollutants. The 1990 Clean Air Act deals more strictly with large sources than small ones, but EPA must regulate small sources of hazardous air pollutants as well.

To reduce air toxics pollution, EPA must first identify the toxic pollutants whose release should be reduced. The 1970 Clean Air Act gave EPA authority to list air toxics for regulation and then to regulate the chemicals. The agency listed and regulated seven chemicals through 1990. The 1990 act includes a list of 189 hazardous air pollutants selected by Congress on the basis of potential health and/or environmental hazard; EPA must regulate these listed air toxics. The 1990 act allows EPA to add new chemicals to the list as necessary.

To regulate hazardous air pollutants, EPA must identify categories of sources that release the 189 chemicals listed by Congress in the 1990 Clean Air Act. Categories could be gasoline service stations, electrical repair shops, coal-burning power plants, chemical plants, etc. The air toxics producers are to be identified as major (large) or area (small) sources.

Once the categories of sources are listed, EPA will issue regulations. In some cases, EPA may have to specify exactly how to reduce pollutant releases, but wherever possible, companies will have flexibility to choose how they meet requirements. Sources are to use Maximum Available Control Technology (MACT) to reduce pollutant releases; this is a very high level of pollution control.

EPA must issue regulations for major sources first and must then issue regulations to reduce pollution from small sources, setting priorities for which small sources to tackle first, based on health and environmental hazards, production volume, etc.

If a company wishes to increase the amount of air toxics coming out of an operating plant, the company may choose to offset the increases so that total hazardous air pollutant releases from the plant do not go up. Otherwise, it may choose to install pollution controls to keep pollutants at the required level.

If a company reduces its releases of a hazardous air pollutant by about 90 percent before EPA regulates the chemical, the company will get extra time to finish cleaning up the remaining 10 percent. This early reduction program is expected to result in a speedy reduction of the levels of several important hazardous air pollutants.

Under the 1990 Clean Air Act, EPA is required to study whether and how to reduce hazardous air pollutants from small neighborhood polluters such as auto paint shops, print shops, etc. The agency also will have to look at air toxics pollution after the first round of regulations to see whether the remaining health hazards require further regulatory action.

Cars, trucks, buses, and other mobile sources release large amounts of hazardous air pollutants such as formaldehyde and benzene. Cleaner fuels and engines and making sure that pollution-control devices work should reduce hazardous air pollutants from mobile sources.

The Bhopal tragedy inspired the 1990 Clean Air Act requirement that factories and other businesses develop plans to prevent accidental releases of highly toxic chemicals. The Act establishes the Chemical Safety Board to investigate and report on accidental releases of hazardous air pollutants from industrial plants. The Chemical Safety Board will operate like the National Transportation Safety Board (NTSB), which investigates plane and train crashes.

Mobile Sources (Cars, Trucks, Buses, Off-Road Vehicles, Planes, Etc.). Each of today's cars produces 60 to 80 percent less pollution than cars in the 1960s. More people are using mass transit. Leaded gas is being phased out, resulting in dramatic declines in air levels of lead, a very toxic chemical. Despite this progress, most types of air pollution from mobile sources have not improved significantly.

At Present, the United States. Motor vehicles are responsible for up to half the smog-forming volatile organic compounds (VOCs) and nitrogen oxides (NO_x). Motor vehicles release more than 50 percent of the hazardous air pollutants. Motor vehicles release up to 90 percent of the carbon monoxide found in urban air.

What Went Wrong? More people are driving more cars more miles on more trips. In 1970, Americans traveled 1 trillion miles in motor vehicles, and we are expected to drive more than 4 trillion miles each year by 2001. Many people live far from where they work; in many areas, buses, subways, and commuter trains are not available. Also, most people still drive to work alone, even when van pools, high-occupancy vehicle (HOV) lanes, and other alternatives to one-person-per-car commuting are available. Buses and trucks, which produce a lot of pollution, have not had to clean up their engines and exhaust systems as much as cars. Automobile fuel has become more polluting. As lead was being phased out, gasoline refiners changed gasoline formulas to make up for octane loss, and the changes made gasoline more likely to release smog-forming VOC vapors into the air. Although cars have had pollution-control devices since the 1970s, the devices only had to work for 50,000 miles, whereas a car in the United States is usually driven for 100,000 miles.

The 1990 Clean Air Act takes a comprehensive approach to reducing pollution from motor vehicles. The Act provides for cleaning up fuels, cars, trucks, buses, and other motor vehicles. Auto inspection provisions were included in the law to make sure cars are well maintained. The 1990 law also includes transportation policy changes that can help reduce air pollution.

Cleaner Fuels. It will be very difficult to obtain a significant reduction in pollution from motor vehicles unless fuels are cleaned up. The 1990 Clean Air Act will clean up fuels. The phaseout of lead from gasoline was to be completed by January 1, 1996. Diesel fuel refining must be changed so that the fuels contain less sulfur, which contributes to acid rain and smog.

Gasoline refiners will have to reformulate gasoline sold in the smoggiest areas; this gasoline will contain fewer VOCs such as benzene (which is also a hazardous air pollutant that causes cancer and aplastic anemia, a potentially fatal blood disease). Other polluted areas can ask EPA to include them in the reformulated gasoline marketing program. In some areas, wintertime carbon monoxide (CO) pollution is caused by people starting their cars. In these areas, refiners will have to sell oxyfuel, gasoline with oxygen added to make the fuel burn more efficiently, thereby reducing carbon monoxide release.

All gasoline will have to contain detergents, which, by preventing buildup of engine deposits, keep engines working smoothly and burning fuel cleanly. Low-VOC oxyfuel and detergent gasoline are already sold in several parts of the country. Gas stations in smoggy areas will install vapor-recovery nozzles on gas pumps that will cut down on vapor release. The 1990 Clean Air Act also encourages development and sale of alternative fuels such as alcohols, liquefied petroleum gas (LPG), and natural gas.

Cleaner Cars. The 1990 Clean Air Act requires cars to have under-the-hood systems and dashboard warning lights that check whether pollution-control devices are working properly. Pollution-control devices must work for 100,000 miles, rather than the current 50,000 miles. Auto makers must build some cars that use clean fuels, including alcohol, and that release less pollution from the tailpipe through advanced engine design. Electric cars, which are low-pollution vehicles, also will be built. Since California, especially southern California, has one of the worst smog problems, manufacturers will first sell clean-fuel cars in a pilot project in California. By 1999, at least 500,000 of these clean-fuel cars were to be manufactured for sale in California each year. Other states can require that cars meeting the California standards be sold in their states.

Many companies and government agencies have fleets of cars. Fleet owners in very smoggy areas had to buy the new cleaner cars starting in the late 1990s.

Inspection and Maintenance (I/M) Programs. Under the 1990 Clean Air Act, auto manufacturers will build cleaner cars, and cars will use cleaner fuels. However, to get air pollution down and keep it down, a third program is needed: vehicle inspection and maintenance (I/M), which makes sure cars are being maintained adequately to keep pollution emissions (releases) low. The 1990 Clean Air Act includes very specific requirements for inspection and maintenance programs.

Before the 1990 Clean Air Act went into effect, 70 U.S. cities and several states already had auto emission inspection programs. The 1990 law requires inspection and maintenance programs in more areas: 40 metropolitan areas, including many in the northeastern United States, are required to start emission inspection and maintenance programs.

Some areas that already have inspection and maintenance programs are required to enhance (improve) their emission inspection machines and procedures. Enhanced inspection and maintenance machines and procedures will give a better measurement of the pollution a car releases when it is actually being driven, rather than just sitting parked at the inspection station. Enhanced inspection and maintenance programs may result in changes in where cars are inspected in a local area. Since the enhanced emission inspection and maintenance machines are expensive, some of the private stations now conducting inspection and maintenance programs may not want to buy the enhanced machinery. But the added expense for the new machinery will be more than made up for by air pollution reductions; emission inspection and maintenance programs are expected to have a big payoff in reducing air pollution from cars.

Cleaner Trucks and Buses. Starting with model year 1994, engines for new big diesel trucks will have to be built to reduce particulate (dust, soot) releases by 90 percent. Buses will have to reduce particulate releases even more than trucks. To reduce pollution, companies and governments that own buses or trucks will need to buy new clean models. Small trucks will be cleaned up by requirements similar to those for cars.

Nonroad Vehicles. Locomotives, construction equipment, and even riding lawn mowers may be regulated under the 1990 Clean Air Act. Air pollution from locomotives must be reduced. For the other nonroad vehicles, EPA must issue regulations if a study shows that controls would help cut pollution.

Transportation Policies. The smoggiest metropolitan areas will have to change their transportation policies to discourage unnecessary auto use and to encourage efficient commuting (van pools, HOV lanes, etc.). States carrying out the 1990 Clean Air Act may add surcharges to parking fees.

Acid Rain.   Although incidents of acid rain have been reported extensively, acid snow, acid fog or mist, acid gas, and acid dust have not received much attention. All these "acids" are related air pollutants and can harm public health, cause hazy skies, and damage the environment and property. The 1990 Clean Air Act includes an innovative program to reduce acid air pollutants (all referred to here as *acid rain*).

The acid rain that has received the most attention is caused mainly by pollutants from big coal-burning power plants in the Midwest. These plants burn midwestern and Appalachian coal, some of which contains a lot of sulfur compared with western coal. Sulfur in coal becomes sulfur dioxide (SO_2) when coal is burned. Big power plants burn large quantities of coal, so they release large amounts of sulfur dioxide, as well as NO_x (nitrogen oxides). These are acid chemicals related to two strong acids: sulfuric acid and nitric acid.

The sulfur dioxide and nitrogen oxides released from the midwestern power plants rise high into the air and are carried by winds toward the East Coast of the United States and Canada. When winds blow the acid chemicals into areas where there is wet weather, the acids become part of the rain, snow, or fog. In areas where the weather is dry, the acid chemicals may fall to earth in gases or dusts.

Lakes and streams are normally slightly acid, but acid rain can make them very acid. Very acid conditions can damage plant and animal life. Acid lakes and streams have been found all over the country. For instance, lakes in Acadia National Park on Maine's Mt. Desert Island have been very acidic due to pollution from the Midwest and the East Coast. Streams in Maryland and West Virginia, lakes in the Upper Peninsula of Michigan, and lakes and streams in Florida also have been affected by acid rain. Heavy rainstorms and melting snow can cause temporary increases in acidity in lakes and streams in the eastern and western United States. These temporary increases may last for days or even weeks.

Acid rain has damaged trees in the mountains of Vermont and other states. Red spruce trees at high altitudes appear to be especially sensitive to acid rain. The pollutants that cause acid rain can make the air hazy or foggy; this has occurred in the eastern United States, including some mountain areas popular with vacationers, such as the Great Smokies.

Acid rain does more than environmental damage; it can damage health and property as well. Acid air pollution has been linked to breathing and lung problems in children and in people who have asthma. Even healthy people can have their lungs damaged by acid air pollutants. Acid air pollution also can eat away stone buildings and statues.

Health, environmental, and property damage also can occur when sulfur dioxide pollutes areas close to its source. Sulfur dioxide pollution has been found in towns where paper and wood pulp are processed and in areas close to some power plants. The 1990 Clean Air Act's sulfur dioxide reduction program will complement health-based sulfur dioxide pollution limits already in place to protect the public and the environment from both nearby and distant sources of sulfur dioxide.

The Act takes a new nationwide approach to the acid rain problem. The law sets up a market-based system designed to lower sulfur dioxide pollution levels. Beginning in the year 2000, annual releases of sulfur dioxide will be about 40 percent lower than the 1980 levels. Reducing sulfur dioxide releases should cause a major reduction in acid rain.

Phase I of the acid rain reduction program went into effect in 1995. Big coal-burning boilers in 110 power plants in 21 midwestern, Appalachian, southeastern, and northeastern states will have to reduce releases of sulfur dioxide. In 2000, Phase II of the acid rain program went into effect, further reducing the sulfur dioxide releases from the big coal-burning power plants and covering other smaller polluters. Total sulfur dioxide releases for the country's power plants will be permanently limited to the level set by the Clean Air Act for the year 2000.

Reductions in sulfur dioxide releases will be obtained through a program of emission (release) allowances. EPA will issue allowances to power plants covered by the acid rain program; each allowance is worth 1 ton of sulfur dioxide released from the smokestack. To obtain reductions in sulfur dioxide pollution, allowances are set below the current level of sulfur dioxide releases. Plants may only release as much sulfur dioxide as they have allowances. If a plant expects to release more sulfur dioxide than it has allowances, it has to get more allowances, perhaps by buying them from another power plant that has reduced its sulfur dioxide releases below its number of allowances and therefore has allowances to sell or trade. Allowances also can be bought and sold by "middlemen," such as brokers, or by anyone who wants to take part in the allowances market. Allowances can be traded and sold nationwide. There are stiff penalties for plants that release more pollutants than their allowances cover.

The acid rain program provides bonus allowances to power plants for (among other things) installing clean-coal technology that reduces sulfur dioxide releases, using renewable energy sources (solar, wind, etc.), or encouraging energy conservation by customers so that less power needs to be produced.

All power plants under the acid rain program will have to install continuous emission-monitoring systems (CEMS), machines that keep track of how much sulfur dioxide and nitrogen oxides the plant is releasing. A power plant's program for meeting its sulfur dioxide and nitrogen oxide limit will appear on the plant's permit, which will be filed with the state and EPA.

To cut down on nitrogen oxide pollution, EPA will require power plants to reduce their nitrogen oxide releases and will require reductions in nitrogen oxide releases from new cars. Reducing nitrogen oxide releases will reduce both acid rain and smog formation. The flexible market-based acid rain reduction program is expected to be a model for pollution-control efforts in the United States and other countries.

Repairing the Ozone Layer.   Scientists have found "holes" in the ozone layer high above the earth. The 1990 Clean Air Act has provisions for fixing the holes, but repairs will take a long time.

Ozone in the stratosphere, a layer of the atmosphere 9 to 31 miles above the earth, serves as a protective shield, filtering out harmful sun rays, including a type of sunlight called *ultraviolet B*. Exposure to ultraviolet B has been linked to development of cataracts (eye damage) and skin cancer.

In the mid-1970s, scientists suggested that chlorofluorocarbons (CFCs) could destroy stratospheric ozone. CFCs were widely used then as aerosol propellants in consumer products such as hairsprays and deodorants and for many uses in industry. Because of concern about the possible effects of CFCs on the ozone layer, in 1978 the U.S. government banned CFCs as propellants in aerosol cans.

Since the aerosol ban, scientists have been measuring the ozone layer. A few years ago, an ozone hole was found above Antarctica, including the area of the South Pole. This hole, which has been appearing each year during the Antarctic winter (our summer), is bigger than the continental United States. More recently, ozone thinning has been found in the stratosphere above the northern half of the United States; the hole extends over Canada and up into the Arctic regions (the area of the North Pole). The hole was first found only in winter and spring but more recently has continued into summer. Between 1978 and 1991, there was a 4 to 5 percent loss of ozone in the stratosphere over the United States; this is a significant loss of ozone. Ozone holes also have been found over northern Europe.

What could a thinned-out ozone layer do to people's lives? There could be more skin cancers and cataracts. Scientists are looking into possible harm to agriculture, and there is already some evidence of damage to plant life in Antarctic seas.

Evidence that the ozone layer is dwindling led 93 nations, including the major industrialized nations, to agree to cooperate in reducing production and use of chemicals that destroy the ozone layer. As it became clear that the ozone layer was thinning even more quickly than first thought, the agreement was revised to speed up the phase-out of ozone-destroying chemicals.

Unfortunately, it will be a long time before we see the ozone layer repaired. Because of the ozone-destroying chemicals already in the stratosphere and those which will arrive within the next few years, ozone destruction likely will continue for another 20 years.

The 1990 Clean Air Act sets a schedule for ending production of chemicals that destroy stratospheric ozone. Chemicals that cause the most damage will be phased out first. The phase-out schedule can be speeded up if an earlier end to production of ozone-destroying substances is needed to protect the ozone layer. Table A2.1 listing ozone-destroying chemicals includes "speeded up" phase-out dates that were proposed by EPA in early 1993.

CFCs, halons, HCFCs (hydrochlorofluorocarbons), and other ozone-destroying chemicals were listed by Congress in the 1990 Clean Air Act and must be phased out. The Act also lets EPA list other chemicals that destroy ozone. HCFCs and halons are chemicals much like CFCs. HCFCs may be somewhat less harmful to the ozone layer than are CFCs.

EPA issues allowances to control manufacture of chemicals being phased out. Companies also can sell unused allowances to companies still making the chemicals or can use the allowances, within certain limits, to make a different, less ozone-destroying chemical on the phase-out list.

TABLE A2.1 Ozone-Destroying Chemicals

Name	Use	When U.S. Production ends*
CFCs (chlorofluoro-carbons)	Solvents, aerosol sprays (most spray can uses banned in 1970s), foaming agents in plastic manufacture	January 1, 1996
Halons	Fire extinguishers	January 1, 1994
Carbon tetrachloride	Solvents, chemical manufacture; carbon tetrachloride causes cancer in animals	January 1, 1996
Methyl chloroform (1,1,1-trichloroethene)	Very widely used solvent; in many workplace and consumer solvents, included in auto repair and maintenance products	January 1, 1996
HCFCs (hydro-CFCs)	CFC substitutes, chemicals slightly different from CFCs	January 1, 2003†

*The 1990 Clean Air Act includes a schedule for ending U.S. production of ozone-destroying chemicals and provisions for speeding up the phase-out schedule if that is necessary. The dates in this table are "speeded up" dates proposed by EPA in early 1993.

†Production of the HCFC with the most severe ozone-destroying effects will end by January 1, 2003. Production of the rest of the HCFCs will end by January 1, 2030.

In addition to requiring the phasing out of production of ozone-destroying chemicals, the Clean Air Act takes other steps to protect the ozone layer. The law requires recycling of CFCs and labeling of products containing ozone-destroying chemicals. The 1990 Clean Air Act also encourages the development of "ozone friendly" substitutes for ozone-destroying chemicals.

CFCs from car air conditioners are the biggest single source of ozone-destroying chemicals. By the end of 1993, all car air conditioner systems must be serviced using equipment that recycles CFCs and prevents their release into the air. Larger auto service shops were required to start using this special equipment in January 1992. Only specially trained and certified repair persons will be allowed to buy the small cans of CFCs used in servicing auto air conditioners.

As CFCs and related chemicals are phased out, appliances and industrial processes that now use the chemicals will change. For example, industrial and home refrigerators will be changed to use refrigerants that do not destroy ozone. In the meantime, refrigerator servicing and disposal will have to be done in ways that do not release CFCs. Methyl chloroform, also called 1,1,1-trichloroethane, which will be phased out by 1996, is a very widely used solvent found in products such as automobile brake cleaners (often sold as aerosol sprays) and spot removers used to take greasy stains off fabrics. Replacing methyl chloroform in workplace and consumer products will lead to changes in many products and processes.

As substitutes are developed for ozone-destroying substances, before the chemicals can be produced and sold, EPA must determine that the replacements will be safe for health and the environment.

Consumer Products.   Hair sprays, paints, foam plastic products (such as disposable styrofoam coffee cups), and carburetor and choke sprays all are consumer products that may be regulated under the 1990 Clean Air Act. These products will be regulated to reduce releases of smog-forming VOCs and ozone-destroying chemicals (CFCs and related chemicals).

By May 1993, consumer products containing CFCs and related chemicals identified in the 1990 Clean Air Act as most damaging to the ozone layer must have this label:

WARNING: Contains or manufactured with [name of chemical], a substance which harms public health and the environment by destroying ozone in the upper atmosphere.

All products containing less destructive ozone-destroying chemicals identified in the 1990 act must be labeled by 2015.

Consumers should be aware of product changes and any safety or health problems that may be caused by the new ozone-safe formulations. Material safety data sheets for the products should be read for health and safety information and information on how to use and dispose of the product.

Material safety data sheets are product safety information sheets prepared by manufacturers and marketers. These sheets can be obtained by requesting them from the manufacturer. Some stores, such as hardware stores, may have material safety data sheets on hand for products they sell. The 1990 Clean Air Act orders EPA to study VOC releases from consumer products and report to Congress by 1993 on whether these products should be regulated. If they are to be regulated, EPA is to list the consumer products that account for at least 80 percent of VOC releases and issue regulations for product categories, starting with the worst polluters. Labeling, repackaging, chemical formula changes, fees, or other procedures may be used to reduce VOC releases.

Home Woodstoves.   Woodstoves and fireplace inserts have become very popular in the past 20 years. Although these wood-burning heat suppliers are relatively cheap to operate, they have some disadvantages, including polluting the air. In some areas of the country, wintertime air pollution from wood smoke has become so bad that governments have had to curtail the use of woodstoves and fireplaces under certain weather and pollution conditions.

Wood smoke often contains a lot of particulates (dust, soot) and much higher levels of hazardous air pollutants, including some cancer-causing chemicals, than smoke from oil- or gas-fired furnaces. Steps to clean up wood smoke pollution have included redesigning the burning system in woodstoves; newer woodstoves put out much less pollution than older models.

Under the 1990 act, EPA has issued guidelines for reducing pollution from home wood-burning. These guidelines, which are not requirements, include design information for less polluting stoves and fireplaces.

Everyone in the United States has a role to play to make the Clean Air Act a success. One of the most important things Americans can do is to keep track of how the law is working. There are several ways the public will be able to tell how well the Clean Air Act is working.

EPA, state, regional, and local air pollution control agencies have to issue regulations (rules), give out permits, enforce the Act against violators, and do other things described in the Clean Air Act. Many groups with an interest in how the Clean Air Act works are watching EPA and the other air pollution control agencies. These groups include local and national business and trade organizations (from state associations of dry cleaners to the U.S. Chamber of Commerce), local community organizations (such as neighborhood associations), and local and national environmental and public health organizations (such as the Clean Air Network of the Natural Resources Defense Council and the American Lung Association). Newspapers, radio, and television will report on how the Act is being carried out, both nationally and locally. The public also can contact EPA and state, regional, or local air pollution control agencies to receive information directly on Clean Air Act activities.

The U.S. Congress monitors how federal agencies are carrying out the laws. By contacting state congressional representatives or senators the public can get more information on congressional hearings and reports on how EPA is carrying out the Clean Air Act. The public also can request reports from the U.S. General Accounting Office (GAO), the congressional investigative agency that reviews how EPA carries out the Clean Air Act.

The Common Air Pollutants

Ozone.   Ground-level ozone is the principal component of smog.

Source. Chemical reaction of pollutants; VOCs and NO_x.

Health effects. Breathing problems, reduced lung function, asthma, eye irritation, stuffy nose, reduced resistance to colds and other infections, may speed up aging of lung tissue.

Environmental effects. Ozone can damage plants and trees; smog can cause reduced visibility.

Property damage. Damages rubber, fabrics, etc.

VOCs.   Volatile organic compounds, smog-formers.

Source. VOCs are released from burning fuel (gasoline, oil, wood coal, natural gas, etc.), solvents, paints, glues, and other products used at work or at home. Cars are an important source of VOCs. VOCs include chemicals such as benzene, toluene, methylene chloride, and methyl chloroform.

Health effects. In addition to ozone (smog) effects, many VOCs can cause serious health problems such as cancer and other effects.

Environmental effects. In addition to ozone (smog) effects, some VOCs such as formaldehyde and ethylene may harm plants.

Nitrogen Dioxide. One of the NO_x compounds, a smog-forming chemical.

Source. Burning of gasoline, natural gas, coal, oil, etc. Cars are an important source of NO_2.

Health effects. Lung damage, illnesses of breathing passages and lungs (respiratory system).

Environmental effects. Nitrogen dioxide is an ingredient of acid rain (acid aerosols), which can damage trees and lakes. Acid aerosols can reduce visibility.

Property damage. Acid aerosols can eat away stone used on buildings, statues, monuments, etc.

Carbon Monoxide (CO)

Source. Burning of gasoline, natural gas, coal, oil, etc.

Health effects. Reduces ability of blood to bring oxygen to body cells and tissues; cells and tissues need oxygen to work. Carbon monoxide may be particularly hazardous to people who have heart or circulatory (blood vessel) problems and people who have damaged lungs or breathing passages.

Particulate Matter (PM-10). Dust, smoke, soot.

Source. Burning of wood, diesel, and other fuels; industrial plants; agriculture (plowing, burning off fields); unpaved roads.

Health effects. Nose and throat irritation, lung damage, bronchitis, early death.

Environmental effects. Particulates are the main source of haze that reduces visibility.

Property damage. Ashes, soot, smoke, and dusts can dirty and discolor structures and other property, including clothes and furniture.

Sulfur Dioxide (SO_2)

Source. Burning of coal and oil, especially high-sulfur coal from the eastern United States; industrial processes (paper, metals).

Health effects. Breathing problems; may cause permanent damage to lungs.

Environmental effects. SO_2 is an ingredient in acid rain (acid aerosols), which can damage trees and lakes. Acid aerosols also can reduce visibility.

Property damage. Acid aerosols can eat away stone used in buildings, statues, monuments, etc.

Lead

Source. Leaded gasoline (being phased out), paint (houses, cars), smelters (metal refineries), manufacture of lead storage batteries.

Health effects. Brain and other nervous system damage; children are at special risk. Some lead-containing chemicals cause cancer in animals. Lead causes digestive and other health problems.

Environmental effects. Lead can harm wildlife.

A2.2.4   The Clean Water Act (CWA); 33 USC s/s 121 et seq. (1977)

The Clean Water Act is a 1977 amendment to the Federal Water Pollution Control Act of 1972, which set the basic structure for regulating discharges of pollutants to waters of the United States. The law gave EPA the authority to set effluent standards on an industry basis (technology-based) and continued the requirements to set water quality standards for all contaminants in surface waters. The CWA makes it unlawful for any person to discharge any pollutant from a point source into navigable waters unless a permit (NPDES) is obtained under the Act.

The 1977 amendments focused on toxic pollutants. In 1987, the CWA was reauthorized and again focused on toxic substances, authorized citizen suit provisions, and funded sewage treatment plants (POTWs) under the Construction Grants Program.

The CWA provides for the delegation by EPA of many permitting, administrative, and enforcement aspects of the law to state governments. In states with the authority to implement CWA programs, EPA still retains oversight responsibilities.

A2.2.5   Comprehensive Environmental Response, Compensation and Liability Act (CERCLA or Superfund) 42 USC s/s 9601 et seq. (1980)

The Comprehensive Environmental Response, Compensation and Liability Act (CERCLA), commonly known as *Superfund,* was enacted by Congress on December 11, 1980. This law created a tax on the chemical and petroleum industries and provided broad federal authority to respond directly to releases or threatened releases of hazardous substances that may endanger public health or the environment. Over 5 years, $1.6 billion was collected, and the tax went to a trust fund for cleaning up abandoned or uncontrolled hazardous waste sites. CERCLA

Established prohibitions and requirements concerning closed and abandoned hazardous waste sites.

Provided for liability of persons responsible for releases of hazardous waste at these sites.

Established a trust fund to provide for cleanup when no responsible party could be identified.

The law authorizes two kinds of response actions:

Short-term removals. Actions may be taken to address releases or threatened releases requiring prompt response.

Long-term remedial response actions. Actions that permanently and significantly reduce the dangers associated with releases or threats of releases of hazardous substances that are serious but not immediately life-threatening. These actions can be conducted only at sites listed on EPA's National Priorities List (NPL).

CERCLA also enabled revision of the National Contingency Plan (NCP). The NCP provided the guidelines and procedures needed to respond to releases and threatened releases of hazardous substances, pollutants, or contaminants. The NCP also established the NPL.

CERCLA was amended by the Superfund Amendments and Reauthorization Act (SARA) on October 17, 1986.

A2.2.6 The Emergency Planning and Community Right-to-Know Act (EPCRA); 42 USC 11011 et seq. (1986)

Also known as Title III of SARA, EPCRA was enacted by Congress as the national legislation on community safety. This law was designated to help local communities protect public health, safety, and the environment from chemical hazards.

To implement EPCRA, Congress required each state to appoint a State Emergency Response Commission (SERC). The SERCs were required to divide their states into Emergency Planning Districts and to name a Local Emergency Planning Committee (LEPC) for each district. Broad representation by firefighters, health officials, government and media representatives, community groups, industrial facilities, and emergency managers ensures that all necessary elements of the planning process are represented.

A2.2.7 The Endangered Species Act (ESA); 7 USC 136; 16 USC 460 et seq. (1973)

The Endangered Species Act provides a program for the conservation of threatened and endangered plants and animals and the habitats in which they are found. The U.S. Fish and Wildlife Service (FWS) of the Department of the Interior maintains the list of 632 endangered species (326 are plants) and 190 threatened species (78 are plants).

Species include birds, insects, fish, reptiles, mammals, crustaceans, flowers, grasses, and trees. Anyone can petition FWS to include a species on this list. The law prohibits any action, administrative or real, that results in a "taking" of a listed species or adversely affects habitat. Likewise, import, export, interstate, and foreign commerce of listed species are all prohibited.

EPA's decision to register a pesticide is based in part on the risk of adverse effects on endangered species as well as environmental fate (how a pesticide will affect habitat). Under FIFRA (see below), EPA can issue emergency suspensions to the use of certain pesticides to cancel or restrict their use if an endangered species will be adversely affected. Under a new program, EPA, FWS, and the U.S. Department of Agriculture (USDA) are distributing hundreds of county bulletins that include habitat maps, pesticide use, and other actions required to protect listed species.

A2.2.8   Federal Insecticide, Fungicide and Rodenticide Act (FIFRA); 7 USC s/s 135 et seq. (1972)

The primary focus of FIFRA was to provide federal control of pesticide distribution, sale, and use. EPA was given authority under FIFRA not only to study the consequences of pesticide use but also to require users (farmers, utility companies, and others) to register when purchasing pesticides. Through later amendments to the law, users also must take examinations for certification as applicators of pesticides. All pesticides used in the United States must be registered (licensed) by EPA. Registration ensures that pesticides will be labeled properly and, if in accordance with specifications, will not cause unreasonable harm to the environment.

A2.2.9   Federal Food, Drug and Cosmetic Act (FFDCA) 21 USC 301 et seq., Sec. 321. (U.S. Code as of 01/23/00)

The Federal Food, Drug and Cosmetic Act (FFDCA) provides the definitions of *food* as: "(1) articles used for food or drink for man or other animals, (2) chewing gum, and (3) articles for components of any such article," and *drug* as "(A) articles recognized in the official United States Pharmacopoeia, official Homeopathic Pharmacopoeia of the United States, or official National Formulary, or any supplement to any of them; and (B) articles intended for use in the diagnosis, cure, mitigation, treatment, or prevention of disease in man or other animals; and (C) articles (other than food) intended to affect the structure or any function of the body of man or other animals; and (D) articles intended for use as a component of any article specified in clauses (A), (B), or (C)."

Subchapter IV of the Act provides the definitions and standards for food (Sec. 341) organized under

Sec. 342. Adulterated food
   (a) Poisonous, insanitary, etc. ingredients
   (b) Absence, substitution, or addition of constituents

Subchapter V of the Act provides the regulations on drugs and devices. Parts A and B of this subchapter are organized as follows:

Parts C, D, and E of Subchapter V provide the regulations for electronic product radiation control, dissemination of treatment information, and general provisions relating to drugs and devices, respectively. Other subchapters of the Act
cover cosmetics, general authority, imports and exports, and miscellaneous issues.

A2.2.10   Food Quality Protection Act (FQPA), Public Law 104-170, August 3, 1996

Background. In 1996, Congress unanimously passed landmark pesticide food safety legislation supported by the administration and a broad coalition of environmental, public health, agricultural, and industry groups. President Clinton promptly signed the bill on August 3, 1996, and the Food Quality Protection Act of 1996 became law (P.L. 104-170, formerly known as H.R. 1627).

EPA regulates pesticides under two major federal statutes. Under the Federal Insecticide, Fungicide and Rodenticide Act (FIFRA), EPA registers pesticides for use in the United States and prescribes labeling and other regulatory requirements to prevent unreasonable adverse effects on health or the environment. Under the Federal Food, Drug and Cosmetic Act (FFDCA), EPA establishes tolerances (maximum legally permissible levels) for pesticide residues in food. Tolerances are enforced by the Department of Health and Human Services/Food and Drug Administration (HHS/FDA) for most foods, the U.S. Department of Agriculture/Food Safety and Inspection Service (USDA/FSIS) for meat, poultry, and some egg products, and the U.S. Department of Agriculture/Office of Pest Management Policy.

For over two decades, there have been efforts to update and resolve inconsistencies in the two major pesticide statutes, but consensus on necessary reforms remained elusive. The 1996 law represents a major breakthrough, amending both major pesticide laws to establish a more consistent, protective regulatory scheme, grounded in sound science. It mandates a single, health-based standard for all pesticides in all foods, provides special protections for infants and children, expedites approval of safer pesticides, creates incentives for the development and maintenance of effective crop protection tools for American farmers, and requires periodic reevaluation of pesticide registrations and tolerances to ensure that the scientific data supporting pesticide registrations will remain up to date in the future.

FQPA Amendments to FFDCA Provisions

Previous and Law Practice. EPA is currently addressing some of the high-priority issues identified in the 1993 National Academy of Sciences (NAS) report entitled, "Pesticides in the Diets of Infants and Children." The agency routinely assesses risks by age group, ethnicity, and region when setting tolerances. There was no explicit mandate to do so under previous law, however, and the data available to EPA have been criticized as outdated and inadequate. Lack of funding has prevented implementation of some NAS recommendations.

New Legislation. The new law explicitly requires EPA to address risks to infants and children and to publish a specific safety finding before a tolerance can be established. It also provides for an additional safety factor (tenfold, unless reliable data show that a different factor will be safe) to ensure that tolerances are safe for infants and children and requires collection of better data on food consumption patterns, pesticide residue levels, and pesticide use.

Implications. The potentially greater exposure and/or sensitivity of infants and children will be explicitly taken into account in all tolerance decisions. Placing these specific requirements in the statute will help EPA in its efforts to implement the NAS report and ensure that risks to infants and children are always considered in the future.

Issue: Consumer "Right to Know" Provision

Previous Law and Practice. No comparable law or practice at the federal level.

New Legislation. The new law requires EPA to publish a short pamphlet containing consumer-friendly information on the risks and benefits of pesticides, any tolerances that EPA has established based on benefits considerations, and recommendations for reducing exposure to pesticide residues and maintaining a healthy diet (including foods that could substitute for foods for which tolerances have been established based on benefits). This information would be distributed each year to "large retail grocers for public display (in a manner determined by the grocer)." In addition, petitions for tolerances must include informative summaries that can be published and made publicly available. The law also recognizes a state's right to require warnings or labeling of food that has been treated with pesticides, such as California's Proposition 65.

Implications. EPA must coordinate with USDA and HHS to accomplish this. While the agency has general information materials that describe how to reduce pesticide exposure, developing materials to reach consumers at the supermarket level is a new departure, particularly with respect to substitute foods. Publishing an informative summary of tolerance petitions will open up the tolerance-setting process to more informed public input.

Health Issue: Endocrine Disruptors (Estrogenic Substances)

Previous Law and Practice. No specific provision in law. EPA is in the process of updating its testing requirements for reproductive and developmental effects, which should provide better information about possible endocrine disruption effects. These tests do not provide information about mechanism, however, and EPA has not required testing specifically for endocrine disruption. EPA is working with others in the public and private sectors to develop appropriate testing strategies.

New Legislation. The new law requires the development and implementation of a comprehensive screening program for estrogenic and other endocrine effects within 3 years of enactment.

Implications. EPA must develop a screening program within 2 years of enactment, implement it within 3 years of enactment, and report to Congress within 4 years. This is a very ambitious schedule. Little is known about mechanisms of endocrine disruption and possible synergistic effects. This is a high priority for EPA.

Tolerances/Residues Issue: General Standards for Tolerances

Previous Law and Practice. Previous law generally required EPA to establish tolerances that will "protect the public health." With respect to chemicals that pose carcinogenic risks, EPA used a negligible-risk standard, except in cases where the Delaney clause of the FFDCA applied, as described below. For effects determined to have a "threshold," EPA used safety factors to ensure that lifetime exposure would not exceed a safe level.

New Legislation. The new law requires that tolerances be "safe," defined as "a reasonable certainty that no harm will result from aggregate exposure," including all exposure through the diet and other nonoccupational exposures, including drinking water, for which there is reliable information. It also distinguishes between "threshold" and "nonthreshold" effects, consistent with EPA practice.

Implications. The new law establishes a single, health-based standard for all pesticide residues in all types of food, replacing the sometimes conflicting standards in the old law. There are no differences in the standards applicable to tolerances set for raw and processed foods. Additional provisions ensure coordination with standards and actions under FIFRA for a more consistent regulatory scheme.

Issue: Resolution of the "Delaney Paradox"

Previous Law and Practice. If a pesticide that causes cancer in humans or laboratory animals is concentrated in ready-to-eat processed food at a level greater than the tolerance for the raw agricultural commodity, the Delaney clause of the FFDCA prohibited the setting of a tolerance. This had paradoxical effects in terms of food safety, since alternative pesticides could pose higher (noncancer) risks and EPA allowed the same pesticide in other foods based on a determination that the risk was negligible.

New Legislation. The new law provides that tolerances for pesticide residues in all types of food (raw or processed) will be set under the same provisions of law. The standards apply to all risks, not just cancer risks.

Implications. This legislation eliminates the Delaney paradox. The Delaney clause no longer applies to any tolerances set for pesticide residues in food. Rather, the EPA must determine that tolerances are "safe," defined as "a reasonable certainty that no harm will result from aggregate exposure" to the pesticide. EPA and others will be able to devote resources that have been consumed by Delaney-related activities to higher-priority public health and environmental protection issues.

Issue: Consideration of Pesticide Benefits

Previous Law and Practice. Previous law required EPA to set tolerances "to protect the public health" and to give appropriate consideration "to the necessity for the production of an adequate, wholesome and economical food supply." In practice, economic considerations have not driven tolerance decisions or been the basis for granting tolerances that allow unsafe pesticide residues in food.

New Legislation. In certain narrow circumstances, the new law allows tolerances to remain in effect that would not otherwise meet the safety standard, based on the benefits afforded by the pesticide. Pesticide residues would only be "eligible" for such tolerances if use of the pesticide prevents even greater health risks to consumers or the lack of the pesticide would result in "a significant disruption in domestic production of an adequate, wholesome and economical food supply." Tolerances based on benefits considerations would be subject to a number of limitations on risk and more frequent reassessment than other tolerances. All tolerances would have to be consistent with the special provisions for infants and children.

Implications. This provision narrows the range of circumstances in which benefits may be considered and places limits on the maximum level of risk that could be justified by benefits considerations. It also would apply only to "non-threshold" risks posed by pesticides, e.g., carcinogenic effects for which conservative quantitative risk assessment is appropriate.

Issue: Other Factors to be Considered in Setting Tolerances

Previous Law and Practice. EPA already considers many of the factors in the new law, although they were not all required under the previous law.

New Legislation. The new law requires EPA to consider the validity, completeness, and reliability of available study data; the nature of potential toxic effects and available information on the relationship of study results to human risk; dietary consumption patterns and variations in the sensitivities of major identifiable subpopulations; cumulative and aggregate (dietary and nondietary) effects of exposure to the pesticide and other substances with common mechanisms of toxicity; effects on the endocrine system; and scientifically recognized appropriate safety factors.

These considerations are in addition to the special provisions for infants and children. In assessing potential risks, EPA also may consider exposure to actual residues expected on foods (which are often far lower than tolerances) and the percentage of a crop treated with the pesticide, but these assessments must be reevaluated periodically to ensure that they are still valid. EPA is given new authority to require data under FFDCA if the data cannot otherwise be obtained under FIFRA or the Toxic Substances Control Act. Finally, there must be a practical method for detecting residues in food before a tolerance can be granted.

Implications. As scientific understanding of potential cumulative and aggregate effects advances, it is likely that additional data will be required for EPA decisions, along with more information on subpopulation exposure and risk. In most cases, EPA will be able to use existing FIFRA authority to require this information. The additional data will enhance the scientific basis and protectiveness of pesticide regulations.

Issue: Reevaluation of Existing Tolerances

Previous Practice. EPA has been reassessing tolerances in connection with its ongoing reregistration review of chemicals first registered before November 1984. Pesticides approved after that were not subject to reregistration requirements.

New Legislation. While transitional provisions maintain existing tolerances in place on enactment, the new law requires review of all tolerances on the following schedule: 33 percent within 3 years, 66 percent within 6 years, and 100 percent within 10 years. Priority will be given to pesticides that may pose the greatest risk to public health.

Implications.   All tolerances will be required to meet the new safety standards, which should increase assurance that they are protective of all American consumers, including infants and children. The magnitude of this task is considerable; well over 9000 tolerances are currently in place. EPA will coordinate review schedules under FFDCA and FIFRA to the maximum extent possible.

Issue: International Standards for Pesticide Residue Levels

Previous Law and Practice.   As a matter of policy, EPA considered international standards for maximum residue levels (MRLs) established by the Codex Alimentarius Commission as part of its reregistration tolerance reassessments for chemicals first registered before November 1984. There is no presumption in favor of accepting international MRLs. Under international trade agreements, the United States must be able to explain differences, based on scientific evidence, as necessary to achieve the level of protection the United States has chosen to provide American consumers in accordance with existing laws and standards that apply to both domestically produced and imported foods.

New Legislation.   The new law requires EPA to publish a notice for public comment whenever the agency proposes a tolerance that differs from an established Codex MRL.

Implications.   This requirement furthers the goal of international harmonization of pesticide residue limits to the extent that international MRLs meet U.S. food safety standards.

Issue: National Uniformity of Tolerances

Previous Law and Practice.   Previous law allowed states to set tolerances that were stricter than EPA tolerances, although in practice this was done rarely.

New Legislation.   Generally, the new law preempts states from establishing tolerances that differ from EPA tolerances first established or reassessed after April 25, 1985. States may petition EPA for exemptions to this provision if there are compelling local conditions that justify the exemption.

Implications.   As a practical matter, states have rarely set tolerances that differ from EPA's, and the protective safety standards in the new law probably will decrease the incentive to do so even further. States will continue to have authority to establish tolerances stricter than those EPA may set based on benefits considerations and to require warning or other statements about the presence of pesticide residues (such as those required under California's Proposition 65).

Issue: Residue Monitoring and Civil Penalties

Previous Law and Practice.   No comparable provisions. FDA and USDA monitor pesticide residues as resources permit.

New Legislation.   The new law provides an additional authorization of $12 million for increased FDA monitoring in fiscal years 1997–1999. It also establishes substantial civil penalties for introducing foods with violative pesticide residues into interstate commerce. The penalties do not apply to growers.

Implications. If Congress were to appropriate additional funds under this authority, FDA would be able to increase monitoring. Penalty provisions should be a deterrent to violations.

FQPA Amendments to FIFRA Provisions—Issue: Emergency Suspension Authority

Previous Law and Practice. Under previous law, EPA could not suspend a pesticide's registration unless a proposed notice of intent to cancel the registration had been issued or was issued simultaneously. This could delay suspensions when there is an emergency, e.g., an imminent threat to public health or the environment.

New Legislation. The new law allows EPA to suspend a pesticide registration immediately. A notice of intent to cancel must be issued within 90 days, or the emergency suspension would expire.

Implications. This change will allow EPA to move quickly in situations that warrant immediate and decisive action to prevent serious risks to human health and the environment while preserving the rights of registrants in the cancellation process.

Issue: Extension of Reregistration Fee Authority

Previous Law and Practice. EPA had authority to collect fees to implement the ongoing reregistration program (covering pesticides first registered before November 1984) until September 30, 1997.

New Legislation. The new law extends fee collection authority through September 30, 2001. It provides for the collection of $14 million per year to support the current reregistration program and the expedited processing of applications for substantially similar "me too" pesticides, with an additional $2 million per year to be collected in 1998, 1999, and 2000. It also requires an annual independent audit to ensure that performance goals are met and specifies that tolerance fees will be available to carry out the requirements of the new law.

Implications. If fee authority had been allowed to expire in 1997, EPA would have completed only approximately one-half of its ongoing reregistration program. Many of the pesticides for which review would have been delayed are chemicals used on foods most often eaten by infants and children. The extension of fee authority will help EPA review these pesticides more quickly and keep the current reregistration program on track.

Issue: Minor-Use Pesticides (Including Public Health Uses)

Previous Law and Practice. Minor uses of pesticides are generally defined as uses for which pesticide product sales do not justify the costs of developing and maintaining EPA registrations. In the aggregate, such "minor" crops are very important to a healthy diet and include many fruits and vegetables. Previous law and practice included notice requirements and other efforts to facilitate support of minor uses. For agricultural pesticides, the Interregional Research Project No. 4

(IR-4), administered by USDA in cooperation with state land grant universities, has been working to develop needed data. There was no corresponding program for public health uses, e.g., for pesticides used to control vector-borne diseases.

New Legislation. The new law enhances incentives for the development and maintenance of minor use registrations in a number of ways. These special provisions do not apply, however, if the minor use may pose unreasonable risks or the lack of data would significantly delay EPA decisions. The legislation also establishes a USDA revolving grant program and a program for support of public health pesticides, analogous to the IR-4 program for agricultural uses, to be implemented jointly by the Public Health Service and EPA. The law encourages minor-use registrations through extensions for submitting pesticides residue data, extensions for exclusive use of data, flexibility to waive certain data requirements, and requiring EPA to expedite review of minor-use applications. These incentives are coupled with safeguards to protect the environment.

Implications. While maintaining safeguards for public health and environmental protection, these provisions will enhance efforts to ensure that effective pest-control products are available to growers and for public health purposes.

Issue: Periodic Review of Pesticide Registrations

Previous Law and Practice. Under amendments to FIFRA enacted in 1988, EPA is in the process of conducting reregistration reviews for all pesticides first registered before November 1984 and their associated tolerances, although tolerance reviews were not specifically required by the law. This review would bring the science base supporting registrations and tolerances up to current standards, but on a one-time-only basis. No further periodic reviews were required.

New Legislation. In addition to requiring tolerance reassessments, the new law requires EPA to establish a system for periodic review of all pesticide registrations, aimed at updating them on a 15-year cycle. If new data are needed for these reviews, or for any other review, EPA may require them at any time under FIFRA's "data call-in" authority in Sec. 3(c)(2)(B).

Implications. The goal of this requirement is to ensure that all pesticides continue to meet up-to-date standards for safety testing and public health and environmental protection. EPA retains authority to require data and take action if needed in the interim, but at a minimum, registrations should be updated on a 15-year cycle. Although other provisions of the new law provide for continuing fees to support the reregistration effort through 2001, additional resources may be needed to sustain periodic review efforts beyond that date.

Issue: Review of Antimicrobial Pesticides

Previous Law and Practice. There were no special provisions for antimicrobial pesticides under previous law. EPA and FDA shared responsibilities for some products, and EPA reviewed applications consistent with agency priorities, resources, and the timing of submissions.

New Legislation. The new law amends the definition of pesticide under FIFRA to exclude liquid chemical sterilants, which are to be regulated exclusively by FDA. It also reforms the antimicrobial registration process, with the goal of achieving significantly shorter EPA review times.

Implications. While the review times set out in the new law will be difficult to achieve, EPA believes that they are attainable and will strive to develop a program to meet them. Ending dual regulation and establishing more specific regulatory requirements will benefit manufacturers.

Issue: Expediting Review of Safer Pesticides

Previous Law and Practice. While there were no comparable provisions in previous law, EPA has established policies that give priority to applications for pesticides that appear to meet reduced-risk criteria.

New Legislation. The new law requires EPA to develop criteria for reduced-risk pesticides and expedite review of applications that reasonably appear to meet the criteria.

Implications. These provisions give a statutory mandate for expedited consideration of applications for safer pesticides, thereby enhancing public health and environmental protection. EPA will develop formal criteria and procedures for expedited reviews.

Issue: Maintenance Applicator and Service Technician Training

Previous Law and Practice. No specific comparable provisions.

New Legislation. The new law creates a category of "Maintenance Applicators" and "Service Technicians" to include janitors, sanitation personnel, general maintenance personnel, and grounds maintenance personnel who use or supervise the use of structural or lawn pest-control agents (other than restricted-use pesticides). States are authorized to establish minimum training requirements for such individuals.

Implications. States have explicit authority to require training. EPA is authorized only to make certain that states understand these provisions on minimum training requirements.

A2.2.11 The Freedom of Information Act (FOIA); USC s/s 552 (1966)

The Freedom of Information Act provides specifically that "any person" can make requests for government information. Citizens who make requests are not required to identify themselves or explain why they want the information they have requested. The position of Congress in passing FOIA was that the workings of government are "for and by the people" and that the benefits of government information should be made available to everyone.

All branches of the federal government must adhere to the provisions of FOIA, with certain restrictions for work in progress (early drafts), enforcement confidential information, classified documents, and national security information.

A2.2.12 The Occupational Safety and Health Act (OSHA); 29 USC 651 et seq. (1970)

Congress passed the Occupational and Safety Health Act to ensure worker and workplace safety. Its goal was to make sure employers provide their workers a place of employment free from recognized hazards to safety and health, such as exposure to toxic chemicals, excessive noise levels, mechanical dangers, heat or cold stress, or unsanitary conditions.

In order to establish standards for workplace health and safety, the Act also created the National Institute for Occupational Safety and Health (NIOSH) as the research institution for the Occupational Safety and Health Administration (OSHA). OSHA is a division of the U.S. Department of Labor that oversees the administration of the Act and enforces standards in all 50 states.

A2.2.13 The Oil Pollution Act of 1990 (OPA); 33 USC 2702 to 2761

The Oil Pollution Act (OPA) of 1990 streamlined and strengthened EPA's ability to prevent and respond to catastrophic oil spills. A trust fund financed by a tax on oil is available to clean up spills when the responsible party is incapable or unwilling to do so. The OPA requires oil storage facilities and vessels to submit to the federal government plans detailing how they will respond to large discharges. EPA has published regulations for above-ground storage facilities; the Coast Guard has done so for oil tankers. The OPA also requires the development of area contingency plans to prepare and plan for oil spill response on a regional scale.

A2.2.14 The Pollution Prevention Act (PPA); 42 USC 13101 and 13102, s/s et seq. (1990)

The Pollution Prevention Act focused industry, government, and public attention on reducing the amount of pollution through cost-effective changes in production, operation, and raw materials use. Opportunities for source reduction often are not realized because of existing regulations and the industrial resources required for compliance focus on treatment and disposal. Source reduction is fundamentally different and more desirable than waste management or pollution control.

Pollution prevention also includes other practices that increase efficiency in the use of energy, water, or other natural resources and protect our resource base through conservation. Practices include recycling, source reduction, and sustainable agriculture.

A2.2.15 The Resource Conservation and Recovery Act (RCRA); 42 USC s/s 321 et seq. (1976)

RCRA gave EPA the authority to control hazardous waste from the "cradle to grave." This includes the generation, transportation, treatment, storage, and disposal of hazardous waste. RCRA also set forth a framework for the management of nonhazardous wastes.

The 1986 amendments to RCRA enabled EPA to address environmental problems that could result from underground tanks storing petroleum and other hazardous substances. RCRA focuses only on active and future facilities and does not address abandoned or historical sites (see CERCLA).

The Federal Hazardous and Solid Waste Amendments (HSWA) are the 1984 amendments to RCRA that required phasing out land disposal of hazardous waste. Some of the other mandates of this strict law include increased enforcement authority for EPA, more stringent hazardous waste management standards, and a comprehensive underground storage tank program.

The Resource Conservation and Recovery Act contains, among others, sections on

Sec. 6901. Congressional findings on (a) solid waste, (b) environment and health, (c) materials, (d) energy
    (a) Congressional findings: used oil recycling
Sec. 6902. Objectives and national policy
Sec. 6904. Governmental cooperation
Sec. 6905. Application of chapter and integration with other acts
Sec. 6907. Solid waste management information and guidelines
Sec. 6908. Small town environmental planning
    (a) Agreements with Indian tribes
Sec. 6921. Identification and listing of hazardous waste
Sec. 6922. Standards applicable to generators of hazardous waste
Sec. 6923. Standards applicable to transporters of hazardous waste
Sec. 6924. Standards applicable to owners and operators of hazardous waste treatment, storage, and disposal facilities
Sec. 6925. Permits for treatment, storage, or disposal of hazardous waste
Sec. 6926. Authorized state hazardous waste programs
Sec. 6927. Inspections
Sec. 6928. Federal enforcement
Sec. 6933. Hazardous waste site inventory
Sec. 6934. Monitoring, analysis, and testing
Sec. 6935. Restrictions on recycled oil
Sec. 6937. Inventory of federal agency hazardous waste facilities
Sec. 6938. Export of hazardous wastes
Sec. 6939. Domestic sewage
    (a) Exposure information and health assessments
    (b) Interim control of hazardous waste injection
    (c) Mixed waste inventory reports and plan
Sec. 6944. Criteria for sanitary landfills; sanitary landfills required for all disposal
Sec. 6945. Upgrading of open dumps

Secs. 6951 to 6956. Duties of Secretary of Commerce in Resource and Recovery
Secs. 6961 to 6965. Federal responsibilities
Secs. 6991 to 6991i. Regulation of underground storage tanks
Secs. 6992 to 6992k. Listing, tracking, inspections, and enforcement for medical waste

A2.2.16 The Safe Drinking Water Act (SDWA); 42 USC s/s 300f et seq. (1974)

The Safe Drinking Water Act was established to protect the quality of drinking water in the United States. This law focuses on all waters actually or potentially designed for drinking use, whether from above-ground or underground sources.

The Act authorized EPA to establish safe standards of purity and required all owners or operators of public water systems to comply with primary (health-related) standards. State governments, which assume this power from EPA, also encourage attainment of secondary standards (nuisance-related).

The Safe Drinking Water Act defines, among others, the primary and secondary drinking water regulations and the public water supplies as follows:

(1) The term "primary drinking water regulation" means a regulation which
   (A) applies to public water systems;
   (B) specifies contaminants which, in the judgment of the Administrator, may have any adverse effect on the health of persons;
   (C) specifies for each such contaminant either

   (i) a maximum contaminant level, if, in the judgment of the Administrator, it is economically and technologically feasible to ascertain the level of such contaminant in water in public water systems, or
   (ii) if, in the judgment of the Administrator, it is not economically or technologically feasible to so ascertain the level of such contaminant, each treatment technique known to the Administrator which leads to a reduction in the level of such contaminant sufficient to satisfy the requirements of section 300g–1 of this title; and

   (D) contains criteria and procedures to assure a supply of drinking water which dependably complies with such maximum contaminant levels; including accepted methods for quality control and testing procedures to insure compliance with such levels and to insure proper operation and maintenance of the system, and requirements as to (i) the minimum quality of water which may be taken into the system and (ii) siting for new facilities for public water systems.

At any time after promulgation of a regulation referred to in this paragraph, the Administrator may add equally effective quality control and testing procedures by guidance published in the *Federal Register*. Such procedures shall be treated as an alternative for public water systems to the quality control and testing procedures listed in the regulation.

(2) The term "secondary drinking water regulation" which applies to public water systems and which specifies the maximum contaminant levels which, in the judgment of the Administrator, are requisite to protect the public welfare. Such regulations may apply to any contaminant in drinking water (A) which may adversely affect the odor or appearance of such water and consequently may cause a substantial number of the persons served by the public water system providing such water to discontinue its use, or (B) which may otherwise adversely affect the public welfare. Such regulations may vary accordingly to geographic and other circumstances.

(3) The term "maximum contaminant level" means the maximum permissible level of a contaminant in water which is delivered to any user of a public water system.

(4) Public water system.

   (A) In general. The term "public water system" means a system for the provision to the public of water for human consumption through pipes or other constructed conveyances, if such system has at least fifteen service connections or regularly serves at least twenty-five individuals. Such term includes (i) any collection, treatment, storage, and distribution facilities under control of the operator of such system and used primarily in connection with such system, and (ii) any collection or pretreatment storage facilities not under such control which are used primarily in connection with such system.

   (B) Connections.

      (i) In general. For purposes of subparagraph (A), a connection to a system that delivers water by a constructed conveyance other than a pipe shall not be considered a connection, if

         (I) the water is used exclusively for purposes other than residential uses (consisting of drinking, bathing, and cooking, or other similar uses);

         (II) the Administrator or the State (in the case of a State exercising primary enforcement responsibility for public water systems) determines that alternative water to achieve the equivalent level of public health protection provided by the applicable national primary drinking water regulation is provided for residential or similar uses for drinking and cooking; or

         (III) the Administrator or the State (in the case of a State exercising primary enforcement responsibility for public water systems) determines that the water provided for residential or similar uses for drinking, cooking, and bathing is centrally treated or treated at the point of entry by the provider, a pass-through entity, or the user to achieve the equivalent level of protection provided by the applicable national primary drinking water regulations.

(ii) Irrigation districts. An irrigation district in existence prior to May 18, 1994, that provides primarily agricultural service through a piped water system with only incidental residential or similar use shall not be considered to be a public water system if the system or the residential or similar users of the system comply with subclause (II) or (III) of clause (i).

(C) Transition period. A water supplier that would be a public water system only as a result of modifications made to this paragraph by the Safe Drinking Water Act Amendments of 1996 shall not be considered a public water system for purposes of the Act until the date that is two years after August 6,1996. If a water supplier does not serve 15 service connections [as defined in subparagraphs (A) and (B)] or 25 people at any time after the conclusion of the 2-year period, the water supplier shall not be considered a public water system.

(5) The term "supplier of water" means any person who owns or operates a public water system.

(6) The term "contaminant" means any physical, chemical, biological, or radiological substance or matter in water.

(7) The term "Administrator" means the Administrator of the Environmental Protection Agency.

(8) The term "Agency" means the Environmental Protection Agency.

(9) The term "Council" means the National Drinking Water Advisory Council established under section 300j-5 of this title.

(10) The term "municipality" means a city, town, or other public body created by or pursuant to State law, or an Indian Tribe.

(11) The term "Federal agency" means any department, agency, or instrumentality of the United States.

(12) The term "person" means an individual, corporation, company, association, partnership, State, municipality, or Federal agency (and includes officers, employees, and agents of any corporation, company, association, State, municipality, or Federal agency).

(13) (A) Except as provided in subparagraph (B), the term "State" includes, in addition to the several States, only the District of Columbia, Guam, the Commonwealth of Puerto Rico, the Northern Mariana Islands, the Virgin Islands, American Samoa, and the Trust Territory of the Pacific Islands.

(B) For purposes of section 300j-12 of this title, the term "State" means each of the 50 States, the District of Columbia, and the Commonwealth of Puerto Rico.

(14) The term "Indian Tribe" means any Indian tribe having a Federally recognized governing body carrying out substantial governmental duties and powers over any area. For purposes of section 300j-12 of this title, the term includes any Native village [as defined in section 1602(c) of title 43].

(15) Community water system. The term "community water system" means a public water system that

(A) serves at least 15 service connections used by year-round residents of the area served by the system; or

(B) regularly serves at least 25 year-round residents.

(16) Noncommunity water system. The term "noncommunity water system" means a public water system that is not a community water system.

Section 300j that the preceding act refers to provides assurances of availability of adequate supplies of chemicals necessary for treatment of water.

A2.2.17 The Superfund Amendments and Reauthorization Act (SARA); 42 USC 9601 et seq. (1986)

The Superfund Amendments and Reauthorization Act (SARA) amended the Comprehensive Environmental Response, Compensation and Liability Act (CERCLA) on October 17, 1986. SARA reflected EPA's experience in administering the complex Superfund program during its first 6 years and made several important changes and additions to the program.

SARA

Stressed the importance of permanent remedies and innovative treatment technologies in cleaning up hazardous waste sites.

Required Superfund actions to consider the standards and requirements found in other state and federal environmental laws and regulations.

Provided new enforcement authorities and settlement tools.

Increased state involvement in every phase of the Superfund program.

Increased the focus on human health problems posed by hazardous waste sites.

Encouraged greater citizen participation in making decisions on how sites should be cleaned up.

Increased the size of the trust fund to $8.5 billion.

SARA also required EPA to revise the hazard ranking system (HRS) to ensure that it accurately assessed the relative degree of risk to human health and the environment posed by uncontrolled hazardous waste sites that may be placed on the National Priorities List (NPL).

A2.2.18 The Toxic Substances Control Act (TSCA); 15 USC s/s 2601 et seq. (1976)

The Toxic Substances Control Act (TSCA) of 1976 was enacted by Congress to give EPA the ability to track the 75,000 industrial chemicals currently produced or imported into the United States. EPA repeatedly screens these chemicals and can require reporting or testing of those which may pose an environmental or human health hazard. EPA can ban the manufacture and import of those chemicals that pose an unreasonable risk.

Also, EPA has mechanisms in place to track the thousands of new chemicals that industry develops each year with either unknown or dangerous characteristics. EPA then can control these chemicals as necessary to protect human health and the environment. TSCA supplements other federal statutes, including the Clean Air Act and the Toxic Release Inventory under EPCRA.

The Toxic Substances Control Act contains, among others, articles on

Sec. 2601. Findings, policy, and intent of Congress

Sec. 2603. Testing of chemical substances and mixtures

Sec. 2604. Manufacturing and processing notices

Sec. 2605. Regulation of hazardous chemical substances and mixtures

Sec. 2606. Imminent hazards

Sec. 2607. Reporting and retention of information

Sec. 2608. Relationship to other federal laws

Sec. 2609. Research, development, collection, dissemination, and utilization of data

Sec. 2610. Inspections and subpoenas

Sec. 2611. Exports

Sec. 2612. Entry into customs territory of the United States

Sec. 2614. Prohibited acts

Sec. 2615. Penalties

Sec. 2616. Specific enforcement and seizure

Sec. 2618. Judicial review

Sec. 2619. Citizens' civil actions

Sec. 2620. Citizens' petitions

Sec. 2622. Employee protection

Secs. 2641 to 2656. Asbestos hazard emergency response

Secs. 2661 to 2671. Indoor radon abatement

Secs. 2681 to 2692. Lead exposure reduction

REFERENCES

Abramson, B., and A. J. Finizza. Probabilistic forecasts from probabilistic models case study: The oil market. *International Journal of Forecasting* **11**:63–72 (1995).

Aitchison, J., and J. A. C. Brown. *The Lognormal Distribution (With Special Reference to Its Uses in Economics)*. Cambridge University Press, Cambridge, England, 1999.

Alaska Fish and Game. Special oil spill issue. *Alaska Fish and Game* **21**(4) (July-August 1989).

Al-Bahar, J. F., and K. C. Crandall. Systematic risk management approach for construction projects. *Journal of Construction Engineering and Management* **116**(3): 533–546 (1990).

Asante-Duan, D. K. *Management of Contaminated Site Problems*. Lewis, Boca Raton, FL, 1996.

Bagneschi, L., Pollution prevention: The best-kept secret in loss control. *Risk Management* (July):31–38 (1998).

Bakr, A. A., L. W. Gelhar, A. L. Gutjar, and J. R. MacMillan. Stochastic analysis of spatial variability in subsurface flows: 1. Comparison of one- and three-dimensional flows. *Water Resources Research* **14**(2):263–271 (1978).

Beard, R. T. Bankruptcy and care choice. *RAND Journal of Economics* **21**(4):23–28 (1990).

Benjamin, J. R., and C. A. Cornell. *Probability, Statistics, and Decision for Civil Engineers*. McGraw-Hill, New York, 1970.

Berger, J. O. *Statistical Decision Theory and Bayesian Analysis*. Springer-Verlag, New York, 1985.

Bleistein, N., and R. A. Handelsman. *Asymptotic Expansions of Integrals*. Dover, New York, 1986.

Bogardi, I., L. Duckstein, and F. Szidarovszky. Bayesian analysis of underground flooding. *Water Resources Research* **18**(4):1110–1116 (1982).

Bouchart, F. J.-C., and I. C. Goulter. Is rational decision making appropriate for management of irrigation reservoirs? *Water Resources Planning and Management*, ASCE, **124**(6):301–309 (1998).

Bracewell, R. *The Fourier Transform and Its Applications*. McGraw-Hill, New York, 1965.

Bunn, D. *Applied Decision Analysis*. McGraw-Hill, New York, 1984.

Burt, O. R., and M. C. Stauber. Economic analysis of irrigation in subhumid climate, *American Journal of Agricultural Economics* **53**(1):33–46 (1971).

Canter, L. W., and R. C. Knox. *Ground Water Pollution Control*. Lewis, Chelsea, MI, 1986.

Carrera, J., and S. P. Neuman. Estimation of aquifer parameters under transient and steady state conditions: 1. Maximum likelihood method of incorporating prior information. *Water Resources Research* **22**(2):199–210 (1986a).

Carrera, J., and S. P. Neuman. Estimation of aquifer parameters under transient and steady state conditions: 3. Application to synthetic and field data. *Water Resources Research* **22**(2):228–242 (1986b).

Christakos, G., and D. T. Hristopulos. *Spatiotemporal Environmental Health Modeling: A Tractatus Stochasticus*. Kluwer Academic Publishers, Boston, 1998.

Clemen, R. T. *Making Hard Decisions: An Introduction to Decision Analysis*. Duxbury Press, Brooks/Cole Publishers, Pacific Grove, CA, 1996.

423

Clemen, R. T., and R. Winkler. Limits for the precision and value of information from dependent sources. *Operations Research* **33**:427–442 (1985).

Clifton, P. M., and S. P. Neuman. Effects of Kriging and inverse modeling on conditional simulation of the Avra Valley aquifer in southwestern Arizona. *Water Resources Research* **18**(4):1215–1234 (1982).

Corwin, D. L., K. Loague, and T. R. Ellsworth (eds.). *Assessment of Non-point Source Pollution in the Vadose Zone.* American Geophysical Union, Geophysical Monograph 108, 1999.

Cozzolino, J. M. A simplified utility framework for the analysis of financial risk. *Economic Evaluation Symposium of the Society of Petroleum Engineers,* SPE 6359 (1977a).

Cozzolino, J. M. *Management of Oil and Gas Exploration Risk.* Cozzolino Associates, West Berlin, N.J., 1977b.

Cozzolino, J. M. A new method for measurement and control of exploration risk. *Society of Petroleum Engineers,* AIME, SPE No. 6632, 1978.

Council of Environmental Quality. *Ocean Dumping: A National Policy.* U.S. Government Printing Office, Washington, DC, 1970.

Cushman, J. H. On measurement, scale, and scaling. *Water Resources Research* **22**(2):129–134 (1986).

Dagan, G. *Flow and Transport in Porous Formations.* Springer-Verlag, New York, 1989.

Dagan, G., and S. P. Neuman (eds.). *Subsurface Flow and Transport: A Stochastic Approach.* Cambridge University Press, UNESCO, 1997.

Davis, D. R., and W. M. Dvoranchik. Evaluation of the worth of additional data. *Water Resources Bulletin* **7**(4):700–707 (1971).

Davis, D. R., L. Duckstein, and R. Krzysztofowicz. The worth of hydrologic data for nonoptimal decision making, *Water Resources Research* **15**(6):1733–1742 (1979).

Davis, D. R., C. Kisiel, and L. Duckstein. Bayesian decision theory applied to design in hydrology. *Water Resources Research* **8**(1):33–41 (1972).

DeGrott, M. *Optimal Statistical Decisions.* McGraw-Hill, New York, 1970.

Delhomme, J. P. Spatial variability and uncertainty in groundwater flow parameters: A geostatistical approach. *Water Resources Research* **15**(2):269–280 (1979).

Desbarats, A. J. Spatial averaging of hydraulic conductivity in three-dimensional heterogeneous porous media. *Mathematical Geology* **24**(3):249–267 (1992).

Desbarats, A. J., and R. Dimitrakopoulos. Geostatistical modeling of transmissibility for 2D reservoir studies. SPE Formation Evaluation, 1990, pp. 437–443.

De Souza Porto, M. F., and C. M. de Freitas. Major chemical accidents in industrializing countries: The socio-political amplification of risk. *Risk Analysis* **16**(1): 19–29 (1996).

Deutsch, C. V., and A. G. Journel. *GSLIB: Geostatistical Software Library and User's Guide.* Oxford University Press, New York, 1992.

Diekmann, J. E., and D. W. Featherman. Assessing cost uncertainty: Lessons from environmental restoration projects. *Journal of Construction Engineering and Management,* **124**(6):445–451 (1998).

Dixon, S. A. *Lessons in Professional Liability: DPIC's Loss Prevention Handbook for Environmental Professionals.* DPIC Co., Monterey, CA, 1996.

Dooge, J. C. I. Scales problems in hydrology. In *Reflections on Hydrology, Science and Practice,* N. Buras (ed.). American Geophysical Union, Washington, DC, 1997, pp. 85–143.

Drewnowski, S. Evaluating and managing environmental risk. *Energy Economist* **182**:10–14 (1996).

Duckstein, L., and C. C. Kisiel. Efficiency of hydrologic data collection systems: Role of type I and II errors. *Water Resources Bulletin* **7**(3):592–604 (1971).

Duckstein, L., R. Krzysztofowicz, and D. Davis. To build or not to build: A Bayesian analysis. *Journal of Hydrological Sciences* **5**(1):55–68 (1978).

Dunn, J. H. Environmental liability: New insurance policies can help protect investors against unseen risk. *Journal of Property Management* **62**(6):54–60 (1997).

Edwards, W. (ed.). *Utility Theories: Measurements and Applications.* Kluwer Academic Publishers, Boston, 1992.

Eiser, J. R., and J. van der Pligt. *Attitudes and Decisions.* Routledge, London, 1988.

Fischer, J. N. *Hydrologic Factors in the Selection of Shallow Land Burial for the Disposal of Low-level Radioactive Waste.* U.S. Geological Survey Circular 973, Washington, 1986.

Fishburn, P. C. *Nonlinear Preference and Utility Theory.* Johns Hopkins University Press, Baltimore, 1988.

Fishburn, P. C. Foundations of decision analysis: Along the way. *Management Science* **35**:387–405 (1989).

Flach, G. P., and M. K. Harris. *Integrated Hydrogeological Model of the General Separations Area (U),* Vol. 2: *Groundwater Flow Model (U).* Westinghouse Savannah River Company, Savannah River Site, WSRC-TR-96-0399, Rev. 0, 1997.

Fletcher, C. D., and E. K. Paleologos. *Environmental Risk and Liability Management for Corporations and Consultants,* American Institute of Professional Geologists and Geological Society of America, Boulder, CO, 2000.

Frano, A. J. Pricing hazardous-waste risk revisited. *Journal of Management in Engineering* **7**(4):428–440 (1991).

Freeman, P. K., and H. Kunreuther. *Managing Environmental Risk Through Insurance.* Kluwer Academic Publishers, Boston, 1997.

Freeze, R. A. Groundwater contamination: Technical analysis and social decision making. In *Reflections on Hydrology, Science and Practice,* N. Buras (ed.). American Geophysical Union, Washington, DC, 1997, pp. 148–180.

French, S. *Decision Theory: An Introduction to the Mathematics of Rationality.* Wiley, London, 1986.

Gardner, W. R. Some steady-state solutions of the unsaturated moisture flow equation with application to evaporation from a water table. *Soil Sciences* **85**(4):228–232 (1958).

Garrison, R. H. *Managerial Accounting,* 6th ed. Irwin, Homewood, IL, 1991.

Gelhar, L. W. *Stochastic Subsurface Hydrology.* Prentice-Hall, Upper Saddle River, NJ, 1993.

Glezen, W. H., and I. Lerche. A model of regional fluid flow: Sand concentration factors and effective lateral and vertical permeabilities. *Mathematical Geology* **17**(3):297–315 (1985).

Goldman, D. Estimating expected annual damage for levee retrofits. *Water Resources Planning and Management,* ASCE, **123**(2):89–94, 1997.

Gomez-Hernandez, J. J., and S. M. Gorelick. Effective groundwater model parameter values: Influence of spatial variability of hydraulic conductivity, leakance, and recharge. *Water Resources Research* **25**(3):405–419 (1989).

Gradshteyn, I. S., and I. M. Ryzhik. *Table of Integrals, Series, and Products.* Academic Press, San Diego, 1980.

Graf, W. L. *The Colorado River.* Association of American Geographers, Washington, DC, 1985.

Grasso, D. *Hazardous Site Remediation.* Lewis, Boca Raton, FL, 1993.

Grayson, C. J. *Decisions under Uncertainty: Drilling Decisions by Oil and Gas Operators.* Harvard Business School, Division of Research, Cambridge, MA, 1960.

Herrmann, D. Environmental liability: Abundant insurance key to unpolluted profits. *Corporate Cashflow* **16**(4):20–24 (1995).

Hillel, DW *Fundamentals of Soil Physics,* Academic Press, Orlando, FL, 1980.

Hillier, F. S., and G. J. Lieberman. *Introduction to Operations Research.* Holden-Day, Inc., Oakland, CA, 1980.

Howard, L. S. Risk managers analyze broad business risks, *National Underwriter,* 17–39 (1998).

Howard, R. A., and J. E. Matheson, *The Principles and Applications of Decision Analysis.* Strategic Decisions Group, Palo Alto, CA, 1984.

Holloway, C. A. *Decision Making Under Uncertainty: Models and Choices.* Prentice-Hall, Upper Saddle River, NJ, 1979.

Hughson, D. L., and T.-C. Yeh. A geostatistically based inverse model for three-dimensional variably saturated flow. *Stoch. Hydro. and Hydraul.* **12**:285–298 (1998).

Hunter, R. L., and C. J. Mann. *Techniques for Determining Probabilities of Geologic Events and Processes.* Oxford University Press, New York, 1992.

Isaaks, E. H., and R. M. Srivastava. *An Introduction to Applied Geostatistics.* Oxford University Press, New York, 1989.

James, B. R., J.-P. Gwo, and L. Toran. Risk-cost framework for aquifer remediation design. *Water Resources Planning and Management,* ASCE, **122**(6):414–420 (1996).

Janney, J. R., C. R. Vince, and J. D. Madsen. Claims analysis from risk-retention professional liability group. *Journal of Performance of Constructed Facilities* **10**(3):115–122 (1996).

Jaynes, E. T. Where do we stand on maximum entropy? In *The Maximum Entropy Formalism.* R. D. Levine and M. Tribus (eds.). MIT Press, Cambridge, MA, 1978, pp. 15–118.

Jeljeli, M. N., and J. S. Russell. Coping with uncertainty in environmental construction: Decision-analysis approach. *Journal of Construction Engineering and Management* **121**(4):370–380 (1995).

Kahneman, D., and A. Tversky. Prospect theory: An analysis of decision under risk. *Econometrica* **47**:263–291 (1979).

Kahneman, D., P. Slovic, and A. Tversky. *Judgement Under Uncertainty: Heuristics and Biases.* Cambridge University Press, Cambridge, England, 1982.

Keeney, R. L. *Sitting Energy Facilities.* Academic Press, New York, 1980.

Keeney, R. L., and H. Raiffa. *Decisions with Multiple Objectives: Preferences and Values Tradeoffs.* Cambridge University Press, Cambridge, England, 1993.

Keller, E. A. *Environmental Geology,* 8th ed. Prentice-Hall, Upper Saddle River, NJ, 2000.

Krzysztofowicz, R. Strength of preference and risk attitude in utility measurement. *Organ. Behav. Human Perform.* **31**:88–113 (1983a).

Krzysztofowicz, R. Why should a forecaster and a decision maker use Bayes theorem. *Water Resources Research* **19**(2):327–336 (1983b).

Krzysztofowicz, R. Expected utility, benefit, and loss criteria for seasonal water supply planning. *Water Resources Research* **22**(3):303–312 (1986).

Larson, B. A. Environmental policy based on strict liability: Implications of uncertainty and bankruptcy. *Land Economics* **72**(1):33–42 (1996).

LaValle, I. On cash equivalents and information evaluation in decisions under uncertainty: I. Basic theory. *Journal of the American Statistical Association* **63**:252–276 (1968a).

LaValle, I. On cash equivalents and information evaluation in decisions under uncertainty: II. Incremental information decisions. *Journal of the American Statistical Association* **63**:277–284 (1968b).

LaValle, I. *Fundamentals of Decision Analysis.* Holt, New York, 1978.

Lerche, I. *Geological Risk and Uncertainty in Oil Exploration.* Academic Press, San Diego, 1997.

Lerche, I., and J. A. MacKay. *Economic Risk in Hydrocarbon Exploration.* Academic Press, San Diego, 1999.

Levy, H. Stochastic dominance and expected utility: Survey and analysis. *Management Science* **38**:555–593 (1992).

Lindgren, B. W. *Elements of Decision Theory.* Macmillan, New York, 1974.

Lindley, D. W. *Making Decisions,* 2d ed. Wiley, New York, 1985.

Loaiciga, H. A., and M. A. Marino. Risk analysis for reservoir operation. *Water Resources Research,* **22**(4):483–488 (1986).

MacDonald, J. A., and M. C. Kavanaugh. Superfund: The cleanup standard debate. *Policy and Planning* (Feb.):55–61 (1995).

MacKay, J. A., and I. Lerche. On the influence of uncertainties in estimating risk aversion and working interest. *Energy Explor. Exploit.* **14**:13–46 (1996).

Macsyma, *Symbolic/Numeric/Graphical Mathematics Software: Mathematics and System Reference Manual,* Macsyma 2.1, Arlington, MA, 1996.

Magnuson, E. The poisoning of America. *Time* **116**(12):58–69 (1980).

Malle, K. G. Cleaning up the River Rhine. *Scientific American* **274**:70–75 (1996).

Mantoglou, A., and L. W. Gelhar. Stochastic modeling of large-scale transient unsaturated flow systems. *Water Resources Research* **23**(1):37–46 (1987a).

Mantoglou, A., and L. W. Gelhar. Capillary tension head variance, mean soil moisture content, and effective specific soil moisture capacity of transient unsaturated flow in stratified soils. *Water Resources Research* **23**(1):47–56 (1987b).

Mantoglou, A., and L. W. Gelhar. Effective hydraulic conductivities of transient unsaturated flow in stratified soils. *Water Resources Research* **23**(1):57–68 (1987c).

McCord, M., and R. D. Neufville. Lottery equivalents: Reduction of the certainty effect problem in utility assessment. *Management Science* **32**:56–60 (1986).

McMahon, N., Ruitenbeek, K., Wams, J., and S. Slawson. Cost-effective acquisition in a low oil price environment. *The Leading Edge* **18**:1162–1168 (1999).

Merkl, A., and H. Robinson. Environmental risk management: Take it back from the lawyers and engineers. *The McKinsey Quarterly* **14**(3):150 (1997).

Miller, A. C., and T. R. Rice. Discrete approximations of probability distributions. *Management Science* **29**:352–362 (1983).

Milligan, A. Mergers can bring new environmental liabilities: concern about acquired risks sparks EIL buys. *Business Insurance* **32**(21):3–6 (1998).

Minato, T., and D. B. Ashley. Data-driven analysis of "corporate risk" using historical cost-control data. *Journal of Construction Engineering and Management* **124**(1):42–47 (1998).

Mood, A. M., F. A. Graybill, and D. C. Boes. *Introduction to the Theory of Statistics.* McGraw-Hill, New York, 1974.

Moorhouse, D. C., and R. A. Millet. Identifying causes of failure in providing geotechnical and environmental consulting services. *Journal of Management in Engineering* **10**(3):56–64 (1994).

Morgan, M. G., and M. Henrion. *Uncertainty: A Guide to Dealing with Uncertainty in Quantitative Risk and Policy Analysis.* Cambridge University Press, Cambridge, England, 1990.

Morris, P. W. G., and G. H. Hough. *The Anatomy of Major Projects.* Wiley, New York, 1987.

National Research Council. *Opportunities in the Hydrologic Sciences.* National Academy Press, Washington, 1991.

Ness, A. A contracting for environmental remediation. *Construction Business Review* **2**(March-April):20–75 (1992).

Nyer, E. K. *Practical Techniques for Groundwater and Soil Remediation.* Lewis, Boca Raton, FL, 1992.

Paek, J. H. Pricing the risk of liability associated with environmental clean-up projects, *Transactions of the American Association of Cost Engineers,* CC4.1-CC4.5, 1996.

Paleologos, E. K., and C. Fletcher. Assessing risk retention strategies for environmental project management. *Environmental Geoscience,* **6**(3):130–138 (1999).

Paleologos, E. K., and I. Lerche. Multiple decision-making criteria in the transport and burial of hazardous and radioactive wastes. *Stochastic Environmental Research and Risk Assessment,* **13**(6):381–395 (1999).

Paleologos, E. K., I. Lerche, and Y. Mylopoulos. Optimal use of multiple statistical criteria in Bayesian decision-making. In *Proceedings of the IVth International Conference: Protection and Restoration of the Environment,* Vol. II, Bouris Publishing Co., Kalamaria, Greece, 1998, pp. 863–870.

Paleologos, E. K., S. P. Neuman, and D. Tartakovsky. Effective hydraulic conductivity of bounded, strongly heterogeneous porous media. *Water Resources Research,* **32**(5):1333–1341 (1996).

Parent, E., P. Hubert, B. Bobee, and J. Miquel (eds.). *Statistical and Bayesian Methods in Hydrological Sciences.* UNESCO, IHP-V, Technical Documents in Hydrology No. 20, Paris, France, 1998.

Percival, D. B., and A. T. Walden. *Spectral Analysis for Physical Applications.* Cambridge University Press, Cambridge, England, 1993.

Powell, D. M. Selecting innovative cleanup technologies, EPA resources, *Chemical Engineering Progress,* 33–41 (1994).

Press, W. H., B. P. Flannery, S. A. Teukolsky, and W. T. Vetterling. *Numerical Recipes: The Art of Scientific Computing.* Cambridge University Press, Cambridge, England, 1987.

Raiffa, H. *Decision Analysis: Introductory Lectures on Choices under Uncertainty.* McGraw-Hill, New York, 1997.

Raiffa, H., and R. Schlaifer. *Applied Statistical Decision Theory.* Harvard University Press, Cambridge, MA, 1961.

Rotbart, D. The state of the environment. *Financial Executive* 13(2):24–28 (1997).

Rubin, Y., and J. J. Gomez-Hernandez. A stochastic approach to the problem of upscaling of conductivity in disordered media: Theory and unconditional numerical simulations. *Water Resources Research* 26(4):691–701 (1990).

Sandia Report. *Total-System Performance Assessment for Yucca Mountain–SNL Second Iteration,* Vols. 1 and 2 (TSPA-1993). Sandia National Laboratory Technical Report, SAND93-2675, 1994.

Slack, J. R., J. R. Wallis, and N. C. Matalas. On the value of information to flood frequency analysis. *Water Resources Research* 11(5):629–647 (1975).

Smith, J. E. Moment methods for decision analysis. *Management Science* **39:**340–358 (1993).

Spetzler, C. S., and C. A. Stael von Holstein. Probability encoding in decision analysis. *Management Science* **22:**340–352 (1975).

Stephens, D. B. *Vadose Zone Hydrology.* Lewis, Boca Raton, FL, 1996.

Taylor, R. Environmental insurance helps CFOs shed liability. *National Underwriter: Property and Casualty-Risk and Benefits Management,* Oct. 1998, pp. 8–12.

Telego, D. J. A growing role: Environmental risk management. *Journal of Risk Management* 6(1):19–21 (1998).

Thompson, K. D., J. R. Stedinger, and D. C. Heath. Evaluation and presentation of dam failure and flood risks. *Water Resources Planning and Management,* ASCE, **123**(4):216–227 (1997).

Thomsen, R. O., and I. Lerche. Relative contributions to uncertainties in reserve estimates. *Mar. Pet. Geol.* **14:**65–74 (1997).

Tietenburg, T. H. *Environmental Economics and Policy,* 2d ed. Addison-Wesley, Reading, MA, 1998.

U.S. Department of Energy (DOE). *Total System Performance Assessment—1993: An Evaluation of the Potential Yucca Mountain Repository.* DOE Report No. B00000000-01717-2200-00099-Rev. 01, 1994.

U.S. Environmental Protection Agency (EPA). *Is Your Drinking Water Safe?* EPA Report No. 570-9-91-0005, 1991.

Van der Heijde, P., Y. Bachmat, J. Bredehoeft, B. Andrews, D. Holtz, and S. Sebastian. *Groundwater Management: The Use of Numerical Models.* American Geophysical Union, Water Resources Monograph 5, 1986.

Van Genuchten, M. T. A closed-form equation for predicting the hydraulic conductivity of unsaturated soils. *Soil Science Society of America Journal* **44:**892–898 (1980).

Von Neumann, J., and O. Morgenstern. *Theory of Games and Economic Behavior.* Princeton University Press, Princeton, NJ, 1944.

Von Winterfeldt, D., and W. Edwards. *Decision Analysis and Behavioral Research.* Cambridge University Press, Cambridge, England, 1986.

Voorhees, J., and R. A. Woellner. *International Environmental Risk Management.* Lewis, Boca Raton, FL, 1997.

Wentz, C. A. *Hazardous Waste Management.* McGraw-Hill, New York, 1989.

Westinghouse Savannah River Company. *Savannah River Site Environmental Report for 1996.* M. W. Arnett and A. R. Mamatey (eds.), WSRC-TR-97-0171, 1996.

Whitmore, G. A., and M. C. Findlay. *Stochastic Dominance.* Heath, Lexington, MA, 1978.

Winkler, R. L. *Introduction to Bayesian Inference and Decision.* Holt, New York, 1972.

Woodbury, A. D., and E. A. Sudicky. The geostatistical characteristics of the Borden aquifer. *Water Resources Research* 27(4):533–546 (1991).

Yeh, T.-C. One-dimensional steady state infiltration in heterogeneous soils. *Water Resources Research* 25(10):2149–2158 (1989).

Yeh, T.-C., L. W. Gelhar, and A. L. Gutjahr. Stochastic analysis of unsaturated flow in heterogeneous soils. I. Statistically isotropic media. *Water Resources Research* **21**(4):447–456 (1985a).

Yeh, T.-C., L. W. Gelhar, and A. L. Gutjahr. Stochastic analysis of unsaturated flow in heterogeneous soils: II. Statistically anisotropic media with variable α. *Water Resources Research* **21**(4):457–464 (1985b).

Yeh, T.-C., L. W. Gelhar, and A. L. Gutjahr. Stochastic analysis of unsaturated flow in heterogeneous soils: III. Observations and applications. *Water Resources Research* **21**(4):465–472 (1985c).

Yeh, W. W.-G. Review of parameter identification procedures in groundwater hydrology: The inverse problem. *Water Resources Research* **22**(2):95–108 (1986).

Yeo, K. T. Risks, classification of estimates, and contingency management. *Management in Engineering,* ASCE, **6**(4):458–470 (1990).

Zevas, M. *Elements of Mathematical Analysis.* Athanasopoulos, Athens, Greece, 1973.

INDEX

ABOUT THE AUTHORS

IAN LERCHE is an award-winning researcher and Professor of Geology at the University of South Carolina. His current major research interests are basin analysis, salt, economic risk, and environmental problems. He has published several hundred papers, together with over a dozen books. He is the recipient of numerous awards and honors, including the Levorsen Award of the American Association of Petroleum Geologists. Currently, he sits on several editorial boards and is also technical editor of *Energy Exploration & Exploitation.*

EVAN K. PALEOLOGOS, Ph.D., is an Assistant Professor in the Department of Geological Sciences at the University of South Carolina. With a doctorate in hydrology, he currently specializes in problems of groundwater flow and in the transport of contaminants in porous media. He is also concerned with decision making in environmental projects. The author of over thirty journal articles and one book, he has also served on the editorial board of two scientific journals, *Stochastic Hydrology & Hydraulics* and *Stochastic Environmental Research and Risk Assessment.*

www.ingramcontent.com/pod-product-compliance
Lightning Source LLC
Chambersburg PA
CBHW060744220326
41598CB00022B/2321